The Era of Cloud Consumption

雲消費時代

賴陽——

著

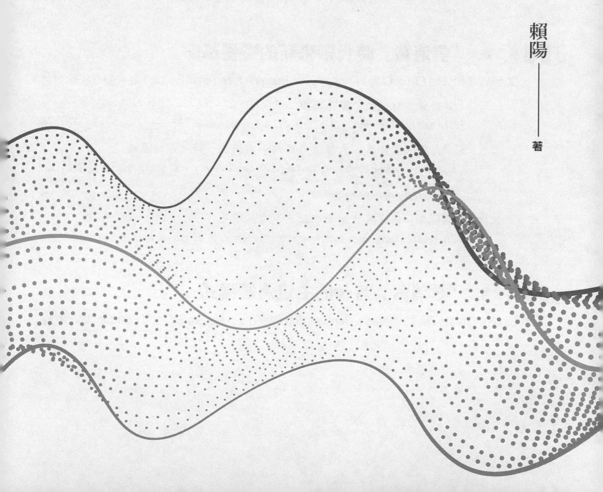

目　錄

第四章　「雲消費」時代交易模式的根本性變化

Chapter Four Fundamental Changes in the Transaction Patterns in the Era of Cloud

Consumption⋯⋯⋯⋯⋯⋯⋯⋯⋯⋯⋯⋯⋯⋯⋯⋯⋯⋯⋯⋯⋯ **238**

第五章　「雲消費」時代傳統零售業的變革

Chapter Five The Change of Traditional Retail Industry in the Era of Cloud Consumption **286**

第一章

商業進入「雲消費」時代

Chapter One
Business Enters the Era of "Cloud Consumption"

一

什麼是「雲消費」？
1. What is the「cloud consumption」?

21 世紀以來，隨著現代資訊互聯技術的飛速發展，商業資訊傳遞突破時空障礙、物流網路逐步實現全通聯，制約消費的一系列障礙正在快速消失，為此，使消費亦得以突破時間、空間的障礙，從而從根本上改變了商業的成長肌理，這使得「雲消費」不僅成為可能，也正在我們眼前發生，成為必然的現實，同時也代表了現代商業的發展走向。如果說，2009 年筆者首次預言「雲消費」時代即將來臨時，「雲消費」還僅僅是對未來的一種判斷，那麼今天，我們已經可以切切實實地感受到「雲消費」給我們生活帶來的全方位的改變。

Since the 21st century, with the rapid development of modern information interconnection technology, commercial information transmission has broken through the obstacles of time and space, and all-pass communication of the logistics network has been gradually coming true. A series of obstacles restricting consumption are rapidly disappearing. For this reason, consumption has also broken through the obstacles of time and space, an event that has fundamentally changed the growth of business. These changes made the 「cloud consumption」 possible; are happening before our eyes and become an inevitable reality; and reflect the developmental trend of modern business. If it was merely a judgment on the future when the author predicted for the first time in 2009 that the era of「cloud consumption」 is coming, then today we can already feel the all-round change in our life brought by the 「cloud consumption」.

什麼是「雲消費」？
What is the "cloud consumption?"

「雲」是指消費者所處的如雲一般無障礙、無邊際、無所不在的消費環境。「雲消費」就是以現代信息互聯技術為基礎，消費者可以透過任意消費終端獲得任何其需要的商品和服務，其接觸的任何有形、無形平臺均能為其提供無縫消費支援。其核心是以消費者需求為主導。

"Cloud" refers to a cloud-like, barrier-free, borderless and omnipresent consumption environment. "Cloud consumption" is based on modern information interconnection technology. Consumers can get any commodity and service they need through any consumption terminal, and any tangible and intangible platform they get in touch with can provide seamless consumption support for them. Its core is oriented towards consumer demand.

當前，我們正面臨一輪真正意義上的零售革命，它打破了零售業業態的界限，打破了虛擬與現實的界限，使有形與無形相融，商流、物流、資訊流交織，改寫了商店的定義，改變了零售的思維。我們認為，正是「雲消費」的發展推進了零售革命，或者說「雲消費」時代帶來新的零售革命。

At present, we are facing a round of real retail revolution, which breaks the boundary of retail industry, breaks the boundary between virtual and reality, makes the tangible and intangible blend and interweaves business, logistics and information flow. It also rewrites the definition of store and changes the thinking of retail. We believe that it is the development of "cloud consumption" that promotes the retail revolution, or the "cloud consumption" era brings a new retail revolution.

「雲消費」在技術層面上表現為三大核心特徵，即體現雲消費發展內涵和容量的「雲內容」，代表「雲消費」終端平臺的「雲終端」，反映「雲消費」的交易形式和支付方式的「雲支付」。

On the technical level, the "cloud consumption" presents three core characteristics, namely "cloud content" that reflects the development connotation and capacity of cloud consumption, "cloud terminal" that represents the terminal platform of "cloud consumption" and "cloud payment" that reflects the transaction form and payment method of the "cloud consumption".

在消費領域，消費突破了傳統店鋪的限制，突破了商品型態的限制，流通的時間障礙、距離障礙、管道環節也已完全突破，呈現出「零時差、零距離、零管道」之趨勢。

In the consumption field, consumption has broken through the restrictions of traditional shops and the restriction of commodity forms, as well as the time barrier, distance barrier and circulation channel, showing the trend of "zero time difference, zero distance and zero channel".

「雲消費」特別強調以消費者需求為主導，即以消費者的個性化需求指

向為導向。未來的消費必將是以消費者為核心的現代消費方式為主導的消費，消費者價值是零售業生存的根基。伴隨互聯網和移動互聯的發展，消費者生活方式正在發生轉型，消費者偏好與消費方式正在發生變化。我們認為，當前社會主流消費群消費模式日益表現出四大基本屬性：消費的體驗化、個人化、社群化和即時化。

In particular, it is emphasized that the "cloud consumption" is dominated by consumer demand, that is, it is oriented by the consumer's personalized demand direction. The future consumption will be dominated by the modern consumption mode with consumers as the core, and consumer value is the foundation for the survival of the retail industry. With the development of the Internet and mobile Internet, consumer lifestyle is undergoing transformation, and consumer preferences and consumption patterns are changing. We believe that the consumption model of the mainstream consumer groups in the current society is increasingly showing four basic attributes: experiencing, personalization, socialization and real-time consumption.

「雲消費」時代，企業資訊成本的獲得方式和信譽評價機制產生根本性變化，同時與現代主流消費模式的變化相適應，使更多小微企業獲得可以快速成長甚至與大企業比肩對話的機會，「新魚吃舊魚」成為可能。

Fundamental changes have taken place in the access method and reputation evaluation mechanism of enterprise information cost in the era of "cloud consumption". Meanwhile, it adapts to the changes of modern mainstream consumption mode and enables more small and micro enterprises to have opportunities to grow rapidly and even have a dialogue with large enterprises. Such thing as 「new-born fish eats old one」 becomes possible now.

隨著「雲消費」時代的來臨，零售業面臨重新洗牌，更多傳統零售業態、零售商家面臨空前的創新變革的壓力。商家必須學會主動應對「雲消費」時代新的零售革命，在新的「雲消費」市場生態環境下找準自己的生態位。

With the advent of the cloud consumption era, the retail industry is facing a reshuffle, and more traditional retail formats and retailers are under unprecedented pressure to innovate. Merchants must learn to take the initiative to respond to the new retail revolution in the era of the 「cloud consumption」 and identify their own niche in the new 「cloud consumption」 market environment.

「雲消費」時代不僅僅是零售業革命的時代，也是全行業產業革命的時代。一切商業資源呈現平臺化整合趨勢，透過行業資源的整合、盈利模式的再造，我們會發現更多的發展機會，創造更多的社會價值。

The era of cloud consumption is not only the era of the retail revolution, but also the era of industrial revolution in the whole industry. All commercial resources show a trend of platform integration. Through the integration of industry resources and the re-creation of profit models, we will find more development opportunities and create more social value.

新的消費時代不僅將徹底改變我們的消費與生活，也將導致提供消費服務的商業面對新的挑戰，傳統的業種業態生存的根基發生動搖，一場新的零售革命將徹底改變流通行業。之所以稱之為「革命」是因為，商業零售業的變化不再是傳統業態的此消彼長，而是整個產業的全面變革和洗牌。未來五年可能現有傳統習慣的零售方式會徹底改變，而新的零售方式會以百倍千倍

速度增長。任何輕視、低估這場革命力量，或沒有跟上革命步伐的企業將面臨淘汰，新生力量快速崛起，未來的零售業將是一個全新的產業格局。

The new era of consumption will not only completely change our consumption and life, but also lead to new challenges to the businesses providing consumption services. The foundation of the survival of the traditional industry is shaken, and a new retail revolution will completely change the circulation industry. The reason why it is called "revolution" is because the change of commercial retail industry is no longer the trade-off of traditional formats, but the overall transformation and reshuffle of the entire industry. Traditional retail methods used now may be changed radically in the next five years, and the new retail methods will grow at a rate of a hundreds and thousands times. Any enterprise that despises or underestimates the revolutionary force, or fails to keep up with the pace of the revolution will face elimination. In the meanwhile new forces will rise rapidly, and the future retail industry will be in a brand new industrial structure.

二

「雲消費」的三大基本特徵
2. Three basic characteristics of the
"cloud consumption"

　　隨著現代資訊互聯技術的飛速發展，使得跨時間、跨空間整合資源成為可能，實體商業與網路商業的界限逐漸消失，產品開發者與銷售者的距離逐步消失，並逐步由量變轉為質變。以消費者需求為主導，我們正面臨一個翻天覆地的新的商業時代——「雲消費」時代。

With the rapid development of modern information interconnection technology, it is possible to integrate resources across time and space. The boundaries between physical business and network business are gradually disappearing, so is the distance between product developers and seller. Such changes are being brought about from quantitative to qualitative. Driven by consumer demand, we are facing an earth-shaking new era of business, the era of 「cloud consumption」.

　　「雲消費」在技術層面具有三大核心特徵：「雲內容」、「雲終端」、「雲支付」。其中雲內容體現「雲消費」的發展內涵和容量，雲終端代表「雲消費」的終端平臺，雲支付反映「雲消費」的交易形式和支付方式。

The 「cloud consumption」 has three core characteristics at the technical level, they are「cloud content」,「cloud terminal」, and「cloud payment」. Cloud content reflects the development connotation and capacity of「cloud consumption」, cloud terminal represents the terminal platform of「cloud consumption」, and cloud payment reflects the transaction form and payment method of「cloud consumption」.

（一）雲內容 Cloud content

1・ 什麼是「雲內容」 What is「cloud content」

消費「雲內容」即在現代資訊技術條件下，消費突破傳統店鋪存儲、面積、陳列限制，突破線上線下、有形與無形的界限，突破商品與服務的界限，一切為了消費者，從消費者出發，消費者所想即所供。

"Cloud content" of consumption means that under the conditions of modern information technology, consumption can now break through the restrictions of storage, area and display of traditional stores, as well as the boundaries of online and offline, tangible and intangible, and the boundaries between goods and services. As is consumer-oriented, it provides exactly what consumers want.

2・ 「雲消費」的四個基本特點
Four basic characteristics of "cloud consumption"

從「雲內容」的基本概念出發，「雲消費」在實踐中表現為四個基本特點：

As for the basic concept of "cloud content", the "cloud consumption" has four basic characteristics in practice:

（1）消費突破傳統店鋪限制

Consumption breaks through the restrictions of traditional stores

「雲消費」時代極大地拓展了零售業的市場空間，這種空間除了為實體零售業提供了地理型態的跨區域發展空間，更主要是體現在虛擬的零售市場的發展。由於虛擬零售市場突破了實體零售業有限的空間市場，其高速發展意味著這一市場具有巨大的容量和體量。

The era of 「cloud consumption」 has greatly expanded the market space of the retail industry. Such space can not only be used for the geographically-oriented cross-regional development for the physical retail industry, but also is mainly reflected in the development of the virtual retail market. As the virtual retail market breaks through the limited market space of the physical retail industry, its rapid development means that this kind of market has huge capacity and volume.[1]

國際上很多零售企業都已經在這方面做了一定的探索，如澳大利亞最大的零售商 Woolworths 就突破實體店鋪的銷售模式，在 2012 年初結合自己的手機應用推出「虛擬」超市，具有搭建成本低、無須門店資源、高度的靈活性、更貼近日常生活的優勢。Woolworths 將產品透過圖片和條碼的方式印刷在展示板上，再將這樣的展示板擺放在人流密集的區域，比如車站附近。消費者

---

1 顧國建. 網路時代零售業格局的變革 [J]. 上海商業 . 2010-07-10   Gu Guojian. The transformation of the retail industry in the Internet age [J]. Shanghai Business. 2010-07-10

在路過或者等車的時候，就可以拿出手機，透過掃描相應的條碼選擇購買需要的產品，產品最快會在第二天送到消費者家裡。

　　Many retailers in the world have already made some explorations in this area. For example, Woolworths, Australia's largest retailer, has broken through the sales model of physical stores. In early 2012, it launched a 「virtual」 supermarket with its own mobile phone application, which was low-cost, highly flexible, had no need for store resources and was closer to everyday life. Woolworths printed the products on the display boards by means of pictures and barcodes, and then placed them in crowded areas, like near the stations. When a consumer is passing by the stations or waiting for the bus, he or she can take out the mobile phone and scan the specific barcode to buy the product he or she wants. At the soonest, the product will be delivered to the consumer's home the next day.

圖 1-1　澳大利亞最大零售商 Woolworths 在地鐵開設「虛擬」超市 Australia's largest retailer Woolworths started the "virtual" supermarket in the subway

　　全球三大零售商之一的 TESCO（樂購）也突破傳統店鋪模式，在韓國地鐵、公交站建設虛擬商場，韓國候車的乘客可以在海報前拍攝商品二維碼購物並透過手機結算，商品會在 24 小時內送達客戶指定的地點。

TESCO, one of the world's three largest retailers, has also broken through the traditional store model. It starts virtual shopping malls in Korea's subway and bus stations. Waiting passengers can scan the QR code printed on the poster to buy goods and pay the bill through mobile phones. Goods will be delivered to the customer's designated location within 24 hours.

（2）消費突破時空限制

Consumption breaks through the restrictions of time and space

「雲消費」時代傳統的時空觀念全面突破，從空間上看，我們不僅面臨傳統的實體市場空間，還面臨由虛擬網路形成的沒有地域界限的網路空間，且該空間範圍內消費活動的各方透過網路彼此發生聯繫。從時間上看，「雲消費」沒有時間間斷，全年全天 24 小時無休。這樣，即使在網路的一頭沒有客服人員，消費者也可以自助購物，而且對於偏遠地區的企業和小生產者、個人，都可以和大城市的企業站在同一條銷售起跑線上，原則上只要網線鋪到哪裡，那裡的人們就可以享受與大城市人們一樣的購物機會，觸摸到網上海量的商品。

In the era of 「cloud consumption」, the traditional concept of time and space has undergone an all-round breakthrough. From the perspective of space, we not only face the traditional physical market space, but also face the cyberspace formed by virtual networks, which has no geographical boundaries. Different parties involved in the consumption activities interrelate within such space through the network. From the perspective of time, "cloud consumption" has no service time break, it runs 24 hours a day throughout the year. In this way, even if there is no customer service staff at the other end of the network, consumers still can self-shop, and for enterprises and small producers and individuals in remote areas, they can stand on the same starting line as the enterprises in big cities do. In principle, as long as the network is accessible, no matter where, people can enjoy the same shopping opportunities as people in big cities do and get vast amount of goods on the Internet.

在「雲消費」的市場環境下，商品之間的交易，除了物流是實體外，商品交易的過程、商品交易的手續都可以是虛擬的。這種交易方式，一方面降低了交易成本，提高了交易效率；另一方面，也增加了競爭的速率和強度，特別是移動網購交易的發展，更使這種虛擬交易呈爆發式增長。2009—2013年5年間中國網路零售市場規模由2009年的2.1%（2786.238億元）增長至2013年的7.8%（18281.64億元），增幅達271%，超過社會消費品零售總額增幅194個百分點。[2]2017年中國網路零售市場交易規模達到7.18萬億元，對社會消費品零售總額增長的貢獻率為37.9%，比2016年又提升了7.6個百分點。[3]2009—2017年9年間，中國網路零售市場規模增長了近25倍！

In the market environment of 「cloud consumption」, besides the fact that logistics is an entity, the process and procedure of commodity trading can both be virtual. This kind of trading method reduces the cost and improves the efficiency

---

2　中國電子商務研究中心發佈《2012年度中國網路零售市場資料監測報告》 2012 China Online Retail Market Data Monitoring Report released by China E-Commerce Research Center

3　2018年1月中國商務部發佈資料 Data of January 2018 released by China's Ministry of Commerce

on the one hand; on the other hand, it also increases the rate and intensity of trading competition, especially with the development of mobile online shopping, virtual trading explodes. China's online retail market grew from 2.1% (278.623 billion yuan) in 2009 to 7.8% (18,281.64 billion yuan) in 2013, an increase of 271%, exceeding the growth of total retail sales of consumer goods by 194 percentage points.2 In 2017, the trading volume of China's online retail market reached 7.18 trillion yuan, contributing 37.9% to the growth of total retail sales of consumer goods, an increase of 7.6 percentage points over 2016.3 In 2009-2017, online retail market in China has grown by nearly 25 times!

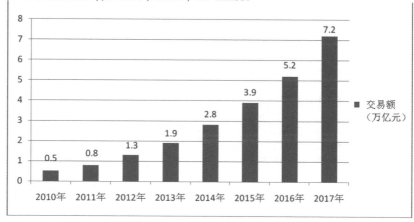

Transaction Volume (Trillion yuan)

圖 1-2 2010-2017 中國網路零售市場交易規模 2010-2017 Transaction scale of China online retail market
資料來源：中國電子商務研究中心 www.100EC.cn
Source: China Electronic Commerce Research Center (www.100EC.cn)

（3）消費突破商品型態限制

Consumption breaks through the restrictions of commodity form

「雲消費」環境下，交易的商品不僅包括所有能夠想像的有形實體商品，

還可以包括所有我們能夠或不能想到的虛擬商品。包括電腦軟體、專業資訊、電子讀物、音樂產品、影視節目、搜索、網路遊戲中的一些產品和線上服務、綜合性服務等。淘寶網上諸如網路遊戲點卡、網遊裝備、QQ 號碼、Q 幣、手機充值卡、IP 卡、網路電話、軟體序號、電子書，網路軟體（如安卓手機軟體、SKYPE 語單軟體等）、功能變數名稱、虛擬空間、網站、搜索服務等網站類產品、電子票（電影票、演出票、火車票、飛機票等）等均可輕鬆採購。

In the 「cloud consumption」, the goods traded include not only all tangible physical goods that can be imagined, but also all virtual goods that we can or cannot think of, including computer software, professional information, electronic books, musical products, films and television programs, search services, some products of online games, online services comprehensive services, etc. In Taobao, the biggest online shopping platform in China, such goods as online game cards, online game equipments, QQ numbers, Q-coins, mobile phone top-up cards, IP cards, Internet phone, CDkeys, e-books, network software (such as Android mobile phone software, SKYPE, etc.), websites products such as domain names, virtual spaces, websites, search services, and e-tickets (movie tickets, show tickets, train tickets, airline tickets, etc.) can be easily purchased.

美國一家名為 TaskRabbits（「任務兔子」）的網站，將任務發佈者（TaskPosters）和「任務兔子」聯繫到一起。任務發佈者透過這個平臺獲得任務兔子的幫助，「任務兔子」完成領取的任務後可以獲得一定報酬。認領任務的「任務兔子」，在認領工作的過程中就有一種與網上的其他人虛擬賽跑的樂趣。認領人參與任務、完成任務之後，可以獲得一定的獎勵，例如完

成 3 項最高可獲得 20 美元報酬，還可以獲得網站獎勵的一根胡蘿蔔。這樣，「任務兔子」不僅將很多人各式各樣奇怪的需求變成虛擬商品，也為很多人及很多人的冗餘時間找到工作機會[4]。中國近年也快速發展起一批任務服務型網站，如「百度眾包」、「閃送」、「UU 跑腿」等，尋求大眾智慧，集約配置社會冗餘資源解決各種現實需求已逐漸成為人們的生活習慣。

A website named TaskRabbits connects the Task Posters with the Task Rabbits. Task Posters obtain the help from Task Rabbits through this platform, and the latter can get paid after completing the task. Task Rabbit that claims the task has a kind of fun to compete with other people on the Internet during the process. After having participated in and completing the task, the claimant can get paid. For example, a maximum reward of 20 dollars and a virtual carrot from the website for completing 3 tasks. In this way, the website not only turns many people's strange needs into virtual goods, but also finds jobs for many people with redundant time.4 In recent years, China has also rapidly developed a number of mission-oriented websites, such as Baidu Crowdsourcing, FlashEx, UU Errands, etc. Seeking public wisdom and intensively deploying social redundant resources to solve various practical needs have gradually become people's life style.

（4）一切以消費者需求為核心

Everything is centered on consumer demand

「雲消費」環境下，由於海量消費資料能夠較為容易地獲得，破解了消

---

4 跑腿網站 TaskRabbitss：將繁重工作變成遊戲 . 搜狐 IT TaskRabbits: Turning heavy work into games. Source: Sohu IT

費者個性化資料分析的首要難題，使真正以消費者為核心成為可能；由於人際關係被重新定義，互聯網將有共同需求的人群聚集在了一起，並且透過知識的分享壯大了消費者的力量；也由於消費者本身購買力的增強，技術應用能力的進一步成熟、經驗的豐富、鑑別能力的增強等，消費者這個概念被重新定義，一切以消費者需求為核心可以真正落到實處。消費者所需就能所有，有需求就能滿足。

In the「cloud consumption」, mass consumption data can be easily obtained, thus the primary problem of consumer personalized data analysis is solved, making it truly possible to focus on consumers. Since interpersonal relationships are redefined, the Internet gathers the people with common needs together and expands the power of consumers through the sharing of knowledge. In addition, because of the increasing purchasing power of consumers, further maturity of technology application capabilities, rich experience and enhanced discriminating ability, the concept of consumer is redefined and the will that everything is centered on consumer demand can be truly implemented. Consumers' needs can be met as long as they have.

（二）雲終端 2.2 Cloud terminal

1・ 什麼是「雲終端」 What is cloud terminal

所謂消費「雲終端」，即凡是消費者接觸的任何店鋪或智慧電子平臺都可以做為提供消費的便捷終端。

The so-called cloud terminal means any store or intelligent electronic platform made for consumers can be used as a convenient terminal for consumption.

雲終端不是割裂的，而是所有消費終端的整合，這意味著零售商將能透過多種終端與顧客互動，包括網站、實體店、服務終端、直郵和目錄、呼叫中心、社交媒體、移動設備、遊戲機、電視、網路家電、上門服務、電子閱讀器等等，這些終端相互整合、相互呼應，成為全方位的行銷力量。如在美國的百貨店下訂單，在國內取貨，在網上下訂單到韓國代購，透過二維碼支付在手機、電腦、網吧、餐館觸控式螢幕、電視機、零售店等任一終端消費……總之，接入即互聯，接入即可任意消費。

Cloud terminals are not separated from each other, but the integration of all consumer terminals, which means retailers will be able to interact with customers through a variety of terminals, including websites, physical stores, service terminals, direct mail and directories, call centers, social media, mobile devices, game consoles, TVs, network appliances, on-site services, e-readers, etc. These terminals integrate and reciprocate among each other, and become an all-round marketing force. For example, you can now place an order in a department store in the United States while pick up the goods in China; place an online order to purchasing agent to buy goods in South Korea, and scan the QR code to pay through any terminal like mobile phones, computers, Internet cafes, touch screens in restaurants, TV sets, retail stores, etc. In brief, access to the Internet means you can consume at will.

2. 「雲終端」體系不斷完善

The system of cloud terminal is constantly improving

隨著現代資訊技術的發展，智慧手機、智慧電視、智慧閱讀器、智慧商店等新型的消費終端不斷出現，使消費者可以透過各種平臺進行消費，進一步體會到「雲終端」無處不在的好處。

With the development of modern information technology, new types of consumer terminals such as smart phones, smart TVs, smart e-readers, and smart stores are constantly appearing, enabling consumers to consume through various platforms and further realize ubiquitous benefits of cloud terminals.

（1）智慧手機消費終端 Smart phone

隨著移動電子商務的發展，智慧手機已迅速成為隨時隨地可購物的移動的微型雲商店。人們可以透過手機來完成整個電子商務流程，根據個人現狀、環境等因素檢索獲得符合需要的資訊，在短時間內便可以做出判斷，然後以手機進行線上付款，在指定時間、指定地點收穫商品。這一流程的最大的特點是，手機做為資訊的接收終端同樣也是資訊的發送終端。這樣，做為資訊收集和分析終端的消費平臺，便可以獲得絕對真實並且即時的消費者資料，從而面對消費者給出符合消費者實際需要的資訊結果；並且經過不斷積累，給出的資訊結果會越來越智慧。

With the development of mobile e-commerce, smart phones have quickly become mobile micro cloud-stores where shopping can be done anytime, anywhere. People can complete the entire e-commerce process through mobile phones. They can search for information that meets the needs according to personal status, environment and other factors, and then make judgments in a short period of time. After that, they can make online payment by mobile phone, and receive the goods at a designated time and place. The biggest feature of the process is that mobile phones are both terminals of receiving and sending information. In this way, as the consumption terminal of information collection and analysis, mobile phones can obtain absolutely true and real-time consumer data, so that they can give the result that meets the actual needs of the consumer. After continuous accumulation, the result given will be more and more intelligent.

（2）智慧電視消費終端 Smart TV

智慧電視擁有龐大的使用者群體，利用有線電視網路，可以開展越來越多的接入式消費。近年發展的雲電視更為我們提供了便捷的電視消費終端。智慧電視是集有線電視、通訊、互聯網三大功能於一體的三網融合業務，消費者不需要自己單獨配置互聯網上的內容，只要透過網路就可以實現社交、辦公、上網、閱讀、遊戲、購物等。

Smart TV has a large user base and can use the cable TV network to carry out more and more access-styled consumption. The development of cloud TV in recent years has provided us with a convenient terminal. Smart TV is a tri-networks integration business that integrates cable TV, communication and the Internet. Consumers do not need to configure the content on the Internet separately. They can socialize, work, surf online, read, play games and shop as long as network is accessible.

智慧電視有五大特點：①具有強大而先進的雲技術運用，可快速回應用戶需求，提供穩定、安全可持續化的雲服務技術，能為客戶帶來更好的體驗

以及交互性使用。②可以實現海量的資源存儲，還能實現遠端控制功能。③透過網線介面智慧電視可以實現 3D、LED、LCD 等觀看效果。④智慧電視在電腦的基礎之上，將技術融入現代化的電視設備裡，可對電視進行升級、維護、資源下載、軟體更新、雙向互動、N 屏互動、物聯生活以及「家庭雲」、「社交雲」、「娛樂雲」、「教育雲」等所有雲端家電的物聯。⑤提供穩定、安全、可持續的個性化線上雲服務。

Smart TV has five characteristics:

1. Because of the powerful and advanced cloud technology, it can respond quickly to users' needs and provide stable, secure and sustainable cloud service, providing better experiences and interactive uses for customers.

2. It has mass resource storage and can also realize remote control.

3. Through the network cable interface, smart TV can achieve 3D, LED, LCD and other viewing effects.

4. Based on computer technology, Smart TV integrates such technology into modern TV equipments in upgrading and maintaining the system, downloading resources, updating software, two-way interaction, N-screen interaction, IoT life and instrumentation of all cloud appliances such as family cloud, social cloud, entertainment cloud, education cloud.

5. It provides a stable, secure, sustainable and personalized online cloud service.

（3）閱讀消費終端 E-reader

電子閱讀器也是消費終端的一種創新，透過電子閱讀消費終端，人們可以隨時隨地讀到想瞭解的內容，暢通獲取，按需使用，同時也可以做為消費

終端，方便地實施消費。亞馬遜早在 2007 年 1 月推出的一款 Kindle 產品就有無線上網功能，可以隨時實地支援網上瀏覽、下載書刊，而不用與電腦連接。亞馬遜還為用戶設立私人圖書館帳戶，只要購買 Kindle，就購買了這本書的終身使用權，用戶可以隨時登錄亞馬遜網站免費下載買過的書籍。這樣，擁有一部 Kindle，就擁有了一個移動圖書館，可以享受隨時隨地閱讀的樂趣。為此，Kindle 成為一個改變全球閱讀習慣、影響億萬讀者的偉大的產品。據亞馬遜公布的資料，僅 2011 年 12 月 Kindle 設備銷量就突破了 400 萬臺，平均每週約 100 萬臺。[5]2009 年 Kindle 銷量突破 100 萬臺，電子書下載數量達到 1200 萬次，為公司帶來約 3.1 億美元的營收[6]。截至 2017 年 4 月 8 日，亞馬遜 (AMZN) 的股價達 894.88 美元，總市值達到了 4270 億美元，已超過美國百思買（Best Buy）、梅西百貨等八大零售商總和。

E-reader is also an innovation of consumer terminals. Through such consumption terminal, people can read the content they want anytime, anywhere smoothly, and use it as needed. It can also be used as a consumer terminal to easily consume. Kindle, introduced by Amazon in January 2007, could connect to the Internet wirelessly, thus supporting online browsing and downloading books at any time without connecting to a computer. Amazon also sets up a private library account for users. As long as they purchase a Kindle, they purchase the book

---

5 亞馬遜：12 月 Kindle 設備銷量超過 400 萬臺 . http://www.techweb.com.cn Amazon: Sales of Kindle devices exceeded 4 million in December. http://www.techweb.com.cn
6 亞馬遜電子書經營模式分析 . 出版發行研究 . 2009 年 6 月 Analysis on business model of Amazon e-book. Publishing research. June 2009

for life. Users can log in to Amazon.com at any time to download books they purchased free. In this way, owning a Kindle means you have a mobile library where you can enjoy reading anytime, anywhere. To this end, Kindle has become a great product that changes global reading habits and affects hundreds of millions of readers. According to data released by Amazon, in December 2011 alone, Kindle sales exceeded 4 million, an average of about 1 million per week.5 In 2009, Kindle sales exceeded 1 million, and the number of e-book downloads reached 12 million, bringing about $310 million in revenue for the company.6 As of April 8, 2017, Amazon (AMZN)'s share price reached 894.88 US dollars, the total market value reached 427 billion US dollars, exceeding that of the eight major retailers in the United States including Best Buy, Macy's.

（4）智慧商店消費終端 Smart store

智慧商店是基於雲計算的應用超市，實現了網站類應用的即點即用，讓網站建設更加容易。消費者不需要具有任何技術基礎，只需要滑鼠點幾下，就可以成功安裝一個應用程式。如當您有淘寶、京東、1號店等多個網店時，無須一個店一個店的分別打單發貨、同步庫存、商品上新、資料監控和分析和 CRM 會員管理，也無須購買伺服器，購買和安裝軟體，只需登錄（支援 PC、iPad、iPhone 等移動設備）智慧商店（雲商店），即可一次操作，同步全網。

Smart Store is a cloud-based application supermarket that enables point-and-click use of website-based applications, making website construction easier. Even unskilled consumers can successfully install an application with just a few mouse clicks. For example, when you have the access to multiple online shopping platforms such as Taobao, Jingdong, and No. 1 shop, you do not need to operate the orders or send out the goods, synchronize inventory, update products, monitor and analyze the data or manage the CRM member one platform after another one. Neither do you need to purchase servers or install software. All you need to do is to log in the smart store (the cloud store) that supports mobile devices such as PC, iPad, iPhone, etc., and then synchronize the whole network at one operation.

（三）雲支付 Cloud payment

1· 什麼是「雲支付」 What is cloud payment

消費「雲支付」指在現代資訊環境下，消費者可以利用任何支付工具（儲蓄卡、信用卡、智慧公交卡、手機儲值卡、支付寶網銀帳戶、消費儲值卡等），無障礙購買商品和服務，現金支付、信用支付、信貸支付一體化，支付便捷安全，且資金互通共用。我們認為，「雲支付」不僅是現代化的支付模式，也代表了現代金融服務創新的方向。

Cloud payment means that in the modern information environment, consumers can use any payment tool (savings card, credit card, smart transportation card, stored value card for mobile phone, Alipay online bank account, stored value card for consumption, etc.) to purchase goods and services barrier-free. Since cash payment, credit payment and loan payment become integrated, payment is convenient and safe, and funds can be shared now. We believe that cloud payment is not only a modern payment model, but also represents the direction of modern financial service innovation.

隨著消費者生活方式的改變，依託互聯網、移動互聯，藉由 3G 網路、雲計算等先進技術，未來傳統支付方式的市場份額將持續降低，貨幣電子化的程度將不斷加深，雲支付將成為支付的必然趨勢。未來支付將存在於雲中，支付方式將實現整合，儲蓄卡、信用卡、智慧公交卡、手機儲值卡、支付寶網銀帳戶、消費儲值卡之間資金可以互通共用，聯合支付；現金支付、信用支付、信貸支付一體化……消費者可以採用一切既便捷又安全的支付手段完成交易。

With the change of consumers' lifestyles, market share of traditional payment methods that rely on mobile network and advanced technologies such as 3G network and cloud computing will continue to decrease in the future, the digitalization of currency will continue to deepen, and cloud payment will become an inevitable trend of payment. Future payment will exist in the cloud, the payment methods will be integrated, that is, money in the savings card, credit card, smart transportation card, stored value card for mobile phone, Alipay online bank account, and stored value card for consumption can be shared. Cash payment, credit payment and loan payment become integrated. Consumers can use all convenient and secure payment methods to complete transactions.

2．「雲支付」市場迅速發展

Market of cloud payment is developing rapidly

在傳統支付方式的基礎上，電子支付、移動支付以及多種支付工具的融合應用，支付方式不斷拓展和創新，「雲支付」市場迅速發展，使消費者可

以利用任何支付工具（儲蓄卡、信用卡、智慧公交卡、手機儲值卡、支付寶網銀帳戶、消費儲值卡等），無障礙購買商品和服務，現金支付、信用支付、信貸支付一體化，支付便捷安全，且資金互通共用。

On the basis of traditional payment methods, with the integration of e-payment, mobile payment and various payment tools, payment methods continue to expand and innovate, market of cloud payment is rapidly developing, thus consumers can use any payment tools (savings card, credit card, smart transportation card, stored value card for mobile phone, Alipay online bank account, stored value card for consumption, etc.), to purchase goods and services barrier-free. Since cash payment, credit payment and loan payment become integrated, payment is convenient and safe, and money can be shared.

（1）電子支付市場規模不斷攀升

The scale of the electronic payment market continues to rise

電子支付源於美國，隨著 Interent 網路的發展，蔓延至世界各國。1998 年 3 月，中國第一筆網上電子交易成功。美國 2013 年度移動支付總額為 300 億美元，其中星巴克以 10 億美元的移動支付規模佔其 3%。據星巴克官方資料，2013 年美國已有超過 1000 萬星巴克消費者使用過移動支付服務，且該服務每週在美國境內的使用次數超過 500 萬次。星巴克移動交易額已佔其店內交易總額的 14%。[7]

E-payment originated in the United States and was spread to countries around the world with the development of the Internet. In March 1998, first online electronic transaction had taken place in China. The total amount of mobile payments in the United States in 2013 was $30 billion, of which Starbucks accounted for 3% with $1 billion. According to the official data, more than 10 million Starbucks consumers in the United States used mobile payment services in 2013, and the service has been used more than 5 million times a week in the United States. Starbucks mobile transactions accounts for 14% of its total in-store transactions.

Choose coffee-Barcode in the screen-Scan with the machine in Starbucks-The amount is deducted from the Starbucks account-More consumptions to become a senior member-More discounts-Free iTune music, e-books and APPs

圖 Figure1-3 星巴克移動支付流程 Process of mobile payment in Starbucks

（2）移動支付市場發展迅猛

The mobile payment market is developing rapidly

全球移動通信協會發佈報告指出，截至 2013 年 6 月底全球的活躍移動

---

7 星巴克移動支付業務大獲成功 全年交易額超 10 億．騰訊科技．2014-02-01 Starbucks mobile payment has achieved great success, the annual transaction volume exceeds $1 billion. Tencent Technology. 2014-02-01

支付帳戶數量超過 6000 萬，共有 5.3 萬戶商家透過移動支付接受付款，1.6 萬個組織使用移動支付做為接受帳單支付或支付薪水的平臺。到 2013 年年底，共在 84 個國家和地區推出 219 項服務，在 52 市場擁有兩個或兩個以上移動支付服務。[8]

According to a report released by the Global Mobile Communications Association, as of the end of June 2013, the number of active mobile payment accounts worldwide exceeded 60 million. A total of 53,000 merchants accepted mobile payment, and 16,000 organizations used mobile payments to pay for the bills or salaries. platform. By the end of 2013, 219 services were launched in 84 countries and regions, and two or more mobile payment services were available in 52 markets.[8]

據工信部提供的資料，截至 2013 年 9 月，中國行動電話用戶總數已突破 12 億戶，其中 3G 用戶佔比達 31%，移動互聯網用戶滲透率達到 67.9%。龐大的移動使用者群體，移動電子商務市場規模的不斷增加，使中國移動支付條件日益成熟。[9]2013 年協力廠商移動支付市場交易規模已突破萬億，達 12197.4 億元，同比增速 707%。[10] 資料顯示，2015 年中國有近 54% 的移動用戶進行過移動端購物，其中超過半數已使用移動購物（服務）應用三年或

8　GSMA：2013 年全球移動支付使用者規模突破 6000 萬 . CCTIME 飛象網 . 2014 年 2 月 28 日 GSMA: The Global Mobile Payment Subscribers Exceeded 60 million in 2013. CCTIME. February 28, 2014

9　中國行動電話用戶總數破 12 億 . 中國資訊產業網——人民郵電報 . 2013 年 10 月 23 日 The Total Number of Mobile Phone Users in China Has Exceeded 1.2 Billion. CNII Paper. October 23, 2013

10　iResearch 艾瑞諮詢 iResearch Consultation

更久，這些用戶中，平均每週使用移動購物（服務）應用一次以上的比例達22.2%。

According to the data provided by the Ministry of Industry and Information Technology, as of September 2013, the total number of mobile phone users in China has exceeded 1.2 billion, of which 3G users accounted for 31%, and mobile Internet users penetration rate reached 67.9%. The huge mobile user group and the increasing scale of the mobile e-commerce market have made China Mobile's payment conditions more and more mature.[9] In 2013, the transaction volume of the third-party mobile payment market has exceeded one trillion, reaching 1,219.74 billion yuan, a year-on-year growth rate of 707%.[10] According to the data, in 2015, nearly 54% of mobile users in China had mobile shopping, and more than half of them had used the mobile shopping (service) application for three years or more. Among these users, 22.2% of them used such service more than once every week.

近年中國移動智慧終端機的普及及其網路化和便攜性優勢為隨時隨地購物創造了條件，在電商平臺的支持和推動下，廣大消費者已逐漸適應並習慣於移動消費體驗。

In recent years, the popularity of China's mobile intelligent terminals and its networking and advantage of portability enabled people to shop anytime, anywhere. With the support and promotion of e-commerce platforms, consumers have gradually adapted and become accustomed to the mobile consumption.

據艾媒諮詢 (iiMedia Research) 發佈的《2015-2016 中國移動電商市場年度報告》，截至 2015 年底，中國移動購物使用者規模達到 3.64 億人，同比

增長 23.8%，中國移動網購市場交易規模達 2.1 萬億元，同比增長 123.8%，（2015 年中國網路購物市場交易規模為 3.8 萬億元，同比增長 36.2%），中國移動網購市場交易規模佔比網路購物市場交易總規模已經超過 55%，且增速遠高於中國網路購物整體增速。2018 年中國移動電商使用者規模將接近 5 億人，交易規模將超過 5 萬億元。移動終端已成為中國消費者線上消費的首選管道。移動購物的普及也加速促進了網購 App、消費金融、快遞 / 物流和移動支付等相應配套服務體系的豐富和完善。

According to Annual Report on 2015-2016 China Mobile E-Commerce Market released by iiMedia Research, as of the end of 2015, the number of China Mobile shopping users reached 364 million, a year-on-year increase of 23.8%. The transaction scale of China Mobile online shopping market reached 2.1 trillion yuan, a year-on-year increase of 123.8%. (In 2015, the transaction scale of China online shopping market in China was 3.8 trillion yuan, a year-on-year increase of 36.2%.) China Mobile online shopping market accounted for more than 55% of the total online shopping market. The growth rate is much higher than the overall growth rate of China online shopping. In 2018, the number of mobile e-commerce users in China will be close to 500 million, and the transaction scale will exceed 5 trillion yuan. Mobile terminals have become the preferred channel for online consumption of Chinese consumers. The popularity of mobile shopping has also accelerated the enrichment and improvement of corresponding supporting service systems such as online shopping Apps, consumer finance, express/logistics and mobile payment.

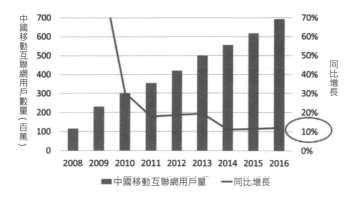

Number of China Mobile Internet Uers (Unit: Million)

Year-on-year growth

灰 Number of China Mobile Internet Uers　深灰 Year-on-year growth

圖 Figure 1-4　中國移動互聯網用戶數量及同比增長，2008 年 -2016 年 China Mobile Internet users and year-on-year growth, 2008-2016
資料來源：CNNIC。統計資料選取截至年底。Source: CNNIC。 The statistics were collected as of the end of the every year.

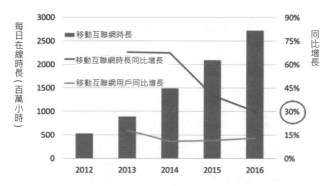

Daily online time(Unit: million hours)

Year-on-year growth

灰 Daily online time 深灰 Year-on-year growth of daily online time

下方淺灰 Year-on-year growth of number of China Mobile Internet Uers

圖 Figure1-5　中國移動互聯網用戶每日線上時長（百萬小時），2012 年 -2016 年 Daily online time of China Mobile Internet users (unit: million hours), 2012-2016
資料來源：高瓴資本基於 ZenithOptimedia 的每日媒體消費時間資料和 QuestMobile 的移動資料估算。Source: Hillhouse Capital's estimation based on ZenithOptimedia's data of daily media consumption time and QuestMobile's mobile data.

隨著移動支付技術的不斷發展，移動支付廣泛應用於各類生活場景，如購買數位產品（鈴聲、新聞、音樂、遊戲等）和實物產品、公共交通（公共汽車、地鐵、計程車等）、票務（電影、演出、展覽等）、公共事業繳費（水、電、煤氣、有線電視等）、現場消費（便利店、超市等）。

With the continuous development of mobile payment technology, the mobile payment is now widely used in various life scenes, such as the purchase of tangible and digital products (ringtones, news, music, games, etc.), public transportation (bus, subway, taxi, etc.), ticketing (movies, performances, exhibitions, etc.), public utilities (water, electricity, gas, cable TV, etc.), on-site consumption (convenience stores, supermarkets, etc.).

（3）多種支付工具的融合應用

Integrated application of multiple payment tools

除此之外，騰訊微信支付、交通卡支付、養老券購物、消費券購物等多種支付工具的融合應用，進一步創新和拓展了支付方式和範圍，使消費者享受到「雲支付」的方便快捷，即可以採用一切既便捷又安全的支付手段完成交易。

In addition, Tencent WeChat payment, transportation card payment, pension voucher shopping, consumption voucher shopping and other payment tools have been integrated to further innovate and expand the payment methods and scope, so that consumers can enjoy the convenience of cloud payment, that is, you can use all the convenient and safe payment methods to complete the transaction.

（3.1）騰訊微信支付 Tencent WeChat payment

騰訊公司的知名移動社交通訊軟體「微信」及協力廠商支付平臺財付通

聯合推出的移動支付創新產品，為廣大微信用戶及商戶提供更優質的支付服務，透過在支付方式上的創新，讓消費者享受到「雲支付」的方便快捷。微信支付主要包括線下掃碼支付、WEB 掃碼支付、公眾號支付三種方式，使用者只需在微信中關聯一張銀行卡，並完成身分認證，即可將裝有微信 APP 的智慧手機變成一個全能錢包，之後即可購買合作商戶的商品及服務，使用者在支付時只需在自己的智慧手機上輸入密碼，無須任何刷卡步驟即可完成支付，整個過程簡便流暢。

Tencent's well-known mobile social communication software "WeChat" and third-party payment platform Tenpay jointly launched mobile payment innovation products to provide better payment services for WeChat users and merchants. With innovation in payment methods, consumers can now enjoy the convenience of cloud payment. WeChat payment mainly includes offline scanning code payment, WEB scanning code payment, and offical account payment. Users only need to associate a bank card with WeChat and complete the identity authentication to change the smart phone installed with WeChat into an all-round wallet. Only after that can consumers purchase the goods and services of the partner merchants. Users only need to enter the password on their smart phones when paying, and the payment can be completed without any card-swapping steps. The whole process is simple and smooth.

據報導，2013 年春節 8 天，參與騰訊微信紅包活動用戶超過 800 萬，超過 4000 萬個紅包被領取，微信支付綁卡用戶數從 2000 萬增長過億。2014 年 1 月 10 日至 2 月 9 日，嘀嘀打車平均日微信支付訂單數 70 萬單，總微信支付訂單約 2100 萬單，補貼總額高達 4 億元。[11] 2016 年微信支付接入線下門

店超過 30 萬家，2017 年中國微信社交服務積累的使用者已達 10 億，2017 年第二季度微信支付交易份額佔中國協力廠商移動市場總規模的 39.12%。[12] 支付寶、微信支付、壹錢包、連連支付等協力廠商移動支付的快速發展，使得中國很多地區已初步實現「無現金生活」理念。

According to reports, in the 8th day of the Spring Festival in 2013, more than 8 million users of Tencent WeChat participated in the red-pocket events, more than 40 million red pockets were opened, and the number of card-associating Wechat users increased from 20 million to more than 100 million. From January 10th to February 9th, 2014, the average number of daily WeChat payment orders of Didi car-hailing was 700,000, and the total number of WeChat payment orders was about 21 million, with a total subsidy of 400 million yuan.[11] In 2016, there were more than 300,000 offline outlets supporting WeChat payment. In 2017, number of users using WeChat social services accumulated 1 billion in China. In the second quarter of 2017, WeChat payment transactions accounted for 39.12% of the total size of China's third-party mobile market.[12] The rapid development of third-party mobile payment such as Alipay, WeChat payment, e-Wallet, and LianLian Pay has made the concept of 「cash-free life」 come true in many areas of China.

（3.2）交通卡支付 Transportation card payment

交通卡支付在小額支付領域具有廣泛的消費基礎、龐大的用戶群體、完

---

11 截至 2014 年 2 月 9 日微信支付嘀嘀打車補貼額達 4 億. 中國電子商務研究中心. 2014 年 02 月 12 日

12 根據 Analysys 易觀發佈的 2017 年度中國協力廠商支付移動支付市場報告整理 Data source: report on 2017 China third-party payment mobile payment market released by Analysys

善的支付環境、成熟的結算平臺、豐富的管理經驗、強大的宣傳能力等優勢，是消費「雲支付」的重要載體。

Traffic card payment has the advantages of extensive consumption base, large user group, perfect payment environment, mature settlement platform, rich managerial experience and strong propaganda ability in micropayment. It is an important carrier of cloud payment.

上海市在交通卡整合支付方面走在全國前列，上海公共交通卡目前支持軌道交通、計程車、輪渡、貨的運輸、高速公路收費、旅遊交通、停車場、咪表、水電煤公用事業付費、加油站、長途汽車客運等，還成功地實現了與常熟公交、無錫公交和部分出租、蘇州公交、杭州部分出租、安徽阜陽公交、廣西南寧海博出租、昆山公交等應用的對接，擁有龐大的用戶群個體和廣泛的使用環境。

Shanghai is in the forefront of the nationwide transportation card payment. The Shanghai Public Transport Card currently supports rail traffic, taxis, ferries, cargo transportation, highway tolls, tourist transportation, parking lots, parking meters, utilities including water, electricity and coal, Gas stations, long-distance passenger transport, etc. It also successfully achieved docking with public transportation in Changshu, Wuxi, Suzhou, Hangzhou, Fuyang in Anhui province, Kunshan and part of cab in Wuxi, Hangzhou, Haibo of Nanning in Guangxi province, etc. It has a huge user group and an extensive applying environment.

（3.3）養老券購物 Pension voucher shopping

養老券購物也是「雲支付」方式的創新。深圳於 2005 年啟動對部分戶籍老人實行貨幣化居家養老補助，2009 年改發現金為發券，老人憑券向定點

服務機構購買服務。後來發放養老券模式也在北京、上海、長沙、成都等地推開。2010 年 1 月 1 日北京市面向 38 萬符合標準的老年人和殘疾人發放養老助殘券，用於購買居家服務。2012 年年底長春更進一步嘗試推出養老卡電子結算，類似於公交 IC 卡，每月政府存入 200 元供老人持卡消費。

Pension voucher shopping is also an innovation in cloud payment. In 2005, Shenzhen initiated the monetized home-based subsidies for part of the aged with registered residence. In 2009, such subsidies were changed into vouchers, and the aged then purchased the services from the designated service agencies. Later, the pension voucher model was also opened in Beijing, Shanghai, Changsha and Chengdu. On January 1, 2010, Beijing issued old-age disabled vouchers for 380,000 eligible elderly and disabled people to purchase residential services. At the end of 2012, Changchun further tried to introduce electronic settlement of pension cards, which were similar to bus IC cards. The government deposits 200 yuan per month for the aged to use card consumption.

（3.4）消費券購物 Consumer voucher shopping

消費券做為實現經濟政策的工具之一，通常用於當經濟不景氣導致民間消費能力大幅衰退時，由政府或企業發放給民眾，做為特定消費的支付憑證，期待藉由增加民眾的購買力與消費慾望的方式以拉動消費、活躍市場。目前也逐漸被用於消費者的消費支付。2009 年 150 家淘寶網店共同聯手，透過淘寶網官方雜誌向消費者派發總值高達 10 億元的電子消費券。這些消費券包括 5 － 8 折不等的優惠券、面值 2 － 300 元不等的代金券以及祕密的 1 元秒殺資訊，消費者可使用這些「消費券」在淘寶網制 (指) 定的商戶中購物。2010 年西城區什剎海文化旅遊節免費發放 1800 萬元消費券，消費券包括電

影票、西單各大商場的購物券、本市各大旅遊景區門票、什剎海酒吧街優惠券等。

As one of the tools for realizing economic policies, consumption vouchers are usually used by the government or enterprises to issue to the public when the economic downturn causes a sharp decline in private consumption power. As the payment credential for specific consumption, it is expected to stimulate consumption and activate the market by increasing the purchasing power and consuming desires of people. It is also gradually being used for ordinary payment now. In 2009, 150 Taobao stores joined forces to distribute electronic coupons worth up to 1 billion yuan to consumers through Taobao's official magazine. These vouchers include coupons ranging from 5 to 8 fold, cash coupons ranging from 2 to 300 yuan, and secret 1 yuan seckilling. Consumers can use these vouchers to shop in merchants designated by Taobao. In 2010, Shichahai Cultural Tourism Festival in Xicheng District issued free coupons worth up to 18 million yuan. The coupons included movie tickets, shopping vouchers for major shopping malls in Xidan, tickets for major tourist attractions in the city, and discount coupons for Shichahai Bar Street.

3．「雲支付」技術不斷創新

Cloud payment technology continues to innovate

隨著消費者支付方式的多元化，雲支付市場迅速發展，雲支付技術也在不斷創新，出現了 NFC（近距離無線通訊）非接觸式支付、二維碼支付、超聲波支付、手機外設刷卡器支付、雲 Key（祕鑰）支付、指紋識別支付等新

型的支付技術，使消費者更加體驗到「雲支付」的好處。

With the diversification of payment methods for consumers, the cloud payment market is developing rapidly, and cloud payment technology is constantly innovating. New payment technologies such as NFC (near field communication) contactless payment, QR code payment, ultrasonic payment, and peripheral card reader for mobile phone, cloud key payment, and fingerprint identification payment appear, all of which enable consumers to experience the benefits of the cloud payment.

（1）NFC（近距離無線通信）非接觸式支付

NFC (near field communication) contactless payment

NFC（近距離無線通訊）非接觸式支付是一種短距高頻的無線電技術，由非接觸式射頻識別（RFID）及互聯互通技術整合演變而來，在單一晶片上結合感應式讀卡器、感應式卡片和點對點的功能，能在短距離內與相容設備進行識別和資料交換。NFC 可以透過結合無線優惠券、會員卡和支付選擇擴展和提升現代購物體驗。消費者可以用個人應用程式掃描產品貨架上的 NFC 標籤，獲得關於該產品更加個性化的資訊。舉個例子，如果某人對堅果過敏，透過掃描產品，他的 NFC 設備就能自動檢測出該產品是否含有堅果並及時做出提醒。透過觸碰 NFC 標籤來獲得資訊、增加到購物籃、獲得優惠券和其他新的用途將對零售業產生越來越大的影響。[13]

NFC (Near Field Communication) contactless payment is a short-range high-frequency radio technology that evolved from the integration of radio frequency identification devices (RFID) and interoperability technologies, combining inductive card readers, inductive cards and peer-to-peer functions on a single chip, which can conduct identification and data exchange with compatible devices over short distances. NFC can extend and enhance the modern shopping experience by combining wireless coupons, membership cards and payment options. Consumers can use a personal application to scan NFC tags on product shelves for more personalized information about the product. For example, if someone is allergic to nuts, by scanning the tag of the product, his NFC device can automatically detect if the product contains nuts and promptly alert. Getting information, adding products into the shopping basket, getting coupons and other new usages, all by scanning NFC tags, will have an ever-increasing impact on the retail industry.[13]

圖 Figure1-6 西方超市中的 NFC 近距離資料交換應用 Application of NFC close-range data exchange in western supermarkets

　　著名的 Google 錢包（Google Wallet）就是使用 NFC 技術，透過在智慧手機和收費終端內植入的 NFC 晶片完成信用卡資訊、折扣券代碼等資料交換，透過智慧手機打造從團購折扣、移動支付到購物積分的「一站式」零

13　NFC 技術有何新玩法．手機之家 What's new in NFC technology? iMobile

售服務。目前 Google 錢包是完全免費開放的平臺，用戶只需在結帳臺支援 PayPass 的終端機即可用手機付帳，它可以讓用戶的手機變成錢包，將塑膠信用卡存儲為手機上的資料，還能加上各種優惠資訊。

The famous Google Wallet uses NFC technology to complete data exchange of credit card information, discount coupon code, etc. through NFC chips embedded in smart phones and charging terminals. Such technology is also used in smart phones to create the "one-stop" retail service from group purchase discounts, mobile payments to shopping points. At present, Google Wallet is completely free and open. Users only need to pay by the mobile phone through the terminal supporting PayPass at the checkout counter. It can turn the user's mobile phone into a wallet and store the plastic credit card as data on the mobile phone with special offers.

國內也有很多 NFC 技術應用的例子，如上海市所有的地鐵站都配備有相應設備，使用者手持具備 NFC 功能的手機，並與銀行卡綁定，在閘機上揮揮手機就可以完成支付，輕鬆進出站。2017 年北京市已普及刷手機坐地鐵，凡是內置 NFC 功能的蘋果 iPhone 以及安卓手機都可以輕鬆刷機進站，從 4 月 29 日起，除西郊線外北京地鐵全路網均實現刷二維碼乘車。

There are also many examples of NFC technology applications in China. For example, all subway stations in Shanghai are equipped with corresponding devices. Users can now use mobile phones with NFC bound with bank cards to complete payment by waving them against the gate machine. In 2017, Beijing has completed the popularization of such way to take the subway. Passenger holding iPhone and Android phone with built-in NFC can easily enter the station. From April 29th, except for the Western Suburban Line, the whole Beijing subway has supported QR Code to buy the tickets now.

（2）二維碼支付 QR code payment

二維碼支付是一種基於帳戶體系搭起來的新一代無線支付方案。在該支付方案下，商家可把帳號、商品價格等交易資訊彙編成一個二維碼，並印刷在各種報紙、雜誌、廣告、圖書等載體上發佈。使用者透過手機用戶端掃描或拍攝二維碼，便可實現與商家支付寶帳戶的支付結算。商家根據支付交易資訊中的使用者收貨、聯繫資料，就可以進行商品配送，完成交易。[14] 早在 2002 年日本的運營商就開始推廣二維碼。目前日本是全球二維碼使用最多的國家，其次是美國。

QR code payment is a new generation wireless payment solution based on the account system. The merchant can compile the transaction information such as the account number and the commodity price into a QR code, and print it on various carriers such as newspapers, magazines, advertisements, books, and the like. User can scan and record the QR code through the mobile app to finish the payment to merchant' s Alipay account. According to the receipt and contact information in the transaction information, merchant can carry out the goods distribution and complete the transaction.[14] As early as 2002, operators in Japan began to promote QR code. At present, Japan used QR code the most in the world, followed by the United States.

---

14 支付寶將推二維碼支付. 騰訊網. 2014 年 3 月 Alipay will Push the QR Code for Payment. Tencent. March 2014

在這方面，國內外很多企業已經進行了大量探索：

2011 年 10 月，支付寶推出中國國內首個推出二維碼支付解決方案，利用手機識讀支付寶二維碼，實現用戶即時支付功能，幫助淘寶電商發展空間從線上向線下延伸。透過該方案，商家可把帳戶、價格等交易資訊編碼成支付寶二維碼，並印刷在各種報紙、雜誌、廣告、圖書等載體上發佈；使用者使用手機掃描支付寶二維碼，便可實現與商戶支付寶帳戶的支付結算，支付方便快捷。

In this regard, many companies at home and abroad have conducted a lot of exploration:

In October 2011, Alipay launched the first domestic QR code payment solution in China, using the mobile phone to read Alipay QR code to realize the instant payment, helping Taobao e-commerce development to extend from online to offline. Through this program, the merchant can encode the account, price and other transaction information into the Alipay QR code, and print it on various newspapers, magazines, advertisements, books and other carriers. User can use the mobile phone to scan the Alipay QR code to finish the payment to merchant' s Alipay account, making the process convenient and quick.

2012 年初，國際線上支付巨頭 PayPal 在全國 WiFi 免費的新加坡地鐵站試驗二維碼閱讀應用支付，此項試驗在新加坡地鐵公司（SMRT）的 15 個地鐵站可允許手機用戶透過拍下商品的二維碼來購物。

In early 2012, international online payment giant PayPal experimented with the QR code reading application payment at the WiFi-free Singapore subway stations. In this trial, 15 stations of the Singapore Mass Rapid Transit (SMRT) allowed passengers to shop by scanning the QR code.

圖 Figure1-7　PayPal 新加坡地鐵站購物 PayPal Shopping in Singapore MRT

　　美國一家知名地理資訊遊戲創業公司 SCVNGR 推出一款基於本地移動支付和獎勵服務應用的 LevelUp，該應用用戶可以透過運行在 iPhone 或者安卓手機上的應用掃描 QR 碼完成支付。iPhone 和安卓手機用戶只需把自己常用的信用卡或者借記卡綁定到 LevelUp 應用中，就可以獲得自己獨一的 QR 二維碼，購物時候只需在公司 1400 個定點合作商戶中掃描即可。交易完成後，使用者會收到一封電子郵件收據。2011 年，SCVNGR 公司透過 LevelUp 生成的交易總額達到 100 萬美元，2012 年用戶的月交易額就突破 100 萬美元，用戶參與度每 5 － 6 週就翻 1 倍。

　　SCVNGR, a well-known geographic information game startup in the United States, launched the LevelUp, a local mobile payment and incentive service app that allows users to scan the QR code to complete payments via an app running on an iPhone or Android phone. iPhone and Android phone users only need to bind their favorite credit card or debit card to the LevelUp app so that they can get their own unique QR code, which they only need to scan in the company's 1400 designated partners to shop. After the transaction is completed, the user will receive an email receipt. In 2011, SCVNGR generated a total of $1 million in transactions through LevelUp. In 2012, the monthly transaction volume exceeded $1 million, and user engagement doubled every 5 to 6 weeks.

圖 Figure1-8　Level Up 二維碼支付 LevelUp QR code payment

（3）超聲波支付 Ultrasonic payment

超聲波支付的技術近似於 NFC 技術，利用超聲波讓手機透過麥克風和揚聲器就能完成一次近場通信，不必依賴專用的晶片，而且使用者體驗一致。兩支手機「碰一碰」，通訊就完成。但是資料傳輸量有限，速度較慢，在對資料加解密傳輸上偏弱。

Ultrasonic payment is similar to NFC, it allows the mobile phone to complete a near field communication through the microphone and speaker, without relying on a dedicated chip, plus the user experience can also be consistent. Put the two mobile phones together just once and the communication is completed. However, the drawbacks are that amount of data transmitted is limited, speed is slow, and encryption and decryption of data come up weak.

2011 年美國公司 Naratte 開發的一種名為「Zoosh」的技術，就是一種典型的超聲波支付技術。該技術利用超聲波進行安全短距離點對點傳輸，可用

來進行移動支付。這種技術和 NFC 擁有一樣的速度，但慢於 WiFi 和藍牙（因聲波遠慢於電磁波）。[15] 該技術對比 NFC 有兩大優點：首先是不需要在硬體上花費額外的費用，不需要另外加裝晶片。只需要對手機的話筒和揚聲器做一些改造，費用不超過 1 美元。第二是不管手機新舊都可使用，對手機性能沒有特殊要求。根據同樣原理，南京音優行資訊技術有限公司開發了一種名為「即付通」的產品，其核心技術被稱為「迅音」，能夠讓具有喇叭、麥克風的終端設備透過聲波進行資料通信，可以順利完成支付。

Zoosh, developed by American company Naratte in 2011, is a typical ultrasonic payment technology. The technology uses ultrasound for secure short-range point-to-point transmission and can be used for mobile payments. It is as quick as NFC, but slower than WiFi and Bluetooth (since sound waves are much slower than electromagnetic waves).[15] This technology has two major advantages over NFC: First, there is no extra expense on the hardware and no additional chips are required. You only need to make some modifications to your phone's microphone and speaker for no more than $1. The second is that whether the phone is old or new if fine and there is no special requirement for the performance of the phone. According to the same theory, Yinyouxing Information Technology Co., Ltd. in Nanjing has developed a product named Gifront. Its core technology is called Fast Sound, which enables terminal devices with speakers and microphones to transmit data through sound waves, thus successfully completes the payment.

---

15 IOTer. 美國開發出低成本 NFC 通訊 Zoosh 技術 . 物聯網線上 . 2011-06-21   IOTer. The United States has developed low-cost NFC Communication Technology Zoosh. IoT online. June 21, 2011

（4）手機外設刷卡器支付 Peripheral card reader for mobile phone

手機刷卡器，類似一款外接讀卡器主要是讀取磁條卡資訊的工具，透過 3.5mm 音訊插孔來傳輸資料的。手機刷卡器本身沒有支付的功能，要有支付通道的軟體來配合才可以有支付、收單的功能。手機刷卡器分為簡易型手機刷卡器、加密手機刷卡器、密碼鍵盤手機刷卡器、EMV 手機讀卡器和 NFC 手機讀卡器。[16]

The peripheral card reader for mobile phone, similar to an external card reader, is mainly a tool to read the information in a magnetic stripe card and transfers data through the 3.5mm audio jack. The card reader itself does not have the function of payment, payment software must be matched to realize such function. The mobile phone card readers are classified into five types, namely the simple type, the encrypted type, the password-keyboard type, the EMV type and the NFC type.[16]

美國一家移動支付公司 Square 公司開發的移動讀卡器就是一種典型的手機外設刷卡器支付方式。該讀卡器配合智慧手機使用，可將信用卡磁條的資訊轉換成音訊，然後 iPhone、安卓的 Square 應用會把音訊再轉換成數位資訊，之後把這些付款資訊用加密的方式傳輸到伺服器端，再返回刷卡是否成功的資訊。透過應用程式匹配刷卡消費，它使得消費者、商家可以在任何地方進行付款和收款，並保存相應的消費資訊，從而大大降低了刷卡消費支付的技術門檻和硬體需求。

16　百度百科 Baidu Encyclopedia

A mobile card reader developed by Square, a US mobile payment company, is a typical peripheral card reader for mobile phone. The card reader is used with a smart phone to convert the information in magnetic strip of the credit card into audio, and then the Square app installed in iPhone and Android would convert the audio into digital information, and then transmit the payment information to the server in an encrypted manner. Finally, it sends out the notification of whether the payment is successful or not. With the app matching the payment information, it enables consumers and merchants to make payments and collections anywhere, and saves corresponding consumption information, thereby greatly reducing the technical threshold and hardware requirements for such payment.

中國深圳盒子支付資訊技術有限公司開發的盒子支付（iboxpay.com）技術類似於 Square，並進行了微創新。該技術致力於讓使用者的各種移動終端變成隨身的 POS 機和閱讀器，從而能夠隨時隨地在用戶的手機上進行信用卡還款、水電、煤氣、電話繳費、手機充值、購物等支付交易。

The iBoxpay, developed by iBoxpay Information Technology Co., Ltd. in Shenzhen, China, is similar to Square and has undergone micro-innovation. The technology is dedicated to making the user's various mobile terminals into portable POS machines and readers, so that payment transactions such as credit card repayment, utility fee, mobile phone recharge and shopping can be performed on the user's mobile phone anytime and anywhere.

（5）雲 Key（祕鑰）支付

雲 Key 支付是使用者可以直接透過網路申請電子帳戶，而這些帳戶透過 AES 加密保存在個人移動設備裡，而鑰匙卻是在雲端，鑰匙採用分散式金鑰。

每次交易時，透過手機去雲端獲取一個臨時鑰匙，當收款方獲取這個鑰匙之後，送給雲端進行解密支付。由於每次獲取的鑰匙不同，你如果截取，即便破解了，也無法使用，因為每次交易的 KEY 都不一樣，所以交易安全性得到保障。

Cloud key payment allows users to apply for electronic accounts directly through the network, and these accounts are stored in personal mobile devices through AES encryption, while the keys are kept in the cloud and are distributed. At each transaction, the mobile phone goes to the cloud to obtain a temporary key. When the payee obtains the key, he sends it to the cloud for decryption. Since the keys acquired each time are different, even if you intercept and crack them, you still cannot use them because the key of each transaction is different. Thus transaction security is guaranteed.

Keypasco（線上支付安全性與雲端身分認證）提供的認證方法是：在用戶名和密碼的雙因素基礎上，增加了可綁定的個人終端設備 ID、地理位置定位，甚至是上線時間，再加上消費者行為分析相關的風險評估機制（可以分析其他使用者是否嘗試透過其他終端登錄使用者帳戶），採取多重因素認證方式提升安全性。為了安全因素與保護客戶隱私，Keypasco 運用雲端資源，透過加密的分散式存放等方式存放使用者登記資訊。用戶只有透過所綁定的一臺或多臺終端，在自己提前設定的地理區域，以自己的身分登錄，所有的訪問和嘗試登錄日誌都會保留在系統中，使用者可以方便查詢。由於這一認證方式基於軟體和雲服務實現，成本極低，在沒有大規模增加成本的前提下提高了整個行業的安全水準。使用者第一次使用某一終端簡單的註冊登錄即

被賦予一個唯一的 Keypasco ID，登錄設備也被賦予一個唯一的設備 ID，兩個 ID 及設備本身在今後的登錄中即被關聯，其後進入系統可設定綁定其他設備、使用區域等。提供該服務的支付機構或網路服務提供機構可以在其中設定自己的風險規則，確定某些情況下的風險閾值和使用要求。可以說，Keypasco 基本在不太改變用戶現有習慣的基礎上，以雲端認證單點登錄的方式，確保了使用者身分的合法性。[17]

Authentication method provided by Keypasco (Online Payment Security and Cloud Identity Authentication), based on the two factors as the user name and password, is to add bindable personal terminal device IDs, geographic location, and even online time plus a risk assessment mechanism related to consumer behavior analysis which can analyze whether other users try to log in to the user account through other terminals. It adopts multiple factors authentication to improve security. For security reasons and to protect user's privacy, Keypasco uses cloud resources to store user registration information through encrypted decentralized storage. Users can log in as their own identity only in the geographical area set by themselves in advance through one or more terminals that have been bound. All access and attempted login logs will remain in the system where users can easily query. Because this authentication method is based on software and cloud services, the cost is extremely low, and the security level of the entire industry is improved without large-scale cost increase. The first time a user registers and logs in with a

---

[17]　線上支付安全性與雲端身分認證 Keypasco.PayCircle 支付圈 . 2013 年 4 月 Online Payment Security and Cloud Identity Authentication Keypasco.PayCircle. April 2013

terminal, a unique Keypasco ID is assigned. The login device is also given a unique device ID. The two IDs and the device itself are associated. In the future login, user can bind the account with other devices, login areas, etc. The payment institution or network service provider can set its own risk rules to determine the risk threshold and usage requirements in certain situations. It can be said that Keypasco basically ensures the legality of the user's identity by means of cloud authentication single login on the basis of not changing the user's existing habits.[17]

（6）指紋識別支付

指紋識別支付，就是將使用者的指紋資訊資料與指定銀行帳戶相互綁定，當用戶購物、消費後，伸出手指在指紋識別終端中掃描，確認是本人後，便可輕鬆完成支付，消費的金額會在對應的銀行帳戶中扣除。[18]

The fingerprint identification payment is to bind the fingerprint information data of the user with the designated bank account. To complete the payment, user only needs to put his or her finger on the fingerprint identification terminal to confirm whether it is the person. The amount of payment will be deducted from the corresponding bank account.[18]

早在 2003 年，美國就有三家公司開始在超市和商場裡推廣它們的指紋支付系統。2006 年，英國成為第一個使用指紋支付技術的歐洲國家。指紋支付之所以傳播如此之快，很大程度上是因為省時，指紋支付的整個過程只需 5 秒鐘就能搞定，比現金和刷卡消費節約 40 秒鐘左右。由於指紋資訊具有獨

---

18   林可. 科技購物之旅, 等你出發 [J]. 經理日報 . 2010-01-03 Lin Ke. Technology Shopping Trip is Waiting for You [J]. Manager Daily. Jan 3, 2010

一無二的特點，因此整個支付過程十分安全便捷。

As early as 2003, three companies in the United States began to promote their fingerprint payment systems in supermarkets and shopping malls. In 2006, the UK became the first European country to use fingerprint payment technology. The reason why fingerprint payment is spreading so quickly is largely because it saves time, as the whole process of fingerprint payment can be completed in only 5 seconds, saving about 40 seconds compared with cash and credit card. Because the fingerprint information has unique characteristics, the entire payment process is very safe and convenient.

中國工商銀行、建設銀行、交通銀行、招商銀行等銀行已相繼成為指紋支付業務的合作銀行。2007 年各家銀行的「指付通」業務已有近 10 萬用戶。2014 年支付寶錢包和華為手機共同推出國內首個指紋支付的標準方案。

Banks such as Industrial and Commercial Bank of China, China Construction Bank, Bank of Communications, and China Merchants Bank have successively become cooperative banks for fingerprint payment services. In 2007, there were already nearly 100,000 PayByFinger users of the aforementioned banks. In 2014, Alipay Wallet and Huawei Mobile launched the first standard solution for fingerprint payment in China.

三

「雲消費」時代是一個全產業革命的時代
3. The era of "cloud consumption" is an era of all-industry revolution

伴隨著現代資訊互聯技術的迅猛發展，互聯網經濟的強勁增長，應對新一輪零售革命的浪潮，我們進入了一個全新的時代──「雲消費」時代，「雲消費」時代也是全產業革命的時代。

With the rapid development of modern information interconnection technology, the strong growth of the Internet economy, and the wave of a new round of retail revolution, we have entered a new era-the era of cloud consumption, which is also the era of all-industry revolution.

（一）消費領域面臨三大核心變化──「零時差、零距離、零管道」
The consumption sector faces three core changes-zero time difference, zero distance and zero channel

在「雲消費」時代，我們面臨著以零售革命為先導的全產業的革命，傳統商業模式正經歷著前所未有的挑戰和變化。這種變化不再是不同產業的此

消彼長，而是整個產業體系的全面變革和洗牌。在消費領域，消費突破了傳統店鋪的限制，突破了商品型態的限制，流通的時間障礙、距離障礙、管道環節障礙也已完全突破，呈現出「零時差、零距離、零管道」之趨勢。

In the era of cloud consumption, we are facing an all-industry revolution led by the retail revolution. The traditional business model is experiencing unprecedented challenges and changes. Such changes are no longer the trade-off between different industries, but the overall transformation and reshuffle of the entire industrial system. In the consumption sector, it has broken through the limitations of traditional stores, commodity forms, and the time barriers, distance barriers, and channel barriers of circulation, showing a trend of zero time difference, zero distance and zero channel.

以「雲終端、雲內容、雲支付」為特徵，虛擬零售市場徹底改變了實體零售業有限的空間市場和時間時差，原則上只要網線鋪到哪裡，借助互聯網和無所不有的移動商店——智慧手機，那裡的人們就可以享受到與大城市人們一樣的購物機會，觸摸到網上海量的商品，24 小時隨時隨地購物已成為很多人的尋常的生活方式。只要有網路和物流支援，西藏阿里的消費者能夠得到與北京、上海消費者同等的購物享受。事實上，2013 年，據支付寶有關統計，中國移動支付排名前十的地區全部位於青海、西藏、內蒙古等邊疆少數民族地區，西藏地區全年的淘寶訂單中，有 29.1％就是透過手機支付完成的。

[19]

---

[19] 支付寶年度對帳單：人均網上總支出超出萬元 . 新浪科技 .2014-01-13 Alipay annual statement of account: total online spending per capita exceeds 10,000 yuan. Sina Tech. Jan 13, 2014

Characterized by cloud terminal, cloud content and cloud payment, the virtual retail market has completely changed the limited space and time difference of the physical retail industry. In principle, as long as the network cable is laid, with the help of the Internet and smart phones,the omnipotent mobile store, people can enjoy the same shopping opportunities as people in big cities, and get access to vast amount of goods on the Internet. Shopping anytime and anywhere has become an ordinary lifestyle for many people. As long as there is network and logistics support, Ali consumers in Tibet can enjoy shopping as those in Beijing and Shanghai do. In fact, in 2013, according to statistics of Alipay, the top ten regions of mobile payments in China were all located in frontier minority areas such as Qinghai, Tibet, Inner Mongolia, etc. In the Tibet region, 29.1% of the annual Taobao orders were completed through mobile payment.[19]

在「雲消費」時代，資訊成本成為交易成本中比重最大的成本。人們在資訊面前完全平等，巨無霸企業與小微企業在資訊面前完全平等，資訊獲得方式更為透明，資訊的評價機制更為透明，在網路交易中交易雙方從洽談、簽約到訂貨、支付等，均可透過互聯網完成，交易過程完全虛擬化，交易價格更為透明，傳統的層層加碼的中間管道已越來越缺乏生存空間，流通的管道環節趨於消失。

In the era of cloud consumption, information cost has become the largest proportion of transaction costs. People are completely equal in the face of information. Big Mac companies and small and micro enterprises are completely equal in information. The way to obtain the information is more transparent, so is its evaluation mechanism. In online transactions, both parties negotiate, sign, order, and pay via the Internet, making the transaction process completely virtualized and the transaction price more transparent. This way, the traditional intermediate channels that were raised without restrictions have become increasingly lacking in living space, and the circulation channels tend to disappear.

（二）傳統零售業賴以生存的根基面臨革命
The foundation of the traditional retail industry is facing a revolution

「雲消費」時代也是傳統零售業革故鼎新的大革命時代。大量資料顯示，近年中國實體零售商業績普遍下滑。2010-2014 年，中國社會消費品零售總額增幅連續五年呈下降趨勢，連鎖百強企業擴張步伐不斷放緩，門店增幅在 5 年內減少 5 成以上。[20] 據有關調查，2014 年全國主要零售企業 (包括百貨、超市) 關閉門店超過 201 家，比上年增長 470% 以上，過去「一鋪難求」的商業街已經普遍出現招租難現象。2015 年上半年，據抽樣調查，全國 60% 的百貨商場業績處於下滑狀態。全國主要零售企業 (包括百貨、超市) 關閉門店已達 121 家。[21] 萬達百貨、瑪莎百貨、天虹商場、金鷹集團、百盛百貨、

---

20 資料來源：德勤中國與中國連鎖經營協會聯合. 中國連鎖零售企業經營狀況分析報 .2014-2015 Source: Deloitte China and China Chain Operation Association. Analysis Report on China Chain Retail Business Operation. 2014-2015

21 贏商網 .2015 年上半年全國 60% 的百貨商場業績處於下滑狀態 .http://hb.winshang. com　Winshang.com. 60% of department stores in the country were in decline in the first half of 2015. http://hb.winshang.com

華堂商場等知名商家均已開始大幅調整門店。

The era of 「cloud consumption」 is also the era of the great revolution of the traditional retail industry. A large amount of data indicates that the performance of physical retailers in china has generally declined in recent years. In 2010-2014, the growth rate of total retail sales of consumer goods in China showed a downward trend for five consecutive years. The pace of expansion of top 100 chain enterprises continued to slow down, and the growth rate of outlets decreased by more than 50% in five years.[20] According to relevant surveys, in 2014, the country's major retail enterprises (including department stores and supermarkets) closed more than 201 stores, an increase of more than 470% over the previous year. In the past, it was hard to rent even one shop on the commercial street. But now the contrary is the case. In the first half of 2015, according to a sample survey, 60% of department stores in the country were in decline. The number of closed stores in major retail enterprises (including department stores and supermarkets) has reached 121.[21] Wanda Department Store, Marks & Spencer, Rainbow Department Store, Golden Eagle Group, Parkson Department Store, Huatang Shopping Mall and other well-known businesses have begun to significantly adjust their outlets.

與此同時，中國 B2C 網路零售市場加速發展。據易觀資料包告，2016 年中國網上零售市場規模達 4.97 萬億元，同比增長 29.6%。其中，B2C 市場交易達 2.73 萬億元，同比增長 36%。交易平臺化是中國 B2C 網路零售市場的基本特徵。2014 年，天貓、京東、蘇寧易購、唯品會、亞馬遜中國、1 號店、聚美優品等 10 大交易平臺佔據了 B2C 網路零售市場份額的 93%。天貓依然穩居第一，佔全國市場 59.3% 的份額。淘寶總成交額達 1.172 萬億人民幣，

其中天貓總成交額為 5050 億人民幣。天貓所佔份額較 2013 年增加了 9.2%，顯示出大平臺商的強勢地位。[22]

At the same time, B2C network retail market in China has accelerated. According to Analysys Data, the online retail market in China reached 4.97 trillion yuan in 2016, a year-on-year increase of 29.6%. Among them, B2C market transactions reached 2.73 trillion yuan, a year-on-year increase of 36%. Trading platform is the basic feature of B2C online retail market in China. In 2014, 10 major trading platforms including Tmall, Jingdong, Suning, Vipshop, Amazon China, No. 1 Store, JUMEI accounted for 93% of the B2C network retail market share. Tmall is still ranked first, accounting for 59.3% of the national market. Taobao's total turnover reached 1.172 trillion yuan, of which Tmall's total turnover was 505 billion yuan. Tmall's share increased by 9.2% compared with 2013, showing the strong position of the big platform.

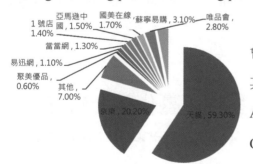

天貓、京東、蘇寧易購、唯品會、亞馬遜中國、1 號店、聚美優品、其他 Tmall, Jingdong, Suning, Vipshop, Amazon China, No. 1 Store, JUMEI, Others

圖 Figure1-9 2014 年中國 B2C 網路購物交易市場份額佔比圖 Ratio map of 2014 China B2C online shopping transaction market share
資料來源：中國電子商務研究中心 www.100EC.cn. Source: China E-Commerce Research Center www.100EC.cn.

---

[22] 資料來源：聯商網中國電子商務研究中心 2014 年中國電商十強榜單 天貓佔 59% 份額．http://www.askci.com/news/chanye/2015/04/13/953471rxg.shtml. Source: 2014 China e-commerce top ten list by China E-Commerce Research Center on Linkshop.com, Tmall accounted for 59% share.

隨著零售革命的不斷深入，消費的便捷性、選擇的多樣性已不再是商業競爭的優勢條件，而成為基礎條件。傳統實體店貼近消費者的管道價值不再是優勢，傳統百貨店、購物中心所代表的傳統零售業選擇的多樣化和一站購足的價值被徹底弱化。隨著更多的管道商作用被交易平臺替代，管道資源壟斷優勢被徹底打破，傳統零售業賴以生存的根基面臨革命。

With the deepening of the retail revolution, the convenience of consumption and the diversity of choices are no longer the advantages of commercial competition, but the basic conditions. The value of traditional physical stores that are close to consumers is no longer an advantage. The diversification of traditional retail choices represented by traditional department stores and shopping centers and the value of one-stop shopping are completely weakened. As more distributors are replaced by trading platforms, the monopoly advantage of channel resources has been completely broken, and the foundation of the traditional retail industry is facing a revolution.

（三）一切商業資源呈現平臺化整合趨勢
All commercial resources show a trend of platform-based integration

「雲消費」時代，隨著互聯網技術的發展，平臺化成為主流，不僅是零售交易，資金、物流、商譽等等一切商業資源均呈現平臺化整合趨勢，交易撮合（淘寶、天貓）平臺化，資金流（支付寶、財付通）平臺化、物流（菜鳥）平臺化，商譽評價（點讚星級制度與旺旺紀錄）平臺化，資本融通（眾籌）

平臺化等等。以 2013 年 6 月 13 日上線的餘額寶為例，經過短短 5 個月，該平臺投資帳戶數已經接近 3000 萬戶，規模超過 1000 億元，[23] 相當於國內全部 78 個貨幣基金總規模的近 20%。到 2014 年年末，餘額寶規模已達到 5789 億元，[24] 約佔中國貨幣基金總規模的 27.6%。可以說，平臺意味著整合，代表了更低的成本、更優的效率、更廣闊的市場和更好的消費體驗。在中國城市社區商業領域，當前大量湧現的社區 O2O 探索，本質就是透過平臺化整合資源，突破社區生活服務最後一公里的瓶頸，在滿足民生，提升民生服務品質的同時，讓參與服務的各方獲得收益。

In the era of cloud consumption, with the development of Internet technology, platformization has become the mainstream. Not only retail transactions, but also all the commercial resources such as capital, logistics, goodwill, etc., have been presenting a platform-based integration trend, including trading match (Taobao, Tmall), capital flow (Alipay, Tenpay), logistics (Cainiao Supply), goodwill evaluation (star rating system and AliWanwan chat history), capital financing (kick starter) platform, etc. Taking the Yu' E Bao released on June 13, 2013 as an example, after just 5 months, the number of investment accounts of the platform has reached nearly 30 million, and the scale exceeds 100 billion yuan,[23] which is equivalent to 20% of the total size of all 78 domestic money funds. By the end of 2014, Yu' E Bao had reached 578.9 billion yuan,[24] accounting for 27.6% of the total

---

23 新浪財經 . 餘額寶問世 5 個月規模破千億 使用者數近 3000 萬戶 .2013-11-14. http://finance.sina.com.cn/money/fund/20131114/161217324359.shtml
24 網易財經 . 餘額寶 2014 年末規模達 5789 億元 .2015-01-05. http://money.163.com/15/0105/11/AF6LUQ0100253B0H.html

size of domestic money funds. It can be said that the platform means integration, representing lower cost, better efficiency, a broader market and a better consumer experience. In the urban commercial community of China, the current emergence of a large number of community O2O explorations is to integrate resources through the platform, break through the bottleneck of the last mile of community life services, and satisfy the people's livelihood and improve the quality of people's livelihood services. At the same time, the parties involved in the service will all gain benefits.

一定意義上，平臺即管道。

In a certain sense, the platform is the channel.

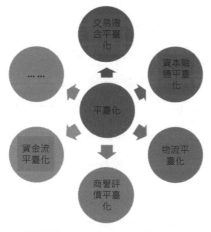

交易撮合平臺化，資金流平臺化、物流平臺化，商譽評價平臺化，資本融通平臺化

Trading match platform, capital flow platform, logistics platform, goodwill evaluation platform, capital financing platform

圖 Figure1-10　一切商業資源呈平臺化整合趨勢 All commercial resources are in a trend of platform-based integration

隨著平臺化成為商業主流，交易成本進一步呈現出趨零化態勢，同時由於 P2P（peer to peer 點對點）交易逐步成為交易主流，也帶來一切商業的眾包化趨勢，我們認為，眾籌和眾包是一切產業的未來。以阿里眾包為例，這個阿里巴巴旗下的預就業和兼職平臺面世僅 1 個月，用戶數就突破了 10 萬人，其「雲客服」任務 5 年間吸納了 32000 名大學生為阿里巴巴提供線上服務，月活躍服務人數達 6500 人，每年為近 2500 萬人次的會員提供幫助。[25]

As platformization becomes the mainstream of the business, transaction costs are further trending toward zero. At the same time, P2P (peer to peer) transactions are gradually becoming the mainstream of transactions, and also bring about the trend of all-business crowdsourcing, which we believe is the future of all industries. Taking Ali Crowdsourcing as an example, this pre-employment and part-time platform of Alibaba has started for only one month and the number of users has exceeded 100,000. Its task, the cloud customer service, has absorbed 32,000 college students to provide online services for Alibaba in 5 years, with monthly active service number reaching 6,500, providing assistance to nearly 25 million members every year.

---

[25] 阿里眾包官網. http://zhongbao.alibaba.com/jury/zhongbao/about.htm?spm=0.0.0.0.TaifGl　AliCrowdsourcing Official Website. http://zhongbao.alibaba.com/jury/zhongbao/about.htm?spm=0.0.0.0.TaifGl

（四）消費者價值是零售業生存的根基
Consumer value is the foundation of retail survival

商業生存價值的在於能為消費者提供價值，「雲消費」時代，零售業生存的根基是消費者價值。The value of business survival lies in providing value to consumers. In the era of cloud consumption, the foundation of retail survival is consumer value.

「雲消費」時代，與社會主流消費模式和生活方式相適應，消費呈現出體驗化、個人化、社群化、定位化的趨勢。體驗化是指消費者更關注消費過程的滿足，願意透過參與互動，進入情景化的環境，感受獨特的氛圍，體驗獨特的文化，而不僅僅是獲得直接的商品和服務；個人化表現為商業智慧和雲資料的發展使個人化訂製不再是少數人的特權，每個消費者都能在不額外付出成本的情況下，享受量身訂製的、專屬的商品和服務，享受消費的尊崇感、自豪感，實現真正意義上的個性化消費；社群化是指在雲消費時代，每個人都處於一個又一個，一環套一環的資訊輻射圈中，驢友圈、社區鄰里圈、家長圈、同事圈、親友圈、粉絲圈等等，每個人都可能被他人影響，每個人又都可能影響他人，隨著QQ、微博、微信等網路平臺的擴散式傳播，網路大V、意見領袖、明星、論壇達人，身邊的時尚達人等等都以他們的消費愛好、偏好，自然地帶動消費，引領時尚。消費者的消費中更加看重社群認同，消費意見與消費結果透過意見領袖或「群」友推薦而透過QQ、微信等網路平臺擴散式傳播達成消費意向，引領消費潮流；定位化是指消費者透過智慧終端機定位消費目標，以參與互動、移動搜索等滿足消費者隨時隨地消費需

求，特別強調消費的專屬性和移動性。大量精準的網路地圖搜索服務的出現，為消費的定位化提供了基礎。

In the era of cloud consumption, in line with the mainstream consumption patterns and lifestyles of the society, consumption has shown a trend of experiencing, personalization, socialization and positioning. Experiencing refers to consumers paying more attention to the satisfaction of the consumption process. They are willing to participate in the contextualized environment, feel the unique atmosphere, and experience the unique culture, not just get the goods and services directly. Personalization means that the development of business intelligent and cloud data makes itself no longer a privilege of a few people, nowadays each consumer can enjoy tailor-made, exclusive goods and services without any additional cost, and enjoy the sense of identity and pride of consumption, realizing personalized consumption in the true sense. Socialization means that in the era of cloud consumption, everyone is in the information circles that are linked with one another, such as the circles of tour pals, neighborhood, parents, colleagues, relatives and fans, etc. Everyone may be affected by others and in the same time, affect others. With the spread of QQ, Weibo, WeChat and other network platforms, big Vs(the verified people who have many followers on Sina Weibo), opinion leaders, celebrities, forum leaders, fashionistas take advantage of their hobbies and preferences to naturally drive consumption and lead fashion. Consumers are more valued by the community recognition, their consumption opinions and results are affected by opinion leaders or other group members from QQ, WeChat and other online platforms, leading the consumption trends. Positioning means that consumers decide the target through smart terminals. By participating in interactions and using mobile searches, their demand can be met anytime, anywhere. It is particularly emphasized on the specificity and mobility of consumption. The emergence of a large number of accurate web map search services provides the basis for the positioning of consumption.

在「雲消費」時代，消費者在實體商業的消費已經從購物滿足，轉向享受新奇好玩的體驗、感受品質文化的認同，傳統的零售賣場轉變為族群社交的場所和家庭生活的空間，商家必須從坪效導向、品牌導向，向消費者體驗導向轉型。商業經營必須突破傳統業種業態的條條框框的束縛，圍繞提供消費者價值重構核心價值和盈利模式。傳統的百貨店和購物中心必須在賣場的體驗化、功能的社交化、服務的個人化、運營的智慧化、傳播的口碑化、流程的致簡化方面做出更大的努力。

In the era of cloud consumption, consumption in physical business has shifted from shopping satisfaction to enjoying new and fun experiences and feeling the recognition of quality culture. Traditional retail stores have turned into group social and family living spaces. Merchants must transform themselves from area effectiveness-oriented and brand-oriented to consumer experience-oriented. Business operations must break through the constraints of the traditional industry, and rebuild the core values and profit models around consumer value. Traditional department stores and shopping centers must make greater efforts in the experiencing of the store, the socialization of functions, the personalization of services, the intelligence of operations, the word-of-mouth of communication, and the simplification of processes.

可以說，「雲消費」時代是一個革命性的時代。商業流通方式將在整個產業鏈和價值鏈上發生革命性變化，商業消費方式、流通盈利模式等都將發生重構和變化。大至批發和代理經銷管道、零售端的各類業態，小到社區居民的一頓早餐、賣廢品的方式等等都會發生變化。這些變化就在人們每一天的生活中，這些變化將不斷改變人們的生活，讓人們生活得更美好。

It can be said that the era of cloud consumption is a revolutionary era. The commercial circulation mode will undergo revolutionary changes in the entire industrial chain and value chain, and commercial consumption methods and circulation profit models will be reconstructed and changed. Many things from the wholesale, agency distribution channels and the various formats of the retail side to breakfast of community residents and the way to sell waste will change. These changes are in people's daily lives and will constantly change people's lives and make people live better.

「雲消費」時代帶來新的零售革命

Chapter Two
The Era of Cloud Consumption Brings A New Retail Revolution

「雲消費」時代的新零售革命
1. The new retail revolution in the era of cloud consumption

（一）什麼是零售革命 What is retail revolution

1 · 基本概念 Basic conception

　　革命，從本質上說是新的替代舊的，是顛覆性的質變。毋庸置疑，隨著現代資訊技術的不斷推進、電子商務的縱深發展，特別是「雲消費」的發展，我們當前正面臨一場覆蓋全球、烈度空前的零售革命。

　　The revolution, in essence, is a new alternative to the old one, which can be seen as a subversive qualitative change. Undoubtedly, with the continuous advancement of modern information technology and the in-depth development of e-commerce, especially the development of cloud consumption, we are currently facing a global retail revolution with unprecedented intensity.

　　從學術角度看，零售革命是指零售業發生的新舊形式主輔換位變化，內在動力的擴張與延伸。[26] 當前我們正經歷的這場以電子商務為表徵的零售革

26　李飛．迎接中國多管道零售革命的風暴 [J]. 北京工商大學學報 · 2012（3）. Li Fei. Meeting the Storm of China's Multi-channel Retail Revolution[J]. Journal of Beijing Technology and Business University, 2012(3).

命，不是簡單的業態形式、交易形式的革命，而是從生產到供應，再到終端，最後到消費，整個產業相關的供應鏈和價值鏈均發生了翻天覆地的，且是不可逆轉的變化。

From an academic point of view, the retail revolution refers to the change of the old and new forms of the main and auxiliary changes in the retail industry, and the expansion and extension of the internal dynamics. [26]The retail revolution characterized by e-commerce that we are currently experiencing is not just a simple change of business and transaction form, but an earth-shaking and irreversible change of the supply chain and value chain related to the entire industry, from production to supply, to terminal, and finally to consumption.

2 · 歷史上發生的零售革命

The retail revolutions that took place in history

目前，從業態發展的角度，國內外不同的專家學者對零售發展史上究竟有幾次零售革命有不同的看法。

第一，三次說。這是最為普通的說法，即百貨商店、連鎖商店和超級市場。也有人認為是百貨商店、超級市場和自動售貨機。[27]

第二，四次說。法國一些零售專家認為歷史上共有四次零售革命，即共發生了百貨商店、一價商店、連鎖商店和超級市場四次革命。[28]

第三，八次說。這種觀點進行了更加全面的分析，認為人類社會已經歷

---

27 李飛 . 零售革命 [M]. 北京：經濟管理出版社，2003. Li Fei. Retail Revolution [M]. Beijing: Economic Management Press, 2003.

28 李飛 . 零售革命 [M]. 北京：經濟管理出版社，2003. Li Fei. Retail Revolution [M]. Beijing: Economic Management Press, 2003.

了八次零售革命，即百貨商店、一價商店、連鎖商店、超級市場、購物中心、自動售貨機、步行商業街、網上商店等八次零售革命。**29**

At present, from the perspective of the development of the industry, different experts and scholars at home and abroad have different views on how many times of retail revolutions took place in the history of retail development.

First, the three-times theory. This is the most common theory, indicating retail revolutions took place in department stores, chain stores and supermarkets. Others argue that they took place in department stores, supermarkets and vending machines.**27**

Second, the four-times theory. Some retail experts in France believe that there have been four retail revolutions in history, respectively in department stores, one-price stores, chain stores and supermarkets.**28**

Third, the eight-times theory. This theory did a more comprehensive analysis, arguing that human society has seen eight retail revolutions, namely in department stores, one-price stores, chain stores, supermarkets, shopping centers, vending machines, pedestrian malls, online stores, etc. **29**

在多種提法中，學界普遍認同的歷史上發生過的零售革命包括百貨商店、連鎖商店和超級市場三次革命。百貨商店是經營商品擴充方面的革命，連鎖商店是組織型態方面的革命，超級市場是自我服務方面的革命，雖然革新點不同，但每次革命都是新的零售形式形成、發展和成熟的過程。

Among these theories, it has been widely recognized in the academic world

---

29 李飛. 零售革命 [M]. 北京：經濟管理出版社，2003. Li Fei. Retail Revolution [M]. Beijing: Economic Management Press, 2003.

that three revolutions took place in history-in department stores, chain stores, and supermarkets. The one in department stores was the expansion of commodities; the one in chain stores was about the organizational form; and the one in supermarkets was related to self-service. Despite of different innovations, each revolution was the part of the process of formation, development and maturity of new forms of retail.

以百貨商店發展為標誌的零售革命開始於 1852 年世界上第一家百貨商店——波馬舍百貨商店的誕生。隨後美國梅西百貨店、德國爾拉海姆、黑爾曼和奇茨百貨店、英國哈樂德百貨店相繼開業，在全世界掀起百貨商店的高潮。百貨商店誕生後，在生產供應端，傳統作坊式的生產行銷一體化的模式（前店後廠）被工廠和商店的分離所替代，工廠利用新技術高效地大批量生產。在零售端，百貨商店構建起類似於「博物館」式的終端場所，消費者能夠在琳瑯滿目的商品中進行挑選。而在消費端，城市的發展聚集了大量的消費者，他們（特別是女性）把百貨商店當作樂園，遊逛其中成為現代城市人的一種生活方式。

The retail revolution marked by the development of department stores began in 1852 with the birth of the first department store in the world, the Le Bon marché. Subsequently, Macy's in the US, Ellaheim, Hermann and Chitz in Germany and Harrods in the UK started one by one, raising a climax of department stores around the world. After the emergence of the department stores, the traditional workshop-style production and marketing integration model, on the production supply side, was replaced by the separation of factories and stores, and the factory utilized new technologies to efficiently mass-produce. On the retail side, department stores were built into museum-like styles where consumers can choose from a wide range of items. On the consumer side, the development of the city has gathered a large number of consumers, who regarded the department store as a paradise, especially women. Wandering around in the department stores becomes a kind of lifestyle of modern city dwellers.

以連鎖商店發展為標誌的零售革命始於 19 世紀中葉紐約的大美茶葉公司，之後連鎖公司在美國得到迅速發展；同時在歐洲各國也先後產生了一大批連鎖商店，第二次世界大戰後達到高潮，形成了世界範圍的連鎖革命。連鎖商店誕生後，生產端的標準化生產方式延續到零售端，工業化的生產得以進一步強化，只要生產出產品就能透過連鎖店的觸角鋪向任何一個角落。同時生產端逐漸依附於零售商，在議價能力中漸處下風。在零售端，連鎖商店能夠透過成本低廉的標準化複製迅速產生規模效益。在消費端，由於連鎖商店主要經營的是大眾化商品，店鋪又多數開在交通便捷之地或居民社區附近，縮短了消費者購物的空間距離和所用時間，方便了人們的購買。

The retail revolution marked by the development of chain stores began in the mid-19th century in New York's Great American Tea Company, after which the chain companies developed rapidly in the United States. At the same time, a large number of chain stores appeared in European countries and reached the climax after the Second World War, forming a worldwide chain revolution. After the emergence of the chain stores, the standardized production mode at the production end extended to the retail end, and the industrialized production was further strengthened. As long as the product was produced, it could be paved to any corner through the chain stores. At the same time, the production side gradually attached to the retailer and became disadvantaged in the bargaining. On the retail side, chain stores are able to quickly generate economies of scale through cost-effective, standardized replication. On the consumer side, since the chain stores mainly operate popular goods, most of the shops are located in areas with convenient transportation or near neighborhoods, which reduces the distance from the stores and time spent by consumers, making shopping convenient.

以超級市場發展為標誌的零售革命誕生於 20 世紀 30 年代，超級市場在美國得到發展；第二次世界大戰以後，超級市場迅速在歐洲和日本等地蔓延，標誌著世界性的超級市場革命的爆發。在生產端，超級市場進一步掌控了勞動密集型的生產商；在零售端，現代電腦化的收銀系統、訂貨系統、核算系統，科學地管理經營的各個環節，同時應用了現代化的交通手段，使流通速度和週轉效率大為提高。在消費端，開架銷售、自我服務的銷售方式，革命性地改變了消費體驗。

The retail revolution marked by the development of the supermarket was started in the 1930s, the supermarket developed in the United States then. After the Second World War, the supermarket quickly spread in Europe and Japan, marking the outbreak of world-wide supermarket revolution. At the production end, the supermarket further controls the labor-intensive producers and at the retail end, the modern computerized cashier system, ordering system and accounting system manage all aspects of the operation in a scientific manner, and in the same time, apply modern means of transportation to greatly improve the circulation speed and turnover efficiency. At the consumer end, the open-shelf sales and self-service sales have revolutionized the consumer experience.

（二）「雲消費」時代的新零售革命
The new retail revolution in the era of cloud consumption

　　無論既往歷史上發生了幾次零售革命，均具有共同點，即歷次零售革命均產生新的零售業態，並且拓展零售企業的價值鏈。而我們正面臨的「雲消費」時代的新的零售革命，則完全打破了零售業內百貨、超市、3C、家電、建材家具、便利店等業態的界限，打破了虛擬與現實的界限，使有形與無形相融，商流、物流、資訊流交織，改寫了傳統零售的定義。可以說，它重構了零售的概念，改變了零售的思維和遊戲規則，它既是生產方式的全面變革，也是生產力的根本性變革，因此，可以說我們正經歷著一輪真正意義上的零售革命。

　　No matter how many retail revolutions have occurred in the past, they all have something in common, that is, all retail revolutions had produced new retail formats and expanded the value chain of retail enterprises. And the new retail revolution in the era of cloud consumption that we are facing completely breaks the boundaries of department stores, supermarkets, 3C, home appliances, building materials and furniture, and convenience stores, etc. in retail industry, breaking the boundaries between virtual and reality, making the tangible and the intangible merge, business flow, logistics, and information flow intertwined, rewriting the definition of traditional retail. It can be said that it reconstructs the concept of retail and changes the thinking of retail and the rules of the game. It is both a comprehensive change in production mode and a fundamental change in productivity. Therefore, we can say that we are experiencing a round of real retail revolution.

有學者認為，本輪零售革命是一場「電商革命」，突出表現在網路商店業態的誕生。電商銷售是非在場、遠距離銷售，不再如其他零售模式顧客與銷售者在同一時空完成購買過程，是對面對面交付的其他傳統零售方式的最大顛覆。電商銷售是實體零售模式難以企及的超大規模銷售，顧客可以隨時、隨地在千萬種以上的商品中搜尋、選擇、比價，所謂長尾效應在電商銷售模式中才能夠充分體現。[30]

Some scholars believe that this round of retail revolution is an e-commerce revolution, which is highlighted by the emergence of the online store format. E-commerce sales are not-on-the-spot and long-distance sales. Unlike other retail models, customers and sellers don' t need to complete the purchase process at the same time and space, which is the biggest subversion of other traditional retail methods requiring face-to-face delivery. E-commerce sales are ultra-large-scale sales that are difficult to achieve in the physical retail model. Customers can search, select, and compare prices in more than tens of thousands of products anytime, anywhere. The so-called long tail effect can be fully reflected in the e-commerce sales model.[30]

顯然，本輪零售革命始於電商，但其意義和價值遠不只於電商業態的誕生和發展。它不僅產生電商，也深刻觸動了傳統零售商，形成全行業的革命浪潮，線上線下結合，今後將不再有電商和實體零售商的嚴格界限。同時，

30　陸小華. 零售業的六次革命 新一輪零售革命來臨 [N]. 中國經營報，2013-03-31. Lu Xiaohua. Six revolutions in the retail industry A new round of retail revolution is coming [N]. China Business Journal. Mar 31, 2013.

它深刻觸動零售業全產業鏈、全供應鏈，帶來一系列根本性變化：在生產端，生產商可以透過資料的回饋及時並準確地掌握消費者回饋，「長尾」商品不再是雞肋，生產商的決策更加高效、準確；在物流端，物聯網技術重新定義了傳統倉儲物流的概念；在零售端，傳統零售業逐漸從實體店鋪向虛擬空間延伸，再到向全時間、全管道的移動虛擬空間滲透的轉變；在交易端，傳統面對面的交易方式早已改寫，支付可以隨時隨地發生，支付寶等電子支付工具甚至已成為有強勁競爭力的金融工具。同時，移動電子商務也使得零售商對於物流、庫存、補貨的處理更加智慧、便捷和高效。更重要的是，消費者的力量空前強大，推動「雲消費」發展，成為零售革命的主導力量。正如顏豔春先生所提出的：「如今消費者已經成為世界的中心，他們透過移動互聯網、社交網路和物聯網即時連接起來，正在形成一股巨大的、看不見卻能真實感受到的力量，每一個消費者都可能成為零售革命的發起者。」[31]

Obviously, this round of retail revolution began with e-commerce, but its significance and value goes far beyond the emergence and development of the e-commerce format. It not only gives birth to e-commerce, but also deeply affects the traditional retailers, forming a revolutionary wave of the whole industry, combining online and offline, and there will be no strict boundaries between e-commerce and physical retailers in the future. At the same time, it deeply affects the entire industry chain and supply chain of the retail industry, bringing a series of fundamental changes. At the production end, producers can timely and accurately grasp consumer feedback through data feedback, and long-tail products are no longer like chicken ribs, the manufacturer's decision can be more efficient and accurate. At the logistics end, the Internet of Things technology redefines the concept of traditional warehousing logistics. At the retail end, the traditional retail industry gradually extends from physical stores to virtual space, then to the full-time, full-channel mobile virtual space. At the transaction end, traditional face-to-face transaction methods have long been rewritten, payments can occur anytime, anywhere and electronic payment tools such as Alipay have become a highly competitive financial instrument. At the same time, mobile e-commerce also makes retailers more intelligent, convenient and efficient in handling logistics, inventory and replenishment. More importantly, the power of consumers is unprecedentedly strong, promoting the development of cloud consumption and becoming the leading force in the retail revolution. As Mr. Yan Yanchun puts it, 「Today, consumers have become the center of the world. They are connected in real time through the mobile Internet, social networks and the Internet of Things. They are forming a power that is huge, invisible but can be truly felt. Every consumer is likely to be the initiator of the retail revolution."[31]

---

31   顏豔春. 第三次零售革命 [N]. 中國商界 · 2014-01-22. Yan Yanchun. The third retail revolution [N]. Business China, 2014-01-22.

二

電子商務、雲計算、大數據應用為「雲消費」時代創造基礎

2. E-commerce, cloud computing, and big data applications create the foundation for the era of cloud consumption

（一）電子商務的迅猛發展推進「雲消費」時代進程

The rapid development of e-commerce promotes the process of cloud consumption

1. 電子商務的基本概念及現實意義

The basic concept and practical significance of e-commerce

電子商務的發端可以追溯到 20 世紀 70 年代，現代含意的電子商務是在 20 世紀 90 年代中期，Internet 實現商用以後在美國出現的，1996 年前後，美國學術界才正式提出了電子商務的概念。

The origin of e-commerce can be traced back to the 1970s. E-commerce of modern meaning emerged in the United States in the mid-1990s after the Internet was commercialized. Around 1996, the American academic community officially proposed the concept of e-commerce.

通常意義上，電子商務是指在互聯網（Internet）、企業內部網（Intranet）和增值網（VAN，Value Added Network）上以電子交易方式進行交易活動和相關服務活動，是傳統商業活動各環節的電子化、網路化，是利用微電腦技術和網路通信技術進行的商務活動。各國政府、學者、企業界人士根據自己所處的地位和對電子商務參與的角度和程度的不同，給出了許多不同的定義，但其關鍵依然是依靠著電子設備和網路技術進行的商業模式。隨著電子商務的高速發展，它已不僅僅包括其購物的主要內涵，還包括物流配送等附帶服務。

Generally speaking, e-commerce refers to trading activities and related service activities on the Internet, Intranet and Value Added Network (VAN), which are the digitalization and networking of all aspects of traditional business activities, and are business activities using microcomputer technology and network communication technology. Governments, scholars, and business people have given many different definitions depending on their positions, and degree and extent of participation in e-commerce, but the key is still the business model based on electronic devices and network technologies. With the rapid development of e-commerce, it no longer only includes the main connotation of its shopping, but also the incidental services such as logistics and distribution.

電子商務發展 10 餘年間，取得了突飛猛進的發展。根據業內有關分析，

完成了從 1.0 到 4.0 的時代進階。在最開始的 1.0 時代，電子商務發展的核心在於資訊展示，即網站展示產品；2.0 時代則為簡單的線上交易，即線上展示、線下交易；3.0 時代，開始進入社會化電子商務的全新時期，線上展示線上交易。如今電子商務已經進入 4.0 時代，打破時空限制，不分線上線下交易，任何人都是交易的發起者，任何人也都可以是接受者。

Over the past 10 years, e-commerce has achieved rapid development. According to the relevant analysis in the industry, it has advanced from the 1.0 era to the 4.0. In the initial 1.0 era, the core of e-commerce development lies in the information display, that is, websites displayed products; in the 2.0 era, it was simple online transaction, that is, online display and offline transaction; in the 3.0 era, it began to enter a new era of social e-commerce, that is, online display and online transaction. Nowadays, e-commerce has entered the 4.0 era, breaking the time and space restrictions. Anyone can be the originator and the recipient of the transaction regardless of online and offline transactions.

電子商務的發展，其現實意義主要表現在兩方面：①推進零售商內部組織重組。電子商務的出現不可避免地會代替零售商原有的一部分管道和資訊源，並對零售商的企業組織造成重大影響，迫使零售商進行組織的重整。如業務人員與銷售人員減少、企業組織層次減少、企業管理幅度增大、零售門店數量減少、虛擬門市和虛擬部門等企業內外部虛擬組織盛行。②帶來零距離的消費體驗。電子商務在資訊技術的催化下打破了零售市場時空界限，為消費者帶來更加便利的購買模式，可以實現「一對一」的精準行銷，可以實現「一對多」的市場空間拓展。消費者從「進店購物」變為「坐家購物」，

足不出戶，便能輕鬆在網上完成過去要花費大量時間和精力的購物過程。

The development of e-commerce is mainly reflected in two aspects: a. Promoting the internal restructuring of retailers. The emergence of e-commerce will inevitably replace some of the retailers' original channels and information sources, and have a significant impact on the retailer's corporate organization, forcing retailers to reform their organizations. Such impact includes the reduction of business personnel and sales staff, the reduction of corporate organization level, the increase of enterprise management, the reduction of the number of retail stores, the prevalence of internal and external virtual organizations like virtual stores and virtual departments. B. Bringing a zero-distance consumer experience. Under the catalysis of information technology, e-commerce breaks the time and space boundaries of the retail market, and provides consumers with a more convenient purchase model. It can achieve 「one-to-one」 precision marketing and achieve 「one-to-many」 market space expansion. 「Shopping in the store」 has changed into 「shopping at home」, consumers can easily complete the shopping process that took a lot of time and effort on the Internet without leaving home.

2 · 全球電子商務發展迅猛 Global e-commerce is growing rapidly

當前，全球電子商務市場發展迅猛，呈現出美國、歐盟、亞洲「三足鼎立」的局面。美國是世界最早發展電子商務的國家，同時也是電子商務發展最為成熟的國家，一直引領全球電子商務的發展，是全球電子商務的成熟發達地區。歐盟電子商務的發展起步較美國晚，但發展速度快，成為全球電子商務較為領先的地區。亞洲做為電子商務發展的新秀，市場潛力較大，是全

球電子商務的持續發展地區。

At present, the global e-commerce market is developing rapidly, showing the situation of tripartite confrontation among the United States, the European Union and Asia. The United States is the first country in the world to develop e-commerce, and it is also the country with the most mature e-commerce development. It has always led the development of global e-commerce and is a mature and developed region of e-commerce in the world. The development of e-commerce in the EU started later than in the United States, but is rapid and has become a leading region of global e-commerce. As a newcomer of the development of e-commerce, Asia has a large market potential and is a region of continuous development of global e-commerce.

據全球市場研究公司 eMarketer 統計，2012 年全球網上購物人數達 9.036 億，亞太網購者人數 3.911 億，成為全球範圍內網購人均最多的地區。全球人均網購消費額為 1243 美元，其中亞太 850 美元、北美 2221 美元、西歐 1738 美元。全球 B2B 電子商務交易一直佔據主導地位，2007 年全球 B2B 交易額達到 8.3 萬億美元。2012 年全球 B2C 電子商務銷售額突破 1 萬億美元，其中北美地區佔 33.5%，在全球電子商務市場上位居首位，美國 B2C 電子商務銷售額高達 3434.3 億美元，亞洲地區 B2C 電子商務銷售額增幅最大，達 3324.6 億美元。[32] 2016 年中國電子商務市場交易規模已達 20.2 萬億元。

According to the statistics by eMarketer, a global market research company, the number of online shopping consumers in the world reached 903.6 million in 2012, among which the number in Asia-Pacific region was 391.1 million, making it the most online shopping area in the world. The global per-capita online shopping expenditure is 1,243 US dollars, including 850 US dollars in Asia-Pacific region, 2,221 US dollars in North America, and 1,738 US dollars in Western Europe. Global B2B e-commerce transactions have always dominated. In 2007, global B2B transactions reached 8.3 trillion US dollars. In 2012, global B2C e-commerce sales exceeded 1 trillion US dollars, including 33.5% in North America, ranking first in the global e-commerce markets. B2C e-commerce sales in US reached US$343.43 billion. B2C e-commerce sales growth in Asia is the largest, reaching $332.46 billion.[32] In 2016, the scale of China's e-commerce market transactions reached 20.2 trillion yuan.

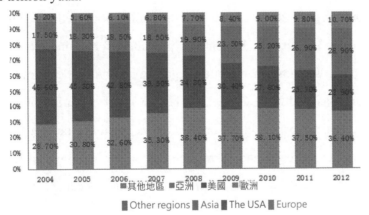

圖 Figure2-1 2004--2012 年全球網路購物市場交易規模地區分佈情況 Regional distribution of global online shopping market transactions in 2004-2012
資料來源：eMarketer 全球網上購物市場統計 Source: Statistics on Global Online Shopping Market by eMarketer

---

32 無忌. 2012 年全球 B2C 電商銷售額突破 1 萬億美元 [J/OL]. 騰訊科技. Wu Ji. Global B2C e-commerce sales exceeded 1 trillion US dollars in 2012 [J/OL]. Tencent Tech.

（1）電子商務在美國 E-commerce in the US

電子商務起源於美國，1995 年美國 B2C 電子商務興起，傑夫·貝索斯在西雅圖成立網上銷售書籍的公司——亞馬遜書店，開創了網上零售的先河。隨後雅虎、美國線上等知名網站開始網上銷售業務，傳統的零售企業也紛紛開設網上業務，如美國第一大零售業連鎖企業沃爾瑪集團，第二大連鎖企業西爾斯集團在 1999 年都已經開展網上零售業務。2000 年亞馬遜書店已經擁有 1800 萬名顧客，年銷售額達到 31 億美元，2003 年首度盈利，成為全球 B2C 電子商務的一面旗幟。

The e-commerce originated in the United States. In 1995, B2C e-commerce emerged in the country. Jeff Bezos established the online book selling company, Amazon Bookstore, in Seattle, which created a precedent for online retail. Then Yahoo, AOL and other well-known websites began to sell online, and traditional retail companies such as Wal-Mart Group, the largest retail chain in the US and Sears Group, the second largest chain, all opened online retail business in 1999. In 2000, Amazon Bookstore already had 18 million customers, with annual sales of 3.1 billion US dollars. It started to generate profit in 2003 and became a banner for global B2C e-commerce.

2003 年美國 B2C 電子商務市場交易額為 960 億美元，2004 年為 1250 億美元，2005 年美國 B2C 電子商務市場交易額達到 1600 億美元，增長率達到 28%。2005-2010 年複合增長率為 24%。[33]2012 年，美國零售電子商務銷售增長了 15%，其速度是傳統零售業的 7 倍。根據美國零售聯合會的統計，2013

33　2005 年中國 B2C 電子商務簡版報告 [R]. 艾瑞網，2006--06--02. Brief report on China B2C e-commerce in 2005[R]. iResearch. Jun 2, 2006.

年美國消費者線上購物的比例大幅提升，感恩節當日有超過 1/4 的消費者選擇線上購物；而在「黑色星期五」當日比例更是高達 47.1%。美國電子商務調查公司康姆斯克的調查資料顯示，2013 年感恩節和「黑色星期五」，主要零售商透過 PC 端的線上銷售額分別達到了 7.66 億美元和 11.98 億美元，與 2012 年同期相比，分別增長了 21% 和 15%，刷新了歷史紀錄。2013 年美國「網路星期一」的線上銷售額再創新高，達成 20 億美元。IBM 資料分析顯示，「網路星期一」線上銷售額較 2012 年同期增長 17.5%。[34]

In 2003, the US B2C e-commerce market transaction volume was US$96 billion, and in 2004 it was US$125 billion. In 2005, it reached US$160 billion, with a growth rate of 28%. The compound growth rate from 2005 to 2010 was 24%.[33] In 2012, US retail e-commerce sales increased by 15%, which is seven times faster than traditional retail. According to the statistics made by the American Retail Federation, the proportion of online shopping for consumers in the US increased significantly in 2013. More than a quarter of consumers chose to shop online on Thanksgiving Day, and the proportion of that on the Black Friday was as high as 47.1%. According to survey data from the US e-commerce research firm Commsk, in 2013 Thanksgiving and the Black Friday, major retailers achieved online sales of $766 million and $1.198 billion respectively through PC-side. Compared with the same period in 2012, it has increased by 21% and 15% respectively, setting a new record. In 2013, online sales on the US Cyber Monday reached a new high, reaching $2 billion. According to IBM data analysis, online sales on Cyber Monday

---

[34] 網購衝擊波蔓延至全球. 美國黑色星期五被電商侵蝕 [J/OL]. 北京商報. Online shopping shockwave spread to the world. American Black Friday was eroded by e-commerce [J/OL]. Beijing Business Today.

increased by 17.5% compared with the same period in 2012.[34]

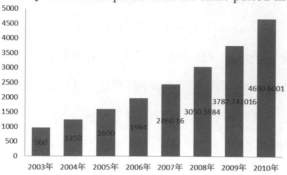

圖 Figure2-2　美國 2003--2010 年 B2C 市場交易情況 B2C market transactions in the US in 2003 to 2010

表 Table 2-1 美國知名商業集團的電子商務年交易情況

Annual e-commerce transactions of well-known American business groups

企業名稱 Company name	2011 年排名 Ranking in 2011	2010 年排名 Ranking in 2010	2009 年排名 Ranking in 2009	2011 年銷售額 (美元)Sales in 2011(US dollar)
Amazon.com Inc.	1	1	1	48080000000
Staples Inc.	2	2	2	10600000000
Apple Inc.	3	3	4	6660000000
Walmart.com.	4	6	6	4900000000
Del Inc.	5	4	3	4609728000
Office Dcpot Inc.	6	5	5	4100000000
Liberty Interactive Corp	7	8	11	3760000000
Sears Holdings Corp	8	7	8	3204577000
Netflix Inc	9	13	14	3000100000
CDW Corp	10	10	9	2950000000
Best Buy Co.	11	11	10	2901497618
OfficeMax Inc.	12	9	7	2901497618
Newegg Inc.	13	12	12	2700000000
Macy' s Inc	14	17	20	2748850000
W.W.Gralnger Inc	15	15	19	2187000000

資料來源：相關資料整理 Source: Relevant data

美國 B2B 電子商務從 B2C 模式起步，但很快成為主流，現在幾乎所有

美國大企業都在使用電子商務。美國電子商務的先行者亞馬遜公司 2011 年實現銷售額 480 億美元，2012 年實現銷售收入 610 億美金，約佔美國電子商務交易額的 20% 左右。2012 年亞馬遜在歐洲網路銷售總額達到 161.1 億美元，比 2011 年增加了 19.3%。[35] 2017 年亞馬遜全年淨收入為 30 億美元，合每股 6.15 美元，全年收入為 1779 億美元，同比增長 31%。[36]

American B2B e-commerce started from the B2C model, but it quickly became mainstream. Now almost all major US companies are using e-commerce. Amazon, the pioneer of e-commerce in the United States, achieved sales of $48 billion in 2011 and achieved sales of $61 billion in 2012, accounting for about 20% of US e-commerce transactions. In 2012, Amazon's total network sales in Europe reached US$16.11 billion, an increase of 19.3% over 2011.[35] In 2017, Amazon's annual net income was $3 billion, or $6.15 per share, and the annual revenue was $177.9 billion, an increase of 31%.[36]

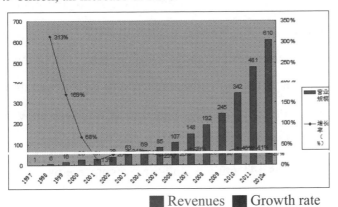

■ Revenues ■ Growth rate

圖 Figure2-3 1997--2012 年亞馬遜電子商務營收情況 Amazon e-commerce revenues from 1997 to 2012
資料來源：艾瑞諮詢報告 Source: Consulting report from iResearch

根據美國製造商協會 2001 年的一項調查，當時 80% 的美國製造商已經擁有自己的網站，電子商務交易的使用率為 32%，採購商的電子商務利用率為 38%。2002-2005 年 4 年間，美國 B2B 電子商務市場交易額始終佔據全球 B2B 市場交易額的 50% 以上，2005 年達到了 2.6 萬億美元。[37] 其中美國國內交易量佔全球市場交易量的 50% 以上。2010 年美國製造業 B2B 滲透率在 40% 以上，批發業 B2B 滲透率在 20% 以上，而零售業 B2C 滲透率則在 5% 左右。2011 年美國網路購物人數是 1.41 億人，電子商務使用率 71.2%。2012 年美國電子商務銷售額達 2255.4 億美元。[38]

According to a survey by the American Manufacturers Association in 2001, 80% of US manufacturers had their own websites, usage rate of e-commerce transactions was 32%, and utilization factor of e-commerce for buyers reached 38%. In the four years from 2002 to 2005, the US B2B e-commerce market transaction volume has always accounted for more than 50% of the global B2B market transaction volume, reaching US$2.6 trillion in 2005.[37] Tthe domestic transaction volume in the United States accounted for more than 50% of the global market transaction volume. In 2010, the US manufacturing B2B penetration rate was above 40%, the wholesale industry B2B penetration rate was above 20%, and the retail B2C penetration rate was around 5%. In 2011, the number of online shopper in the United States was 141 million, and the e-commerce usage rate was

35　中國行業研究網 China Industry Research Network
36　亞馬遜 2017 年度業績報告 Annual report on Amazon performance results in 2017
37　艾瑞市場諮詢 . iResearch marketing consultation.
38　中國行業研究網 . China IRN.

71.2%. In 2012, US e-commerce sales reached US$225.54 billion.[38]

近年在新興的移動資訊獲取模式下，美國移動電子商務快速發展，據公開資料，2016 年美國移動支付市場規模達到 1120 億美元，同比增長 39%，這一規模資料相較於中國協力廠商移動支付金額（5.5 萬億美元）相差近 50 倍，顯示了中國大陸移動支付的普及程度。[39]

In recent years, under the emerging mobile information acquisition model, US mobile e-commerce has developed rapidly. According to public data, the US mobile payment market reached US$112 billion in 2016, a year-on-year increase of 39%. Compared to the third-party mobile payment ($5.5 trillion) in China, this scale is nearly 50 times lower, showing the great popularity of mobile payments in mainland China.[39]

（2）電子商務在歐洲 E-commerce in Europe

歐洲電子商務 20 世紀末遠落後於美國，2000 年後開始加速，2003 年以後進入快速增長通道，2005 年歐洲電子商務市場的收入規模為 519 億歐元，[40] 2006 年西歐 B2C 電子商務銷售額較 2005 年增長 36%，其中英國 B2C 電子商務銷售額位居西歐各國之首，為 479 億美元，佔據 50% 的市場份額，德國以 239 億美元位居第二。[41] 2007 年英國零售業電子商務銷售額突破 253 億美元。

39　張川 . 為什麼美國移動支付比中國落後 . 億邦動力網 .2017-09-11 Zhang Chuan. Why US mobile payments are behind China. Ebrun.com. Sept 11, 2017

40　iResearch 統計。 Statistics by iResearch

41　eMarketer 統計資料顯示。

E-commerce in Europe lagged far behind the United States at the end of the 20th century. It started to accelerate after 2000 and entered the fast-growing channel after 2003. In 2005, e-commerce market in Europe generated € 51.9 billion in revenue.[40] In 2006, B2C e-commerce sales in Western Europe increased 36% compared with 2005, of which B2C e-commerce sales in the UK ranked first in Western European countries, with $47.9 billion, accounting for 50% of the market share. Germany ranked second with $23.9 billion.[41] In 2007, retail e-commerce sales in the UK exceeded $25.3 billion.

根據線民網上平均消費金額統計分析顯示，2006 年 9 月在西歐國家中挪威線民網上平均消費金額最高，達到 1406 歐元，其次是英國線民，平均消費金額為 1210 歐元，同時英國線民的平均購物數量為 18 件，在西歐國家最多。根據 eMarketer 公司預測，2005-2010 年，法國、德國、義大利、西班牙、英國的互聯網用戶數保持 4% 以上的年複合增長率，寬頻用戶滲透率保持 15% 以上的年複合增長率。2006 年西歐線上零售交易額達到 970 億美元，比上年度增長 37%，到 2010 年以前將保持高於 15% 的年增長率，2010 年線上消費市場規模達到 2345 億美元。[42] 據全球電子商務報告顯示，2012 年歐洲電子商務增長 16.62%，達到 3022 億美元。[43]

According to the statistical analysis of the average online consumption of netizens, the average online consumption of Norwegian netizens in Western Europe

---

42　互聯網：分享豐盛的 B2B 電子商務蛋糕. 平安證券. Internet: Sharing a bounteous B2B e-commerce cake. Ping An Securities.

43　歐洲電商增長 16.62% 發展速度超過美國 [J/OL]. 億邦動力網. E-commerce in Europe increased by 16.62%, faster than the United States [J/OL]. Ebrun.com

in September 2006 was the highest, reaching € 1,406, followed by British netizens, with an average consumption of € 1,210 and an average number of goods bought by them was 18, the most in Western Europe. According to eMarketer's forecast, in 2005-2010, the number of Internet users in France, Germany, Italy, Spain, and the UK maintained a compound annual growth rate of more than 4%, and the penetration rate of broadband users maintained a compound annual growth rate of more than 15%. In 2006, online retail transactions in Western Europe reached US$97 billion, an increase of 37% over the previous year. By 2010, it would maintain an annual growth rate of more than 15%. In 2010, the online consumer market reached US$234.5 billion. [42]According to the global e-commerce report, in 2012, e-commerce in Europe grew by 16.62%, reaching $302.2 billion.[43]

表 Table 2-2 2006 年西歐 5 國 B2C 電子商務銷售額 B2C e-commerce sales in five countries in Western Europe in 2006

國家 Country	交易額 （億美元） Transaction volume(100 million US dollars)	市場份額 (%)Market share(%)	2006--2010 年複合增長率 (%)2006-2010 Compound growth rate(%)	淨利潤率 %
英國 The Uk	479	49.60	21.80	3.79
德國 Germany	239	24.70	27.20	4.53
法國 France	160	16.60	27.20	5.13
義大利 Italy	50	5.20	32.90	5.44
西班牙 Spain	38	3.90	30.90	7.72

注：包括網上旅行、訂票、數位產品下載。Note: Includes online travel, booking, digital product downloads.
資料來源：2007.1iresearch.Source: Jan 2007, iResearch.

（3）電子商務在日本、韓國 E-commerce in Japan and Korea

亞太地區電子商務發展迅速，以中國、日本、韓國等國為代表。

日本 B2B 電子商務起步於 1995 年，1998 年日本 B2B 電子商務交易額為 8.6 萬億日元，不到美國的一半，B2C 電子商務交易額為 650 億日元，不足美國的 1/35。1998-2004 年，日本電子商務交易額的年均增長率超過 50%；2004 年 B2B 電子商務交易規模為 102.7 萬億日元，B2C 電子商務規模為 5.6 萬億日元，分別相當於 1998 年的 12 倍和 80 倍[44]。2012 年日本 B2C 市場規模為 9.5 萬億日元，較上年度增長 12.5%，電子商務率為 3.11%。[45]

E-commerce in the Asia-Pacific region has developed rapidly, represented by China, Japan, South Korea and other countries.

Japan's B2B e-commerce started in 1995. In 1998, Japan's B2B e-commerce transaction volume was 8.6 trillion yen, less than half of that in the US, and B2C e-commerce transaction volume was 65 billion yen, less than 1/35 of the US. In 1998-2004, the average annual growth rate of e-commerce transactions in Japan exceeded 50%. In 2004, the scale of B2B e-commerce transactions was 102.7 trillion yen, and the scale of B2C e-commerce was 5.6 trillion yen, equivalent to 12 times and 80 times as 1998 respectively.[44] In 2012, the scale of the Japanese B2C market was 9.5 trillion yen, an increase of 12.5% over the previous year, and the e-commerce rate was 3.11%.[45]

---

[44] 日本電子商務的發展階段及現狀分析 [J/OL]. 中國電子商務研究中心, 2012. Analysis of the Development Stage and Current Situation of Japanese e-commerce [J/OL]. China Electronic Commerce Research Center, 2012.

[45] 2012 年日本 B2C 市場規模為 9.5130 萬億日元 [J/OL]. 中國電子商務研究中心, 2014. The Scale of the Japanese B2C market in 2012 was 9.5130 Trillion Yen [J/OL]. China Electronic Commerce Research Center, 2014.

表 Table 2-3 1998 - 2004 年日本 B2B 及 B2C 電子商務市場規模 1998-2004
Japan B2B and B2C e-commerce market scale

年份 Year	B2B 市場交易額（萬億日元）B2B market transaction volume (Trillion yen)	年增長率 (%) Annual growth rate(%)	B2C 市場交易額（萬億日元）B2C market transaction volume (Trillion yen)	年增長率 (%) Annual growth rate(%)
1998	8.6		0.0646	
1999	12.3	43.0	0.336	420.1
2000	21.6	75.6	0.824	145.2
2001	34.0	57.4	1.484	80.0
2002	46.3	36.1	2.685	80.9
2003	77.4	67.2	4.244	58.1
2004	102.7	32.7	5.643	32.9

資料來源：「Reality and Market Size Research on E-Commerce，FY 2004，」
Ministry of Economy，Trade and Industry，Next-Generation E-Commerce Promotion Council，and NTT Data Management Research Center.
Source:「Reality and Market Size Research on E-Commerce，FY 2004，」
Ministry of Economy，Trade and Industry，Next-Generation E-Commerce Promotion Council，and NTT Data Management Research Center.

　　據韓國統計廳的「2000 年電子商務企業統計調查」統計，2000 年韓國電子商務總規模達 57.6 萬億韓元，遠遠超過 17 萬億韓元的預測值。在所有電子商務交易中，企業之間的 B2B 的規模為 52.3 萬億韓元，B2C 電子商務規模為 7337 億韓元，而面向海外出口的電子商務規模為 4.4 萬億韓元。調查顯示，在韓國網上銷售產品的行業中，製造業所佔比重最大，為 93%。其中，冶金行業為 38.1%，電器行業為 21.1%，汽車行業為 16.3%，電子零部件行業為 11.4%。[46] 韓國 2006 年電子商務交易額達 414 萬億韓元，2007 年 517 萬億韓元，2008 年 630 萬億韓元，2009 年 672 萬億韓元，2010 年 824 萬億韓元。2011 年韓國電子商務交易額達到 999 萬億韓元，同比增加 21.2%，其中，

B2B 增加 22.1%，B2C 增加 15.7%，B2G 增加 10.6%，C2C 增加 14.8%。[47]

According to the statistics of the 2010 E-commerce Enterprise Survey by the Korea National Statistics Office, the total scale of e-commerce in Korea in 2000 reached 57.6 trillion won, far exceeding the forecast of 17 trillion won. Of all e-commerce transactions, the scale of B2B between enterprises was 52.3 trillion won, B2C e-commerce was 733.7 billion won, e-commerce for overseas exports was 4.4 trillion won. According to the survey, manufacturing accounted for the largest share of the online sales of products in Korea, at 93%. Among them, the metallurgical industry was 38.1%, the electrical industry was 21.1%, the automotive industry was 16.3% and the electronic parts industry was 11.4%.[46] The amount of e-commerce transactions in Korea reached 414 trillion won in 2006, 517 trillion won in 2007, 630 trillion won in 2008, 672 trillion won in 2009, and 824 trillion won in 2010. In 2011, the amount of e-commerce transactions in Korea reached 999 trillion won, a year-on-year growth of 21.2%. Among them, B2B increased by 22.1%, B2C increased by 15.7%, B2G increased by 10.6% and C2C increased by 14.8%.[47]

（4）電子商務在中國 E-commerce in China

中國電子商務的發展始於 20 世紀 90 年代初，1997 年中國化工資訊網正

---

46 韓國電子商務發展現狀分析 [J/OL]. 創業網, 2007. Status Analysis of E-commerce Development in Korea [J/OL]. Cye.com.cn. 2007.

47 韓國電子商務交易額連年遞增 [J/OL]. 韓聯社, 2012-02-24. The Amount of E-commerce Transactions in Korea Has Increased Year by Year [J/OL]. Yonhap News Agency, Feb 24, 2012.

式在互聯網上提供服務，開拓了網路化工的先河，是全國第一個介入行業網站服務的國有機構。

The development of e-commerce in China began in the early 1990s. In 1997, China National Chemical Information Center officially provided services on the Internet, pioneered the network chemical industry, and was the first state-owned institution in the country to intervene in industry website services.

1997 年 12 月中國化工網（英文版）上線，成為國內第一家垂直 B2B 電子商務商業網站。

1998 年 10 月美商網（又名「相逢中國」）得到多家美國知名 VC（奉獻投資 Venture Capital 的英文縮寫）千萬美金投資，是最早進入中國 B2B 電子商務市場的海外網站，首開 B2B 電子商務先河。

In December 1997, China National Chemical Information Center (English version) went online and became the first vertical B2B e-commerce commercial website in China.

In October 1998, the USBusinessNetworks (also known as MeetChina) received $10 million worth investment from a number of well-known VCs (short for Venture Capital) in the United States, and was the first overseas website to enter the Chinese B2B e-commerce market, pioneering in B2B electronic business.

2000 年 5 月卓越網成立，為中國早期 B2C 網站之一。

2003 年 12 月，慧聰網（08292-HK）香港創業板上市，為國內 B2B 電子商務首家上市公司。

2005 年中國電子商務交易額達到 7400 億，其中 B2B 電子商務市場交易

額達到了 6500 億元，佔中國整個電子商務市場交易額的 95%。

2006 年電商網站從最初的 B2B 企業對接、C2C 模式的零售平臺，逐漸發展到大型商務網站直接對消費者銷售的 B2C 網路商城模式。

In May 2000, Joyo.com was established as one of the earliest B2C websites in China.

In December 2003, HC International Inc (08292-HK) was listed on the Hong Kong Growth Enterprise Market and was the first listed company in China's B2B e-commerce.

In 2005, China's e-commerce transaction volume reached ¥740 billion, of which B2B e-commerce market transaction volume reached ¥650 billion, accounting for 95% of China's entire e-commerce market transaction volume.

In 2006, the e-commerce website gradually evolved from the original B2B enterprise docking and C2C model retail platform to the B2C e-shop model where large-scale business websites directly sell to consumers.

2009 － 2010 年團購模式迅速崛起，電子商務的發展由商品種類的變革，過渡到了電商模式的變革。

2010 年中國電子商務全年交易額達 4.5 萬億元，增長幅度是 GDP 增幅的 5 倍。中小企業網上交易和網路行銷利用率超過 4 成，相當於社會消費品零售總額的 3.3%，[48] 網購市場金額達 5231 億元。[49]

48　電子商務發展的新趨勢－雲計算於（與）數位技術相結合 [J/OL]. 中國服務外包網，2011 － 10 － 31. The new trend of e-commerce development － combination of cloud computing and digital technology [J/OL]. Chinasourcing, Oct 31, 2011.

49　黃婕 . 1 號店：燒錢邏輯不可持續 [J/OL].21 世紀經濟報導，2011. Huang Jie. No. 1Store: The logic of burning money is not sustainable [J/OL]. 21Jingji, 2011.

In 2009-2010, the group buying model rose rapidly. The development of e-commerce changed from the transformation of commodity types to the transformation of e-commerce model.

In 2010, the annual transaction volume of e-commerce in China reached ¥4.5 trillion, a growth rate of five times as that of GDP. The utilization rate of online transactions and network marketing of SMEs exceeded 40%, equivalent to 3.3% of the total retail sales of consumer goods,[48] and the online shopping market amounted to ¥523.1 billion.[49]

2011 年中國網購市場交易規模近 8000 億元，佔到社會消費品零售總額的 4.3%。[50] 2011 年中國網路購物人數為 1.94 億人，電子商務使用率 37.8%，[51] 電子商務市場規模達到 7736 億元人民幣。[52]

In 2011, the online shopping market in China was nearly ¥800 billion, accounting for 4.3% of the total retail sales of consumer goods.[50] In 2011, the number of online shoppers in China was 194 million, the e-commerce usage rate was 37.8%, and the e-commerce market reached ¥773.6 billion.[52]

2012 年中國線民人數為 5.65 億，電子商務市場整體交易規模為 8.1 萬億元，較 2011 年增長 27.9%，佔中國 GDP 的比例超過 15%，獨立運營電子商務企業超過 4 萬家。其中 B2B 電子商務市場規模達 6.61 萬億元，佔到電子

---

50 2011 年互聯網經濟核心資料 [R]. 艾瑞諮詢 , 2011. 2011 Internet Economy Core Data [R]. iResearch, 2011.

51 日本報告：中美日三國電商最新趨勢 [J/OL]. i 天下網商 , 2012. Japan's Report: The latest trends in e-commerce in China, the US and Japan [J/OL]. i.NET.com, 2012.

52 日本報告：中美日三國電商最新趨勢 [J/OL]. i 天下網商 , 2012. Japan's Report: The latest trends in e-commerce in China, the US and Japan [J/OL]. i.NET.com, 2012.

商務總交易規模的 81.6%。中小企業 B2B 電子商務佔電商總額 53.3%，規模以上企業 B2B 佔電商總額 28.3%；網路購物交易規模達到 18% 以上，其中零售市場份額達到 16.0%；線上旅遊交易規模佔比為 2.1%；B2C 交易規模達 3869.9 億元，佔整體網路購物市場交易規模的比重達到 29.7%，其餘 70.3% 為 C2C 市場交易份額；天貓以 56.7% 的市場份額位居 B2C 電子商務市場第一。

In 2012, the number of Internet users in China was 565 million. The overall transaction scale of the e-commerce market was ¥8.1 trillion, an increase of 27.9% compared with 2011, accounting for more than 15% of China's GDP, and there were more than 40,000 independent e-commerce companies. The B2B e-commerce market reached ¥6.61 trillion, accounting for 81.6% of the total e-commerce transaction volume. The scale of B2B e-commerce of middle and small-sized enterprises accounted for 53.3% of total e-commerce, and above-scale enterprises for 28.3%. Online shopping transactions reached 18%, of which retail market share reached 16.0%. Online travel transactions accounted for 2.1%. The B2C transaction volume reached ¥386.99 billion, accounting for 29.7% of the total online shopping market transaction volume, and the remaining 70.3% was the C2C market trading share. Tmall ranked first in the B2C e-commerce market with a market share of 56.7%.

2013 年中國電子商務交易總額超過 10 萬億元，其中網路零售交易額大約 1.85 萬億元，資料顯示中國已成為世界上最大的網路零售市場。商務部在「十二五」規劃電子商務發展指導意見中指出，到 2015 年中國網路零售額將達到社會消費品零售總額的 9% 以上。

In 2013, China's e-commerce transactions totaled more than ¥10 trillion, of which online retail transactions amounted to about ¥1.85 trillion. The data showed that China had become the world's largest online retail market. In the 「Twelfth Five-Year Plan」 e-commerce development guidance, the Ministry of Commerce pointed out that by 2015, China's online retail sales will reach more than 9% of the total retail sales of consumer goods.

據國家統計局電子商務交易平臺調查資料，2017 年中國電子商務交易額達 29.16 萬億元，同比增長 11.7%。全國網上零售額達 7.18 萬億元，同比增長 32.2%，約佔全球的 50%。[53]

According to the survey data of e-commerce transaction platform of the National Bureau of Statistics, the volume of e-commerce transactions in China reached ¥29.16 trillion in 2017, a year-on-year increase of 11.7%. The national online retail sales amounted to ¥7.18 trillion, a year-on-year increase of 32.2%, accounting for about 50% of that in the world.[53]

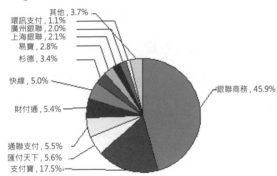

其他 Others / 環訊支付 IPS / 上海銀聯 UnionPay in Shanghai / 易寶 Yeepay 衫德 Sandpay 快線（錢）99bill / 財付通 Tenpay / 通聯支付 Allinpay / 匯付天下 ChinaPnR / 支付寶 Alipay

圖 Figure2-4　2011 年第四季度中國前 10 名 B2C 網站總收入市場份額 Market share of total revenue of China's top 10 B2C websites in the fourth quarter of 2011
資料來源：新線上零售創新與轉型 Source: New online retail innovation and transformation

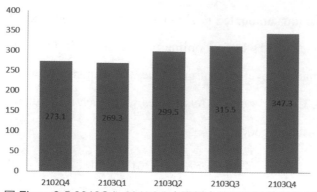

圖 Figure2-5 2012Q4--2013Q4 中國網上銀行交易規模（萬億人民幣）
2012Q4--2013Q4 China online banking transaction scale (trillion yuan)
資料來源：易觀國際、易觀智庫、中國互聯網商情 Source: Analysys International,
Analysys Think Tank, Chinasqbg.com

　　近年中國企業電子商務發展迅速。2012 年當當網實現營業收入 51.9 億元，近 4 年銷售收入年均增長 53.25%，新增註冊用戶近千萬。淘寶網自 2003 年成立以來發展迅速，銷售額從 2003 年的 2271 萬元增長至 2012 年的 10007 億元。2013 年 11 月，支付寶錢包用戶數接近 1 億。2013 年天貓商城「雙 11」活動創造了單日 350.19 億元的銷售奇蹟，相當於中國全國日均社會零售總額的 5 成。「雙 11」當日，支付寶交易筆數達 1.88 億筆，其中手機支付達到 4518 萬筆。支付寶一天的交易筆數已是全國日均刷卡總量的 5 倍。2015 年天貓雙 11 的 24 小時「全球賣」資料顯示，截至北京時間 11 月 12 日 16 點，速賣通雙 11 跨境出口共產生 2124 萬筆訂單，覆蓋到 214 個國家和地區，當天除中國外全球市場中每百人就有一人逛了速賣通。[54]

53　2017 年全國網上零售額約佔全球 50% 電子商務不斷催生新業態 . 中國經濟網 . 2018-06-05 In 2017, the National Online Retail Sales Accounted for about 50% of the World, E-commerce Continues to Generate New Business. China Economic Net. June 5, 2018

54　新浪科技 . 阿里速賣通雙 11 全球 214 個國家地區成交 2124 萬筆 .2015 年 11 月 13 日　Sina Technology. AliExpress' Number of Transactions at Double 11 in Globally 214 Countries and Regions Reached 21.24 Million. November 13, 2015

In recent years, Chinese enterprises have developed rapidly in e-commerce. In 2012, Dangdang achieved an operating income of ¥5.19 billion, with an average annual growth rate of 53.25% in sales in the past four years, and nearly 10 million new registered users. Since its establishment in 2003, Taobao has developed rapidly, with sales increasing from ¥22.71 million in 2003 to ¥1000.7 billion in 2012. In November 2013, the number of Alipay wallet users was close to 100 million. In 2013, Tmall's「Double Eleven」event created a sales miracle of ¥35.019 billion per day, equivalent to 50% of the total daily social retail sales in China. On the day of 「Double Eleven」, the number of Alipay transactions reached 188 million, of which 45.18 million were paid by mobile phone. The number of transactions in Alipay a day is five times that of the national average credit-card comsumption. According to data of 2015 Tmall Double Eleven's 24-hour 「Global Selling」, as of 16:00 Beijing time on November 12th, AliExpress Double Eleven cross-border exports generated a total of 21.24 million orders covering 214 countries and regions. One in 100 people in the global market except China had visited AliExpress.

表 Table 2-4 2013 年中國部分 B2C 購物網站銷售規模排名

序號 No.	B2C 網路購物 網站名稱 B2C e-shopping website	銷售規模（萬 元）Sales scale (ten thousand yuan)	經營範圍 Business scope	類型
1	天貓 Tmall	22000000	綜合百貨 Composite department store	平臺型 Platform-based
2	京東 JDcom	11000000	綜合百貨 Composite department store	自營為主 Self-support- based

3	小米 MIUI	3160000	手機為主 Cell phone-based	自營 Self-support
4	蘇寧易購 Suning E-commerce	2189000	綜合百貨 Composite department store	自營為主 Self-support-based
5	亞馬遜中國 Amazon China	1460000	綜合百貨 Composite department store	自營為主 Self-support-based
6	易迅網 Yixun	1200000	綜合百貨 Composite department store	自營為主 Self-support-based
7	1 號店 No.1 Store	1154000	綜合百貨 Composite department store	自營為主 Self-support-based
8	唯品會 Vipshop	1045000	名品折扣 Famous product discounts	自營 Self-support
9	QQ 網購 QQ Online Shopping	885000	綜合百貨 Composite department store	平臺型 Platform-based
10	凡客 Vancl	850000	服裝服飾 Clothes	垂直型 Vertical type
11	當當網 Dangdang	802020	綜合百貨 Composite department store	平臺型 Platform-based
12	聚美優品 Jumei	600000	化妝品 Cosmetics	自營 Self-support
13	國美線上（含庫巴網）GOME(Coo8 included)	325720	綜合百貨 Composite department store	自營為主 Self-support-based
14	樂蜂網 (Lafaso)	250000	化妝品 Cosmetics	自營 Self-support
15	迪信通移動生活商城 Dixintong mobile shopping mall	200000	手機 Cell phones	自營 Self-support
16	我買網 WoMaiCom	130000	食品 Food	自營為主 Self-support-based
17	宏圖三胞.慧買網 HISAP100	100000	3C	自營 Self-support
18	銀泰網 Yintai.com	85000	時尚百貨 Fashion house	自營為主 Self-support-based

| 19 | 優購網 Yougou.com | 80000 | 鞋帽箱包 Shoes, hats and bags | 自營 Self-support |
| 20 | 夢芭莎 Moonbasa | 63609 | 服裝服飾 Clothes | 自營為主 Self-support-based |

資料來源：中國連鎖經營協會 Source: China Chain Management Association

表 Table 2-5 當當網 2009-2012 年營業收入情況 2009-2012 Operating revenue of Dangdang

年份 Year	營業收入（億元） Operating revenue(one hundred million yuan)	收入同比增長 (%) Year-on-year growth	毛利率 (%)Gross profit rate(%)	淨虧損率 (%) Net loss rate(%)
2012 年	51.90	44	13.90	7.60
2011 年	36.00	59	13.80	10.50
2010 年	22.80	58	22.20	0
2009 年	14.58	56	20.00	-3.00

資料來源：中國連鎖經營協會 Source: CCFA

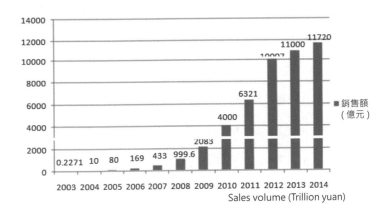

圖 Figure 2-6 淘寶網 2003--2014 年年銷售額情況 2003—2014 Taobao annual sales volume
資料來源：中國連鎖經營協會 Source: CCFA

（二）雲計算為「雲消費」提供技術支援
Cloud computing provides technical support for the cloud consumption

　　雲計算是繼 20 世紀 80 年代大型電腦到用戶端 - 伺服器的大轉變之後資訊技術領域的又一大巨變，描述了一種基於互聯網的新的 IT 服務增加、使用和交付模式。1999 年 IBM 提出透過一個網站向企業提供企業級的應用的概念，使雲計算從理論走向商用，這在雲計算的發展史上具有里程碑的意義。[55] 2007 年 6 月戴爾公司發佈的第 1 季度財報裡提到組建新的戴爾資料中心解決方案部門 Dell Data Center Solution Division，提供戴爾的雲計算 Cloud Computing 服務和設計模型，使客戶能夠根據他們的實際需求優化 IT 系統架構，這是雲計算這一名稱最早見於商用。[56] 而真正將雲計算名稱叫響的是亞馬遜 EC2 產品和 Google-IBM 分散式運算專案。2006 年亞馬遜 Amazon 公司開發了 EC2 產品，命名為 Elastic Computing Cloud（彈性計算雲），並取得了商業上的成功應用，這是目前公認最早的雲計算產品。2006 年，谷歌推出了 Google 101 計畫，並正式提出雲（Cloud) 的概念和理論。隨後，亞馬遜、微軟、惠普、雅虎、英特爾、IBM 等公司紛紛宣布了自己的雲計畫。2007 年 10 月初，Google 和 IBM 聯合與 6 所大學簽署協定，提供在大型分散式運算

---

55 廉琪 . 雲計算在移動學習中的應用探究 [J]. 中國資訊技術教育，2010（20）. Lian Qi. Application of Cloud Computing in Mobile Learning[J]. China Information Technology Education, 2010(20).

56 劉黎明 . 雲計算起源探析 [J]. 電信網技術，2010（9）. Liu Liming. Analysis of the Origin of Cloud Computing[J]. Telecommunications Network Technology, 2010(9).

系統上開發軟體的課程和支援服務，說明學生和研究人員獲得開發網路級應用軟體的經驗。「這種相對新的平行計算（有時也稱雲計算）」明確將雲計算做為一個新概念提出。此後由於 IBM 和 Google 公司在資訊科技領域的影響力，越來越多的媒體、公司、技術人員開始追逐雲計算，甚至將很多 IT 創新都放入雲計算概念中。[57]2010 年 5 月 21 日第二屆中國雲計算大會的媒體見面會上，時任微軟公司全球資深副總裁張亞勤提出微軟未來計畫將 80% 的資源和研發力量都投入到雲計算專案中。[58] 2012 年 3 月份的百度開發者大會上，百度公司正式發佈百度雲戰略，推出百度開發者中心網站。2012 年 10 月，推出僅 2 個月的百度雲個人用戶量已突破 1000 萬。2013 年 9 月，百度雲用戶破億，2016 年百度雲用戶數已達到 4 億。

Cloud computing was another big change in the information technology field after the client-server transformation of large-scale computers in the 1980s. It described a new Internet-based IT service addition, usage and delivery model. In 1999, IBM proposed a concept of providing enterprises with the enterprise-level application through a website, making cloud computing from theory to commercial, which was a milestone in the history of cloud computing. In June 2007, Dell' s first quarter earnings report mentioned the formation of the Dell Data Center Solution Division, which provided Dell' s Cloud Computing services and design models to enable customers to optimize the IT system framework according to actual needs, which was the first time that cloud computing was seen in commercial use. What

57　雲計算的起源 . 中國雲計算網 . The Origin of Cloud Computing. Yjs.cnelc.com.
58　微軟未來五年將把 80% 資源投入雲計算 . 網易科技報導 . Microsoft will Invest 80% of Its Resources in Cloud Computing in the Next Five Years. Netease Technology.

really made the cloud computing well known was the Amazon EC2 product and the Google-IBM distributed computing project. In 2006, Amazon developed the EC2 product, named Elastic Computing Cloud, and achieved commercial success. This was the earliest cloud computing product. In 2006, Google launched the Google 101 program and formally proposed the concept and theory of cloud. Subsequently, Amazon, Microsoft, Hewlett-Packard, Yahoo, Intel, IBM and other companies announced their own cloud plans. In early October 2007, Google and IBM signed an agreement with six universities to provide courses and support services for developing software on large distributed computing systems, helping students and researchers gain experience in developing network-level applications. Such relatively new parallel computing (sometimes also called cloud computing) clearly proposed the cloud computing as a new concept. Since then, due to the influence of IBM and Google in the field of information technology, more and more media, companies, and technicians have begun to chase cloud computing, and even put many IT innovations into the cloud computing concept. At the media meeting of the second China Cloud Computing Conference on May 21, 2010, Zhang Yaqin, senior vice president of Microsoft Corporation then, proposed that Microsoft planned to invest 80% of its resources and R&D efforts in cloud computing projects. At the Baidu Developer Conference in March 2012, Baidu officially released Baidu Cloud Strategy and launched the Baidu Developer Center website. In October 2012, the number of personal users of Baidu Cloud launched for just 2 months had exceeded 10 million. In September 2013, the number of Baidu Cloud users broke 100 million. In 2016, the number of Baidu Cloud users reached 400 million.

1·「雲計算」的基本概念 The basic concept of cloud computing

雲計算是資訊技術領域的一大熱點，雲計算並不是一門新的技術，而是一種基於用戶外部的資料中心為使用者提供各種服務的新型計算模式。20 世紀 60 年代，麥卡錫（John McCarthy）就提出了把計算能力做為一種像水和電一樣的公用事業提供給用戶，這是雲計算思想的最早體現。1997 年，南加州大學教授 Ramnath K. Chellappa 提出雲計算的第一個學術定義，認為計算的邊界可以不是技術侷限，而是經濟合理性。關於雲計算的定義，儘管各研究機構和專家從各個不同的角度、不同層面給予了不同的定義，但其核心思想是一致的，就是使計算無處不在，或者說是使資源最大化地共用和分享。從商業應用的角度來看，雲計算旨在把多個成本相對較低的計算實體整合成一個具有強大計算能力的完美系統，並借助 SaaS（Software- as- a- Service）、PaaS（Platform- as- a- Service）、IaaS（Infrastructure- as- a-Service）等先進的商業模式把強大的計算能力分佈到終端使用者手中。

Cloud computing is a hotspot in the field of information technology. Cloud computing is not a new technology, but a new computing model that provides users with various services based on data centers outside the user. In the 1960s, John McCarthy proposed to provide computing power as a utility like water and electricity to users. This was the earliest manifestation of cloud computing. In 1997, Ramnath K. Chellappa, a professor of University of Southern California proposed the first academic definition of cloud computing, arguing that the boundaries of computing could not be technical limitations, but economic rationality. Regarding the definition of cloud computing, although various research institutions and experts have given different definitions from different angles and levels, the core idea is the same, that is, to make computing everywhere, or to maximize the sharing of resources. From the point of view of business application, cloud computing aims to integrate multiple relatively low-cost computing entities into a perfect system with powerful computing power. And with the help of advanced business models like SaaS (Software-as-a-Service), PaaS (Platform-as-a-Service) and IaaS (Infrastructure-as-a-Service), it distributes powerful computing power to end users.

知識連結：Knowledge link:

關於雲計算的定義，不同研究機構和專家從各個不同的角度、不同層面給予了不同的定義。

Regarding the definition of cloud computing, different research institutions and experts have given different definitions from different angles and levels.

雲計算在維基百科上名為 Cloud Computing，被解釋為是分散式運算技術的一種，其最基本的概念，是透過網路將龐大的計算處理常式自動分拆成無數個較小的副程式，再交由多部伺服器所組成的龐大系統經搜尋、計算分析之後將處理結果回傳給用戶。

Cloud computing, as we can see on Wikipedia, is interpreted as a kind of distributed computing technology. The most basic concept is to automatically split a huge computational processor into a myriad of smaller subroutines through the network. Then after being searched, calculated and analyzed by the huge system composed of multiple servers, the processing result is transmitted back to the user.

狹義的雲計算是指 IT 基礎設施的交付和使用模式，指透過網路以按需、

易擴展的方式獲得所需的資源（硬體、平臺、軟體）。提供資源的網路被稱為「雲」。「雲」中的資源在使用者看來是可以無限擴展的，並且可以隨時獲取、按需使用、隨時擴展、按使用付費，如同像使用水電一樣使用 IT 基礎設施。之所以稱之為雲，主要是因為它在某些方面具有現實中雲的特徵：雲一般都很大，其規模可以動態伸縮且邊界是模糊的；雲在空中飄忽不定，無法也不需要確定它的具體位置，但它確實存在於某處。

Narrowly defined cloud computing refers to the delivery and usage model of IT infrastructure, which means that the required resources (hardware, platform, software) are obtained through the network in an on-demand and scalable manner. The network that provides the resources is called the cloud. The resources in the cloud can be expanded in the user's view and can be acquired at any time, used on demand, expanded at any time, and paid for by use, just like using IT infrastructure like water and electricity. The reason why it is called cloud is mainly because it has the characteristics of cloud in reality: the cloud is generally large and can be dynamically scaled, and the boundary of it is blurred. Cloud is erratic in the air, and its specific location cannot or doesn not have to be determined, but it does exist somewhere.

廣義的雲計算是指服務的交付和使用模式，指透過網路以按需、易擴展的方式獲得所需的服務。這種服務可以是 IT 和軟體、平臺相關的服務，也可以是任意其他的服務。所有這些網路服務可以理解為網路資源，眾多資源形成所謂「資源池」。這種資源池我們通常稱為「雲」。可以說「雲」是一些可以自我維護和管理的虛擬計算資源，通常為一些大型伺服器集群，包括計

算伺服器、存儲伺服器、寬頻資源等等。雲計算將所有的計算資源集中起來，並由軟體實現自動管理，無須人為參與。這使得應用提供者無須為繁瑣的細節而煩惱，能夠更加專注於自己的業務，有利於創新和降低成本。有人打了個比方：這就好比是從古老的單臺發電機模式轉向了電廠集中供電的模式。它意味著計算能力也可以做為一種商品進行流通，就像煤氣、水電一樣，取用方便，費用低廉。最大的不同在於，它是透過互聯網進行傳輸的。

Broadly defined cloud computing refers to the delivery and use models of services, which means that the required services are obtained on-demand and easily expanded through the network. The services can be related to IT, software and platform, or any other services. All of these network services can be understood as network resources, and many resources form a so-called resource pool. This kind of resource pool is often referred to as the cloud. It can be said that cloud is virtual computing resources that can be self-maintained and managed, usually a cluster of large servers, including computing servers, storage servers, broadband resources and so on. Cloud computing centralizes all computing resources and is automatically managed by software without human intervention. This makes it unnecessary for application providers to worry about cumbersome details and enables them to focus more on business, doing good to the innovation and reducing costs. Someone made an analogy: this is like switching from the old single generator mode to the centralized power supply mode of the power plant. It means that computing power can also be circulated as a commodity, just like gas and water-easy to access and the cost is low. The biggest difference is that cloud computing is transmitted via the Internet.

技術層面的雲計算是指虛擬化（Virtualization）、效用計算（Utility Computing）、網格計算（Grid Computing）、海量分散式存儲技術、資料管理技術、分散式資源管理技術、雲計算平臺管理技術，甚至是綠色節能技術形成的一個網路應用框架。

The cloud computing at the technical level refers to a network application framework formed by Virtualization, Utility Computing, Grid Computing, massive distributed storage technology, data management technology, distributed resource management technology, and cloud computing platform management technology and even green energy-saving technology.

正如 Google CEO 埃里克•施密特博士所指出的，雲計算與傳統的以 PC 為中心的計算不同，它把計算和資料分佈在大量的分散式運算機上，這使得計算力和存儲獲得了很強的擴張能力，並方便了使用者透過多種接入方式方便地接入網路獲得應用和服務。其重要特徵是開放式的，不會有一個企業能控制和壟斷它。

As Dr. Eric Schmidt, Google CEO of Google pointed out, cloud computing is different from traditional PC-centric computing in that it distributes computing and data across a large number of distributed computers, which gives computing power and storage strong expansion capability and facilitates users to easily obtain applications and services through multiple access methods. Its important feature is that it is open-ended, and no company can control and monopolize it.

2 · 「雲計算」的基本特徵 The basic features of cloud computing

雲計算既是一種技術，也是一種服務，是一種商業模式。其基本特徵包括：

Cloud computing is both a technology and a service. It is a business model. Its basic features include:

（1）以網路為中心 It is network-centric

雲計算的元件和整體架構由網路連接在一起並存在於網路中，同時透過網路向使用者提供服務。

The components and overall structure of the cloud computing are connected by the network and exist in the network, while providing services to users through the network.

（2）以服務為提供方式 It meets users' needs with service

有別於傳統的一次性買斷統一規格的有形產品，雲服務的提供者可以從一片大雲中切割、組合或塑造出各種型態特徵的雲以滿足不同使用者的多層次個性化需求。

Different from the traditional one-time buyout of tangible products of unified specifications, cloud service providers can cut, combine or shape clouds of various morphological features from a large cloud to meet the multilevel personalized needs of different users.

（3）資源的池化和透明化 Pooling and transparency of resources

對雲服務的提供者而言，各種底層資源（計算/存儲/網路/邏輯資源等）的異構性（如果存在某種異構性）被遮罩，邊界被打破，所有資源可以被統

一管理、調度，成為所謂的「資源池」，從而為使用者提供按需服務；對用戶而言，這些資源是透明的、無限大的，用戶無須瞭解資源池複雜的內部結構、實現方法和地理分佈等，只需要關心自己的需求是否得到滿足。

For cloud service providers, the heterogeneity (if exists) of various underlying resources (computing/storage/network/logical resources, etc.) is blocked, boundaries are broken, and all resources can be managed and scheduled uniformly, becoming a so-called "resource pool" to provide users with on-demand services. For users, these resources are transparent and infinite, users do not need to understand the complex internal structure, implementation methods and geographical distribution of resource pools, only need to care if their needs are met.

（4）高擴展和高可靠性 High scalability and reliability

雲計算快速靈活高效安全地滿足海量用戶的海量需求，有大容量、高彈性、安全可靠、快速回應的底層技術架構。[59]

Cloud computing is fast, flexible, efficient and secure to meet the massive demand of massive users. It has an underlying technical structure with high-capacity, high-elasticity, security, and rapid response.[59]

（5）雲計算的產生和發展是需求推動、技術進步以及商業模式轉換共同作用的結果 The emergence and development of cloud computing is the result of a combination of driving demand, technological advancement, and business model transformation.

---

59 樊勇兵．陳天．陳楠．理解雲計算 [J]．廣東通信技術．2010（11）：2-6 Fan Yongbing, Chen Tian, Chen Nan. Understanding Cloud Computing[J]. Guangdong Communication Technology. 2010(11): 2-6

用戶需求是雲計算發展的動力，技術進步是雲計算發展的基礎，而商業模式的轉換則是雲計算發展的內在要求，也是用戶需求的外在體現，並且需要技術的發展來推動和實現。

User demand is the driving force of cloud computing development. Technology advancement is the foundation of cloud computing development. The transformation of business model is the inherent requirement of cloud computing development and also the external manifestation of user demand, which needs the development of technology to promote and realize.

知識連結：Knowledge link:

與雲計算的相關概念包括：雲伺服器、雲存儲、雲主機、雲資料庫、雲作業系統（雲管理）、雲安全、雲平臺等。其中雲資料庫、雲平臺、雲伺服器、雲存儲定義如下：

Concepts related to cloud computing include: cloud servers, cloud storage, cloud hosting, cloud databases, cloud operating systems (or cloud management), cloud security, cloud platforms, and so on. The cloud database, cloud platform, cloud server, and cloud storage are defined as follows:

雲資料庫是在 SaaS (software-as-a-service: 軟體即服務) 成為應用趨勢的大背景下發展起來的雲計算技術，它極大地增強了資料庫的存儲能力，消除了人員、硬體、軟體的重複配置，讓軟、硬體升級變得更加容易，同時也虛擬化了許多後端功能。雲資料庫具有高可擴展性、高可用性、採用多組形式和支持資源有效分發等特點。總的來說，雲資料庫是部署和虛擬化在雲計算環境中的資料庫。[60] 使用雲資料庫的使用者不能控制運行著原始資料庫的機

器，也不必瞭解它身在何處。

The cloud database is a cloud computing technology developed in the context of SaaS (software-as-a-service) becoming the trend of application. It greatly enhances the storage capacity of the database and eliminates repeated configuration of personnel, hardware, and software, making software and hardware upgrades easier, also virtualizing many backend functions. Cloud databases are highly scalable, highly available, multi-tenant and support efficient resource distribution. In general, a cloud database is a database that is deployed and virtualized in a cloud computing environment.[60] Users who use a cloud database can't control the machine running the original database, and don't have to know where it is either.

「雲平臺」是基於行業領先技術搭建的，集硬體、軟體、網路基礎設施、資料中心為一體的應用導向性的服務平臺，它將企業的各類資訊化需求按功能拆分成不同的模組，以標準化元件的形式集成在這一平臺之上。「雲平臺」所提供的應用服務均透過互聯網提供給使用者，「雲」中的資源對使用者來說可以隨時獲取、按需使用、隨時擴展、按使用付費。因此，「雲平臺」具有開放性、可擴展性，支持無縫升級，其標準化介面能夠靈活對接多種應用服務，使服務內容能夠不斷擴展延伸。而使用者可以根據自身的需要靈活選用平臺上的各類服務，按使用付費，如同使用水電一樣，無須掌握 IT 技術，即可輕鬆應用。

---

60 Yoon JP. Access control and trustiness for resource management in cloud databases. In: Fiore S, Aloisio G, eds. Proc. of the Int' l. Conf. on Grid and Cloud Database Management. Berlin: Springer-Verlag, 2011. 109   131. [doi: 10.1007/978-3-642-20045-8_6]

Cloud platform is an application-oriented service platform based on industry-leading technology, integrating hardware, software, network infrastructure and data center. It splits various information needs of enterprises into different modules according to functions and integrates them on this platform in the form of standardized components. The application services provided by cloud platform are provided to users via the Internet. The resources in the cloud can be acquired by the user at any time, used on demand, expanded at any time, and paid for by usage. Therefore, cloud platform is open and extensible, supports seamless upgrade, and its standardized interfaces can flexibly connect with multiple application services, so that the service content can be extended constantly. Users can flexibly choose various services on the platform according to their own needs, and pay for it by usage. Just like consuming water and electricity, it can be easily applied without having to master IT technology.

雲伺服器是一種類似 VPS 伺服器的虛擬化技術。是一種基於 WEB 服務，提供可調整雲主機配置的彈性雲技術，整合了計算、存儲與網路資源的 Iaas 服務，具備按需使用和按需即時付費能力的雲主機租用服務。[61]

A cloud server is a virtualization technology similar to the VPS server. It is a web-based service that provides flexible cloud technology that can adjust cloud host configuration, integrating the Iaas service of computing, storage and network resources. It has the cloud host renting service with capabilities of on-demand using

---

61 伺服器租用、VPS、雲伺服器主流對比 . 新網 . 2012 年 Comparison Among Server Rental, VPS, Cloud Server Mainstream. ChinaDNS. 2012

and real-time charging.

雲存儲指透過集群應用、網格技術或分散式檔案系統等功能,將網路中大量各種不同類型的存放裝置透過應用軟體集合起來協同工作,共同對外提供資料存儲和業務訪問功能的一個系統。可以說雲存儲是一個以資料存儲和管理為核心的雲計算系統。例如,雲筆記、雲硬碟、網盤等都是這個概念。

Cloud storage refers to a system that integrates a large number of different types of storage devices in a network through application software through cluster application, grid technology, or distributed file system to jointly provide data storage and service access functions. It can be said that cloud storage is a cloud computing system with data storage and management as its core. For example, cloud notes, cloud drives, and network drives are all based on the concept.

(三)大數據應用使「雲消費」成為可能 Application of big data makes cloud consumption possible

自人類社會產生以來,不同社會經濟發展階段產生不同的發展需求,依賴不同的核心產業資源。我們都知道,農業社會的核心資源是土地;工業社會的核心資源是能源,而資訊化社會的核心資源則是大數據及資料分析方法。

Since the beginning of human society, different stages of social and economic development have produced different development needs and depended on different core industrial resources. We all know that the core resource of agricultural society is land, the core resource of industrial society is energy, and the core resource of information society is big data and data analysis methods.

隨著全球 PC 與智慧終端機的普及，全球資料正以每年超過 50% 的速度爆發式增長。截止到 2012 年，全球資料量已經從 TB（1024GB=1TB）級別躍升到 PB（1024TB=1PB）、EB（1024PB=1EB）乃至 ZB（1024EB=1ZB）級別。國際資料公司（IDC）的研究結果顯示，2008 年全球產生的資料量為 0.49ZB，2009 年的資料量為 0.8ZB，2010 年增長為 1.2ZB，2011 年的數量更是高達 1.82ZB，相當於全球每人產生 200GB 以上的資料。到 2020 年全球資料資料存儲量將超過 40ZB。[62] 而到 2012 年為止，人類生產的所有印刷材料的資料量是 200PB，全人類歷史上說過的所有話的資料量大約是 5EB。IBM 的研究稱，整個人類文明所獲得的全部資料中，有 90% 是過去兩年內產生的。到了 2020 年，全世界所產生的資料規模將達到今天的 44 倍。

With the popularity of PCs and smart devices around the world, global data is exploding at an annual rate of more than 50%. As of 2012, the global data volume has jumped from TB (1024GB = 1TB) level to PB (1024TB = 1PB), EB (1024PB = 1EB) and even ZB (1024EB = 1ZB). Reserch by the International Data Corporation (IDC) show that the amount of data generated globally in 2008 was 0.49ZB, 0.8ZB in 2009, and 1.2ZB in 2010. The number was as high as 1.82ZB in 2011, which was equivalent to each person in the world producing more than 200 GB of data. By 2020, global data storage will exceed 40ZB.[62] By 2012, the amount of data for all printed materials produced by humans was 200 PB, and the amount of data for all words spoken in the history of mankind was about 5 EB. According to research by IBM, 90% of all data obtained by human civilization has been produced in the past two years. By 2020, the world will produce 44 times the size of today's data.

「大數據」的爆發正在得到從企業界到政府層面越來越多的重視，所能帶來的巨大商業價值已經被廣泛認為將引領一場足以匹敵 20 世紀電腦革命的巨大變革。《華爾街日報》發表文章〈科技變革即將引領新的經濟繁榮〉大膽預見：「我們再次處於三場宏大技術變革的開端，他們可能足以匹敵 20 世紀的那場變革，這三場變革的震中都在美國，他們分別是大數據、智慧製造和無線網路革命。」瑞士達沃斯論壇《大數據，大影響》（Big Data, Big Impact）的報告宣稱，資料已經成為一種新的經濟資產類別，就像貨幣或黃金一樣。[63] 可以說，當前大到國家戰略、社會經濟運行的各個層面，小到小微企業的商業決策，都離不開大數據，大數據的迅速增長及相關技術的發展正在帶來全新的商業機遇。

The outbreak of big data is gaining more and more attention from business circle to the government level. The great commercial value it can bring is widely believed to lead to a great change that rivals the computer revolution of the 20th century. The Wall Street Journal published an article Technological change will Lead to new Economic Prosperity, which had boldly foreseen that 「we are once again at the beginning of three grand technological changes, which may be enough to match the change of the 20th century. The epicenters of these three changes are

62　維克托·邁爾·舍恩伯格，肯尼士·庫克耶. 大數據時代 [M]，盛楊燕，周濤，譯. 浙江人民出版社，2013. Viktor Mayer-Schönberger, Kenneth Cukier. Big Data Era [M], Translated by Sheng Yangyan, Zhou Tao. Zhejiang People's Publishing House, 2013.

63　郭建偉，李瑛，杜麗萍，趙桂芬，蔣繼婭. 基於 hadoop 平臺的分散式資料採擷系統研究 [J]. 中國科技資訊，2013（13）. Guo Jianwei, Li Wei, Du Liping, Zhao Guifen, Jiang Jiya. Research on Distributed Data Mining System Based on Hadoop Platform[J].CHINA SCIENCE AND TECHNOLOGY INFORMATION, 2013(13).

all in the United States. They are revolutions of big data, smart manufacturing and wireless network respectively.」 Big Data, Big Impact by the Swiss Davos Forum claims that data has become a new kind of economic asset, just like money and gold.[63] It can be said that currently, from national strategy, and social and economic operations to small and micro enterprises business decisions, are all inseparable from big data. The rapid growth of big data and the development of related technologies are bringing new business opportunities.

1· 大數據的基本概念和特徵
The basic concepts and characteristics of big data

對大數據的定義繁多，主要有幾種代表性提法：

研究大數據的先驅麥肯錫公司在其報告《Big data: The nextfrontier for innovation, competition,and productivity》（《大數據：下一個創新、競爭和生產力的前沿》）中給出的大數據定義是：大數據指的是大小超出常規的資料庫工具獲取、存儲、管理和分析能力的資料集。但它同時強調，並不是說一定要超過特定 TB 值的資料集才能算是大數據。[64]

There are many definitions of big data. The most representative ones are as follows:

The definition of big data given by McKinsey & Company, the pioneer of

---

64　趙國棟 . 大數據時代的歷史機遇：產業變革與資料科學 [M]. 清華大學出版社 . 2013 年 6 月 1 日 Zhao Guodong. Historical Opportunities in the Age of Big Data: Industrial Transformation and Data Science [M]. Tsinghua University Press. June 1, 2013

65　趙國棟 . 大數據時代的歷史機遇：產業變革與資料科學 [M]. 清華大學出版社 . 2013 年 6 月 1 日 Zhao Guodong. Historical Opportunities in the Age of Big Data: Industrial Transformation and Data Science [M]. Tsinghua University Press. June 1, 2013

big data, in its report Big data: The next frontier for innovation, competition, and productivity is that big data refers to data sets that go beyond the capabilities of acquisition, storage, management, and analysis of conventional database tool. But it also emphasizes that it is not that data sets that must exceed a certain TB value can be considered as big data.[64]

維基百科對大數據的定義是：巨量資料 (big data)，或稱大數據，指的是所涉及的資料量規模巨大到無法透過目前主流軟體工具，在合理時間內達到擷取、管理、處理並整理成為幫助企業經營決策更積極目的的資訊。[65]

Wikipedia defines big data as「huge amount of data involved that cannot be reached, managed, processed and organized into information that is more positive for helping make business decisions in a reasonable amount of time through current mainstream software tools.[65]

互聯網週刊提出，"大數據"的概念遠不只大量的資料（TB）和處理大量資料的技術，或者所謂的「4個V」之類的簡單概念，而是涵蓋了人們在大規模資料的基礎上可以做的事情，而這些事情在小規模資料的基礎上是無法實現的。換句話說，大數據讓我們以一種前所未有的方式，透過對海量資料進行分析，獲得有巨大價值的產品和服務，或深刻的洞見，最終形成變革之力。[66]

According to Internet Weekly, the concept of big data goes far beyond the

---

66 維克托•邁爾 - 舍恩伯格（Viktor Mayer-Schönberger）、肯尼士•庫克耶（Kenneth Cukier）、盛楊燕、周濤 . 大數據時代 [M]. 浙江人民出版社 . 2013 年 Viktor Mayer-Schönberger, Kenneth Cukier. Big Data Era [M], Translated by Sheng Yangyan, Zhou Tao. Zhejiang People's Publishing House, 2013.

large amount of data (TB) and the technology that processes large amount of data, or the so-called "four Vs" simple concepts, but covers things that can be done on the basis of large-scale data, which cannot be done on the basis of small-scale data. In other words, big data allows us to obtain products and services of great value, or deep insights, in an unprecedented way, by analyzing massive amount of data, and ultimately forms the power of change.

研究機構 Gartner 對「大數據」的定義是：需要新處理模式才能具有更強的決策力、洞察發現力和流程優化能力的海量、高增長率和多樣化的資訊資產。從資料的類別上看，"大數據"指的是無法使用傳統流程或工具處理或分析的資訊。它定義了那些超出正常處理範圍和大小、迫使用戶採用非傳統處理方法的資料集。

Gartner, a Research institution, defines big data as a massive, high-growth, and diverse information asset that requires new processing models to have greater capabilities of decision making, insight discovery and process optimization. From the category of data, big data refers to information that cannot be processed or analyzed by traditional processes or tools. It defines data sets that are beyond the normal processing range and size, forcing users to adopt non-traditional processing methods.

亞馬遜網路服務（AWS）、大數據科學家 JohnRauser 提出相似的定義：大數據就是任何超過了一臺電腦處理能力的龐大數據量。

基於以上認知，業界將大數據歸結為四個主要特徵，或稱 4 個 V：Volume（大量）、Velocity（高速）、Variety（多樣）、Value（價值）。

Volume 指數據體量巨大，從 TB 級別躍升到 PB 級別；Velocity 指處理速度快，符合 1 秒定律（要求在秒級時間範圍內給出分析結果，否則就失去價值。這個資料要求是大數據處理技術和傳統資料採擷技術的最大區別。）Variety 指資料類型繁多，包括各種形式、各種內容的資訊；Value 指個體資料本身價值密度低，但大數據及資料分析商業價值高。

Amazon Web Services (AWS) and big data scientist John Rauser proposed a similar definition: Big data is any huge amount of data that exceeds the processing power of a computer.

Based on this definition, the industry attributes big data to four main features, or four Vs. They are Volume, Velocity, Variety, and Value. Volume refers to the huge amount of data, jumping from the TB level to the PB level. Velocity means that the processing speed is fast, in line with the One Second Law that requires the analysis result in the second level, otherwise the result would lose value, which is the biggest difference between big data processing technology and traditional data mining technology. Variety refers to a wide variety of data types, including information of various forms and various contents. Value refers to fact that value density of individual data itself is low while commercial value of big data and data analysis is high.

2 · 大數據的三種主要類型 Three main types of big data

根據資料的應用情況不同，大數據可以分為興趣資料、需求資料、交易資料三種主要類型。

According to the application of data, big data can be divided into three main types: interest data, demand data, and transaction data.

（1）興趣數據 Interest data

興趣資料指基於共同的興趣或者關係產生的海量資料。如微信基於朋友、親戚的強關係，微博基於共同的興趣或對同一事物關注的弱關係，它們產生編織的一張強大的社交網路圖譜背後則是海量的使用者資料。

Interest data refers to massive amount of data generated based on common interests or relationships. For example, WeChat is based on the strong relationship between friends and relatives. Weibo is based on the weak relationship of common interest or the attention on the same thing. Massive amount of user data is behind the powerful social network map generated by interest data.

微博的資訊以點對面的形式傳播，在這張基於興趣建立起來的社交網路上，使用者不但可以完成陌生人之間的交流，也可以完成資訊、音樂、電影甚至服務等內容的發現、推薦和送達。透過多樣化的興趣內容與關係，完成其商業化及行銷產業鏈條的建立。據統計，新浪微博每天刷新微博數 1 億條，即產生 1 億條興趣資料。微信則以點對點的對話為主，在基於熟人關係建立起來的社交工具上，微信實現移動通訊、私密生活分享、心情分享等功能，借助 O2O、移動支付建立消費閉環，實現與用戶日常生活的緊密綁定從而保持黏性，同時微信將社交關係鏈注入遊戲，使社交遊戲煥發出全新生機。

Information on Weibo is spread in a point-to-point manner. On this social network based on interest, users can not only communicate with strangers, but also discover, recommend and deliver contents such as messages, music, movies and even services. Through the diversified interests and relationships, it finishes commercialization and establishes the marketing industry chain. According to statistics, Sina Weibo refreshed 100 million microblogs every day, which generated 100 million pieces of interest data. WeChat focuses on peer-to-peer dialogue. On social tools based on acquaintance relationships, WeChat implements functions such as mobile communication, private life sharing and mood sharing. With the help of a closed loop of consumption established by O2O and mobile payment, it achieves tight bindng with users' daily lives, thus maintaining stickiness. In the same time, it injects the social relationship chain into the game, giving social games a new life.

（2）需求資料 Demand data

需求資料指根據需求進行搜尋產生的海量資料。如百度每天回應來自 138 個國家超過數億次的搜索請求。用戶可以透過百度主頁，在瞬間找到相關的搜索結果，這些結果來自於百度超過 10 億的中文網頁資料庫，並且這些網頁的數量每天正以千萬級的速度在增長。

Demand data refers to the massive data generated by searching according to demand. For example, Baidu responds to hundreds of millions of search requests from 138 countries every day. Users can find relevant search results in an instant through Baidu's homepage. These results come from more than 1 billion Chinese webpage databases of Baidu, and the number is growing at a rate of tens of millions every day.

（3）交易資料 Transaction data

交易資料指透過實現交易所產生的海量資料。如淘寶平日每天發出的包

裏超過 1200 萬單，每分鐘成交 8300 多筆訂單。⁶⁷ 2013 年淘寶雙 11「光棍節」支付寶交易額達 350.19 億元，相當於中國日均社會零售總額的 5 成。總成交筆數達到 1.71 億，支付寶實現成功支付 1.88 億筆，最高每分鐘支付 79 萬筆。

Transaction data refers to the massive data generated by the implementation of the transaction. For example, Taobao sent out more than 12 million packages per day and achieved more than 8,300 orders per minute.67 In 2013, on Taobao' s Double Eleven, or the Singles' Day, Alipay transaction amounted to ¥35.019 billion, equivalent to 50% of China's daily average social retail sales. The total number of transactions reached 171 million, and successful payments on Alipay was 188 million, with a maximum of 790,000 per minute.

3‧ 應用大數據的 4 個關鍵能力
Four key capabilities of applying big data

更好地應用好大數據主要需要大容量的存儲、高併發的處理、統計分析、智慧推薦 4 個關鍵能力。

Better application of big data mainly requires four key capabilities: large-capacity storage, high-concurrency processing, statistical analysis, and intelligent recommendation.

（1）大容量的存儲 Large-capacity storage

67　淘寶數據：每分鐘成交 8 千筆 包裹每天超千萬 [J/OL]. 錢江晚報‧2013 年 05 月 08 日

大數據存儲是大數據應用的基礎，它有很多實現方式。EMC Isilon 存儲事業部總經理楊蘭江概括說，大數據存儲應該具有以下一些特性：海量資料存儲能力，可輕鬆管理 PB 級乃至數十 PB 的存儲容量；具有全域命名空間，所有應用可以看到統一的檔案系統視圖；支援標準介面，應用無須修改可直接運行，並提供 API 介面進行物件導向的管理；讀寫性能優異，聚合頻寬高達數 GB 乃至數十 GB；易於管理維護，無須中斷業務即可輕鬆實現動態擴展；基於開放架構，可以運行於任何開放架構的硬體之上；具有多級資料冗餘，支援硬體與軟體冗餘保護，資料具有高可靠性；採用多級存儲備份，可靈活支援 SSD、SAS、SATA 和磁帶庫的統一管理。[68]

Big data storage is the foundation of big data applications, and it has many implementations. Yang Lanjiang, general manager of Storage Division of EMC Isilon, said that big data storage should have the following characteristics. It should:

a) have massive data storage capacity, which can easily manage storage capacity of PB level or even tens of PB.

b) have overall namespace so that all applications share a unified file system view.

c) support standard interface so that application can run directly without modification, and provide API interface for object-oriented management.

d) have excellent read and write performance, with aggregate bandwidth up to

---

68 郭濤. 辨析大數據存儲 [J]. 賽迪網－中國電腦報，2014（6）. Guo Tao. Discrimination of Big Data Storage [J].CCID--China InfoWorld, 2014(6).

several GB or even tens of GB.

e) be easy to manage and maintain so that dynamic expansion can be easily implemented without interrupting business.

f) be based on the open architecture so that it can run on any open- architectured hardware.

g) have multi-level data redundancy which supports hardware and software redundancy protection, ensuring high reliability of data.

h) adopt multi-level storage backup to flexibly support Unified management of SSD, SAS, SATA and tape libraries.

（2）高併發的處理 High-concurrency processing

高併發的處理指同時處理海量資料的能力，如春節期間 12530 網站同時處理數百萬人集中購票、「雙十一」期間淘寶同時處理 350 億元消費額的交易等等。為了達到快速高效的處理大量資料的能力，整個 IT 基礎設施需要進行整體優化設計，應充分考量後臺資料中心的高節能性、高穩定性、高安全性、高可擴展性、高度冗餘、基礎設施建設這六個方面，同時更需要解決大規模節點數的資料中心的部署、高速內部網路的構建、機房散熱以及強大的資料備份等問題。

Highly-concurrency processing refers to the ability to process massive data at the same time. For example, during the Spring Festival, 12530.com simultaneously processed millions of people to purchase tickets in a centralized manner; during the Double Eleven, Taobao handled the transaction worth ¥35 billion at the same time. In order to achieve the ability to process large amount of data quickly and efficiently, the entire IT infrastructure needs to be optimized overall. High energy efficiency, high stability, high security, high scalability, high redundancy and infrastructure construction, all six aspects of the background data center should be fully considered. It also needs to solve the problems of data center deployment of large-scale node, high-speed internal network construction, computer room cooling and powerful data backup.

（3）統計分析 Statistical analysis

統計分析指對海量資料進行統計分析並得出結論的能力。大數據最為核心的就要看對於大量資料的核心分析能力，不僅存在於資料管理策略、資料視覺化與分析能力等方面，從根本上也對資料中心 IT 基礎設施架構甚至機房設計原則等提出了更高的要求。

Statistical analysis refers to the ability to perform statistical analysis and draw conclusions on massive data. The core of big data depends on the core analysis capabilities on massive data. The core not only lies in data management strategies, data visualization and analysis capabilities, but also makes higher requirements on the IT infrastructure framework of data center and even the design principles of the computer room.

（4）智慧推薦 Intelligent recommendation

智慧推薦指根據資料統計分析結果預測以後的行為，從而進行目的性的推薦。如京東、淘寶、1 號店等電子商務網站透過消費的消費資料可以分析出消費者消費習慣，從而定期推送適合消費者的商品。如對於曾經購買過嬰兒用品的新手媽媽集中推送其青睐的品牌嬰兒用品；對於關注孩子學習的家長集中推送多種教輔材料；對於定期購買生鮮產品的家庭主婦集中推送生鮮

新品和打折生鮮產品等。

Intelligent recommendation refers to predicting future behavior based on statistical analysis results of data, thereby making targeted recommendations. E-commerce websites such as Jingdong, Taobao, and No. 1 Store can analyze spending habits of consumers through their consumption data, and regularly push products suitable for consumers. For example, for the novice mothers who have purchased baby products intensively, they will push their favorite branded baby products. For parents who care about children's study, they will push a variety of teaching materials. For housewives who regularly purchase fresh products, they will push new products or fresh products on discount.

4· 大數據的應用 Application of big data

從全球範圍看，「大數據」正在對社會經濟領域的各個層面都造成越來越深刻的影響，大到國家層面，小到小微企業發展，決策行為將日益依賴於資料分析，而不是像過去更多憑藉經驗和直覺。

From a global perspective, big data is exerting more and more profound impacts on all levels of the social economy. From the national level to the development of small and micro enterprises, decision-making will increasingly rely on data analysis rather than experience and intuition in the past.

2013 年 3 月 29 日，美國歐巴馬政府在白宮網站發佈《大數據研究和發展倡議》（Big Data Research and Development Initiative），標誌著美國政府把「大數據」上升到國家戰略的層面。2013 年 7 月，中國上海市科學技術委

員會發佈《上海推進大數據研究與發展三年行動計畫》（2013-2015 年），
同時成立「上海大數據產業技術創新戰略聯盟」。

On March 29, 2013, the Obama administration released the Big Data Research
and Development Initiative on the White House website, marking that big data
rised to the level of national strategy. In July 2013, the Shanghai Municipal Science
and Technology Commission issued the Three-Year Action Plan for Promoting
Big Data Research and Development in Shanghai (2013-2015) and in the same
time, established the Strategic Alliance of Shanghai Big Data Industry Technology
Innovation.

在自然災害、公共衛生、經濟預測等領域，「大數據」的預見能力已經
初顯。2009 年，谷歌公司透過應用大數據成功預測了一輪新的甲型 H1N1 流
感爆發，與官方資料的相關性高達 97%。為此，可以準確判斷流感是從哪裡
傳播出來的，範圍大概有多廣，他們的判斷非常及時，不會像疾控中心一樣
要在流感爆發一兩週之後才可以做到。[69] 2013 年 8 月，在東日本大地震中，
由於受災地區範圍較廣，日本政府無法及時準確地掌握災害資訊。為此，日
本政府決定，透過與民間力量合作的形式，在發生大規模災害時導入「大數
據」，分析手機等龐大的電子資料，以求迅速收集情報，預判受災程度，快
速支援受災地區 [70]。

---

69   維克托·邁爾托·-舍恩伯格.肯尼士·庫克大數據時代 [M].盛楊燕，周濤，譯.浙江人民出
      版社 , 2013.
70   大數據將在未來日本地震預測中起作用 [J/OL].企業網 D1Net. Big Data Will Play A
      Role in Future Earthquake Predictions in Japan [J/OL]. D1Net.

In the fields of natural disasters, public health, and economic forecasting, the foresight ability of big data has already begun to appear. In 2009, Google successfully predicted a new round outbreak of H1N1 flu by applying big data, with a correlation of 97% of the official data. Therefore, it can accurately tell where the flu is from and what the scope probably is. Unlike the CDC, who usually makes the estimation after the flu outbreak for a week or two, big data can give the result very timely.[69] In August 2013, due to the extensive disaster area in the Great East Japan Earthquake, the Japanese government was unable to grasp the disaster information in a timely and accurate manner. To this end, the Japanese government has decided to introduce big data in the form of cooperation with non-government forces to analyze massive electronic data such as mobile phones, so as to quickly collect intelligence, predict the extent of the disaster, and quickly support the affected areas.[70]

在企業層面，越來越多的企業也開始認知到大數據對於人們的重要性。早在 20 世紀 70 年代末沃爾瑪就開始透過挖掘資料來改善自己的供應鏈，目前沃爾瑪已經成為全世界擁有最大數據倉庫的零售企業。20 世紀 7,80 年代耐克和愛迪達也紛紛建立了自己的運動實驗室，用來搜集並研究使用者的雙腳資料，耐克現在所有知名的技術產品都出自於其「運動廚房」（Nike Kitchen）。近兩年十分火爆的 Nike ID 業務就是充分挖掘資料潛力的例子。Nike ID 業務允許消費者基於耐克的一些已有產品進行個性化的改造，消費者可以線上對產品進行改造，選擇自己喜歡的顏色搭配、面料，甚至繡上自己的名字縮寫等，完成自己的設計後，Nike 就能為消費者量身打造一款獨一無

二的運動鞋。透過 Nike ID 業務，Nike 公司不僅能夠瞭解到用戶的喜好，同時這些寶貴的資料對於 Nike 將來研發新品都是重要的參考。[71] 據《麻省理工學院斯隆管理評論》和 IBM 商業價值研究院聯合舉行的 2011 年新智慧企業全球高管調查和研究項目指出，58% 的企業已經將分析技術用於在市場或行業內創造競爭優勢，而 2010 年這一比例僅為 37%。

At the enterprise level, more and more companies are beginning to recognize the importance of big data to people. As early as the late 1970s, Wal-Mart began to improve its supply chain by mining data. At present, Wal-Mart has become the retailer with the largest data warehouse in the world. In the 1970s and 1980s, Nike and Adidas also established their own sports labs to collect and study the user's feet. Nike's current well-known technology products are all from its "Nike Kitchen." The Nike ID business, which has been very popular in the past two years, is an example of fully exploiting the potential of data. The Nike ID business allows consumers to personalize their products based on some of Nike's existing products. Consumers can transform their products online, choose their favorite color combinations, fabrics, and even embroider their own name initials. After having finished design, consumers can get a unique sneaker from Nike. Through the business, Nike can not only understand the user's preferences, but also obtain valuable data for Nike's future research and development of new products.71 According to the 2011 Survey and Research on Global Executive of New Smart Enterprise, which was

---

71 大數據藍海：未來世界的新石油 [J/OL]. 商業價值 . Big Data Blue Ocean: New Oil in the Future World [J/OL]. Business Value.

jointly organized by the MIT Sloan Management Review and the IBM Institute for Business Value, 58% of companies have used analytical technology to create competitive advantage in the market or industry, compared to 37% in 2010.

Figure 2-7   NIKE iD Studio
Consumers are designing their unique sports shoes at NIKE iD Studio Shanghai flagship store

麥當勞收入的 1/3 來自直營店，2/3 來自加盟店，其中，房地產收入佔這部分收入的 90％，因此業內普遍認為麥當勞在很大程度上已變成一家經營房地產的企業。而支持麥當勞能夠精準選址做好這門「房地產生意」的基礎，在於該公司對資料的深入挖掘。[72]

One-third of McDonald's income comes from direct-sale stores, and 2/3 comes from franchise stores. Real estate income accounts for 90% of the latter. Therefore, it is widely believed that McDonald's has become a real estate business. The foundation supporting it to accurately select site and operate business smoothly lies in the its in-depth exploration of data.[72]

---

72   大數據藍海：未來世界的新石油 [J/OL]. 商業價值 . Big Data Blue Ocean: New Oil in the Future World [J/OL]. Business Value.

2014 年，麥當勞攜手百度，開展大數據行銷：「讓我們好在一起」。在 3 分鐘的廣告短片中，我們看到：在陌生城市裡奔波求職的男孩，突然想起少年時那些打羽毛球的小夥伴，如今都不知道走向何方。帶著失落，他走進路邊的麥當勞，點完餐剛坐下，突然有兩位拿著羽毛球拍的同齡男生，問：介意坐在一起嗎？3 分鐘的時間裡傳遞著城市中難得的人與人之間濃濃的愛。

　　In 2014, McDonald's teamed up with Baidu to launch big data marketing, 「Let's get together.」 In the 3-minute commercial video, a boy who was looking for a job in a strange city suddenly remembered little friends whom he played badminton with when he was young, and had no idea where they were. With a feeling of loss, he walked into the McDonald's on the side of the road. As he had just sat down after ordering the meal, suddenly two boys of the same age holding badminton rackets asked, 「do you mind we sitting together?」 We can feel deep love between people in the city from the short video.

　　透過對百度平臺的海量資料分析，百度清晰描繪出「我們」──都是哪些人？他們是──「憧憬未來的畢業生」、「在外打工的年輕人」、「尋找幸福的青年們」、「準備結婚的情侶們」、「供養家庭的爸爸」、「關愛孩子的媽媽」六類人群，結合他們的搜索行為標籤，麥當勞與百度共同打造了「城市森林我們在一起」訂製專題，以專題資料發佈的形式，引發網友對於城市中有千百萬夥伴和自己一樣的關注，繼而觸發讓大家尋找同類，「好在一起」，從而引發共鳴。這既是一次成功的行銷行為，也觸發社會關注，宣導人與人、心與心的交流，引發社會的廣泛認同和思考。[73]

　　Through the analysis of the massive data, Baidu clearly depicts who 「we」 are – graduates looking forward to the future, young people working away from home or looking for happiness, couples ready to get married, fathers who supports the family, mothers who cares for the children. All six kinds of people, combined with their search behavior tags, McDonald's and Baidu jointly created the custom topic of 「We are together in the urban forest」. In the form of releasing special data, it motivated netizens to pay attention to millions of partners in the city and then triggered them to look for the people of the same kind and "get together". This was not only a successful marketing act, but also provoked social attention, advocating the exchange between individuals and touching off the broad recognition and social thinking.[73]

圖 Figure 2-8 麥當勞攜手百度開展大數據行銷「讓我們好在一起」廣告頁面
McDonald's teamed up with Baidu to launch big data marketing "Let's get together"
（ad）

73　麥當勞攜手百度 大數據行銷「讓我們好在一起」[J/OL]. 中國廣告網. McDonald's teamed up with Baidu Big Data Marketing "Let's get together" [J/OL]. CNAD.

74　大數據時代：顛覆傳統 異化核心競爭力 [J/OL]. watchstor.com　The Era of Big Data: Subverting the Traditional and Alienating the Core Competitiveness [J/OL]. watchstor.com

近年來，國際企業巨頭越來越多地在生產經營的各個領域大量採用大數據技術，如富豪集團在卡車產品中安裝感測器和嵌入式 CPU，透過對從剎車到中央門鎖系統等形形色色的車輛使用資訊資料進行分析，不僅可以幫助製造更好的汽車，還可以幫助客戶們獲取更好體驗。美國最大醫藥貿易商 McKesson 公司將先進的分析能力融合到每天處理 200 萬個訂單的供應鏈業務中，並且監督超過 80 億美元的存貨。

In recent years, international corporate giants have increasingly used big data technologies in various areas of production and operation. For example, the Volvo Group installed sensors and embedded CPUs in their trucks. By analyzing a variety of vehicle using data from brakes to central locking systems, it can not only help build better cars, but also help users get a better experience. McKesson, the largest pharmaceutical trader in the United States, combines advanced analytical capabilities into a supply chain business that processes 2 million orders per day and supervises inventory worth more than $8 billion.[74]

國外零售巨頭對於資料資產的重視近年來也影響著國內的商業企業。國內知名互聯網企業凡客誠品 2011 年成立自己的資料中心內部，實現互聯網的系統化和數位化的管理。知名婚戀網站百合網將大數據技術應用於婚戀物件服務，大大提高了配對成功率。[75]

The attention paid by foreign retail giants to data assets has also affected

---

75 大數據時代：顛覆傳統 異化核心競爭力 [J/OL]. watchstor.com The Era of Big Data: Subverting the Traditional and Alienating the Core Competitiveness [J/OL]. watchstor.com domestic commercial enterprises in recent years. VANCL, a famous domestic Internet company, established its own data center in 2011 to realize the systematic and digital management of the Internet. Baihe.com, a well-known dating website, applied big data technology to the services of marriage and love, greatly improving the success rate of pairing.[75]

　　透過大數據的應用，使全球時刻存在並不斷產生的海量資料變得可控、可預測，可以大容量的存儲，高併發的處理，精準地統計分析，從而實現智慧推薦，能夠智慧滿足消費者更為細分、更為個性化的消費需求，使其接觸的任何有形、無形平臺均能為其提供無縫的消費支援，因此也使得「雲消費」成為可能。換言之，大數據的應用是「雲消費」時代產生並快速發展的基礎。

The application of big data enables the massive data that exists and is continuously generated in the world become controllable and predictable. It can be stored in large capacity, processed in high concurrency, and accurately and statistically analyzed to achieve intelligent recommendation and satisfy more subdivided and more personalized demand from consumers, enabling any tangible and intangible platforms that they get in touch with to provide seamless consumer support for them, thus making cloud consumption possible. In other words, the application of big data is the foundation for the rapid development of the era of cloud consumption.

三

「雲消費」時代零售業面臨三大
革命性變化

3. The retail industry faces three revolutionary changes in the era of cloud consumption

我們認為，「雲消費」時代，零售產業面臨三大革命性、根本性也是全域性的變化。

We believe that in the era of cloud consumption, the retail industry faces three revolutionary, fundamental and global changes.

（一）商業貿易的時間和空間障礙逐步消失
Time and space barriers to commercial trade are gradually disappearing

「雲消費」時代商業資訊傳遞無障礙、物流網路全聯通，制約消費的一系列障礙將逐漸消失，使消費突破時間、空間的障礙成為必然。

In the era of cloud consumption, commercial information transmission is barrier-free and the logistics networks are all connected, and a series of obstacles restricting consumption will gradually disappear, making it inevitable for consumers to break through the obstacles of time and space.

時間不再是消費的障礙。消費行為可以隨時發生,任何碎片化的時間都可以用來購物,消費者口袋裡的行動電話成為一家移動的微商店。甚至在夜間,商店打烊的時候,往往還成為購物的最佳時間點。從谷歌移動廣告看,來自平板電腦和智慧手機的搜索請求,於晚上 9 點時迎來高峰。[76]

Time is no longer an obstacle to consumption. Consumption behavior can happen at any time, any fragmented time can be used for shopping, and the mobile phone in the consumer's pocket becomes a mobile micro-shop. Even at night, when the traditional store is closed, it often becomes the best time to shop. From Google's mobile ads, search requests from tablets and smartphones ushered in at 9 pm.[76]

商業資源轉移、跨國銷售將突破國境界限,「海外代購」和「網購中國」正打破傳統貿易格局。

The transfer of commercial resources and cross-border sales will break through the borders of the country. 「Overseas purchasing」 and 「online shopping in China」 are breaking the traditional trade pattern.

2013 年,中國海外代購交易規模達 744 億元,增速超過 30%,尤其奢侈品的海外代購管道已經直逼主流銷售管道。巨大的「海外代購」市場需求催生了越來越多的海外代購賣家。據報導,網上的海外代購賣家約超過10萬家,其中僅化妝品代購就有 5000 多家。[77]

---

76 邵文麗 . 基於社交位置的移動零售服務的研究與設計 . 湖南大學碩士論文 . 2013 年 3 月
Shao Wenli. Research and design of mobile retail service based on social location. Master's thesis of Hunan University. March 2013

In 2013, the scale of China's overseas purchasing transactions reached ¥74.4 billion, a growth rate of more than 30%. In particular, the overseas purchasing channels of luxury goods have been almost equal to the mainstream sales channels. The huge "overseas purchasing" market demand has spawned more and more overseas purchasing sellers. According to reports, there are more than 100,000 overseas purchasing sellers on the Internet, of which 5,000 are cosmetics sellers.[77]

「網購中國」已經成為現時代的商業特徵。據 2013 年 5 月普華永道發佈的一份調查報告，中國消費者的網購頻率領先於全球平均水準。在中國，58% 的受訪者每週至少網購一次，而全球這一比例的平均數為 29%。[78] 據該報告研究，中國消費者比其他國家的消費者更快接受互聯網做為零售管道，有 24% 的中國網購受訪者表示計畫在未來 12 個月內會更頻繁地使用平板電腦或智慧手機上網購物，這一比例較全球平均水準 11% 高出 1 倍以上。[79] 根據普華永道 2013 年對 15 個國家 1.5 萬名消費者的調查，中國消費者網上購物以 62% 受訪者每週至少一次的頻率遠遠領先全球平均水準（21%）。

"Online shopping in China" has become a commercial feature of the modern era. According to a survey released by PricewaterhouseCoopers in May 2013,

---

77 方茜. 2014 年海外代購規模將超千億 市場亟需規範. 通信資訊報. 2014-01-08. Fang Qian. The scale of overseas purchasing will exceed ¥100 billion by 2014. The market is in urgent need of regulation. Communication Information. Jan 8, 2014.

78 普華永道報告. 中國消費者網購頻率領先全球平均水準. 新華網. Report by PricewaterhouseCoopers. Online shopping frequency in China leads the global average. NewsXinhua.

79 普華永道研究報告. 揭祕網購者：多管道零售的 10 個迷 (謎). Research report by PricewaterhouseCoopers. Unearthing online shopper: Ten enigmas of multi-channel retail.

Chinese consumers' online shopping frequency was ahead of the global average. In China, 58% of respondents shopped online at least once a week, while the global average was 29%.78 According to the report, Chinese consumers were more likely to accept the Internet as a retail channel than consumers in other countries, 24% of Chinese online shoppers said they plan to use tablets or smartphones more frequently in the next 12 months to shop online. This ratio is more than double the global average of 11%.79 According to PwC's 2013 survey of 15,000 consumers in 15 countries, 62% of respondents shopped online at least once a week, which was far ahead of the global average of 21%.

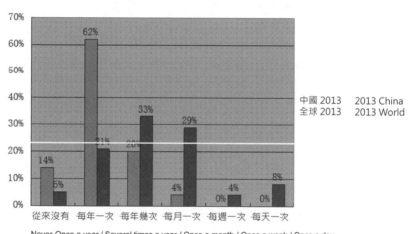

圖 Figure 2-9 全球消費者網購調查 Survey on Global Online Shopping
資料來源：普華永道全球零售調查（2014）Source: 2014 Survey on Global Retail by PricewaterhouseCoopers

網購，讓消費相對單一的大學生群體有了廣闊的消費選擇空間。據報導，2012 年，全國超過 55% 的在校大學生已經加入網上「淘寶」大軍，其中在全國排名第六的福建高校學生的網上消費普及率達到 66%。福州高校學生網上消費人均一年支出達 3000 元，平均每個月上網消費 3.2 次，每月消費 250

元，以萬元／年生活費計算，福州高校學生 30% 的生活費都在網上花掉。⁰⁰

Online shopping has offered much more consumption options to college students who consumed relatively simply in the past. According to reports, in 2012, more than 55% of college students in China had become Taobao shoppers, among which the sixth-ranked Fujian college students' online consumption popularuty rate reached 66%. The average annual expenditure of online students in Fuzhou college students was ¥3,000, with an average of 3.2 times and ¥250 of online consumption per month. Suppose yearly living expense of each student was ¥10,000, then 30% of it was spent online.[80]

即使在相對偏遠的西藏地區，網購消費依然如火如荼。據淘寶網統計，2013 年「雙十一」網購狂潮中，西藏網友創下了單日 4711 萬元的網購消費紀錄。2013 年西藏各地市人均支出由高到低排序依次為拉薩、林芝、日喀則、山南、那曲、阿里、昌都。其中拉薩市的支付寶用戶 2013 年人均網上支出 10790.98 元，而阿里地區則摘下全國無線支付交易筆數第二名。2013 年西藏地區活躍的淘寶賣家有 1216 家，網購也飛速帶動了西藏地區快遞業務的發展，目前西藏除了國內各大型速遞企業十餘家外，還有 DHL 及 UPS 等國際速遞業務，普通國際快遞一週內即可從西藏到達世界各地。[81]

80 鄭曉華．超半數大學生加入網上淘寶大軍 七成用手機購物 [N]．東南快報, 2013-05-15. Zheng Xiaohua. More Than Half of College Students in China Had become Taobao Shoppers. Seventy Percent of Them used Mobile Phones to Shop [N]. Southeast Express, May 15, 2013.

81 西藏活躍淘寶店 1216 家 移動支付全國第二．光明網, 2014-04-09.　There are 1216 Active Taobao Shops in Tibet, Mobile Payment Ranked Nationwide. Gmw.cn, April 9, 2014.

Even in the relatively remote Tibet, online shopping consumption is still in full swing. According to statistics from Taobao, in the 「double Eleven」 online shopping frenzy in 2013, Tibetan netizens set a record of online shopping consumption of ¥47.11 million in one single day. In 2013, the per capita expenditures of cities in Tibet were ranked from Lhasa, Linzhi, Shigatse, Shannan, Nagchu, Ngari, and Qamdo. Among them, the Alipay users in Lhasa City spent ¥10,790.98 per capita online in 2013, while the Ngari area took the second place in the nationwide wireless payment transaction. In 2013, there were 1,216 active Taobao sellers in Tibet. Online shopping also drove the development of express delivery business in Tibet. At present, Tibet has more than ten domestic large-scale express delivery companies, as well as international express delivery services such as DHL and UPS. Express parcels from Tibet can reach all parts of the world within a week.

有理由相信，今後我們某個地區的社會消費品零售額與該地區消費力的相關性將讓位於對全國乃至全球消費的「吸納力」。在可預見的未來，一個個地域的消費孤島將被徹底打破，全國消費平臺一體化正加快形成，全球消費一體化的進程也已經開始。能否佔據「雲消費」的制高點，決定了未來小到一個城市，大到一個國家在世界消費體系中的生態位。

We have reson to believe that in the future, the correlation between the retail sales of social consumer goods and the consumption power in one region will give way to the 「attraction force」 of national and global consumption. In the foreseeable future, the islands of consumption of each region will no longer be separated from each other, the integration of consumer platforms across the country is accelerating, and the process of global consumption integration has begun. Whether it can occupy the commanding heights of cloud consumption determines the future of a small city or a country's niche in the world's consumption system.

（二）實體商業與網路商業界限逐漸消失
Entity business and network business boundaries are gradually disappearing

近年，我們看到更多實體商業與網路商業競合發展的現象。

In recent years, we have seen more competition and development of physical business and online business.

早在 2008 年，淘寶網就開始低調試推線下實體便利店，這些便利店由淘寶賣家加盟開設，定位為「淘寶網特約服務店」，名字統一採用「淘 1 站」。2008 年 11 月，支付寶與拉卡拉電子支付服務有限公司合作，支付寶使用者只需持有銀聯標記的銀行卡，在佈設在連鎖便利店的拉卡拉支付網點，就能給支付寶充值或使用支付寶交易。此舉也意味著，支付寶開始把觸角伸向線下交易，為更方便地拓展實體業務提供基礎。[82]

As early as 2008, Taobao began to push offline physical convenience stores in low profile. These convenience stores were opened by Taobao sellers and

---

82 淘寶低調試水線下實體便利店．網易財經，2009-02-23. Taobao pushed offline physical convenience stores in low profile. Netease Finance, Feb 23, 2009.

positioned as 「Taobao special service stores」. The name was unified as 「Tao Yi Zhan」, or the First Station of Taobao. In November 2008, Alipay cooperated with Lakala E-Payment Services Co., Ltd. Alipay users only needed to hold the UnionPay-labeled bank card to top up Alipay accounts or use Alipay transactions at the Lakala payment outlets in chain convenience stores. The move also meant that Alipay has begun to extend its reach to offline transactions, providing a basis for more convenient expansion of physical business.[82]

2011 年，淘寶在北京試水一家名為「愛蜂潮」的家居實體店，打出「線上實體店＋同城服務體系」的概念，嘗試將線上交易和落地服務一體化。[83]

In 2011, Taobao tested a home-furnishing store called 「Ai Fengchao」 in Beijing and launched the concept of 「online store + city-wide service system」 to try to integrate online transactions and landing services.[83]

2016 年 10 月馬雲提出「新零售」概念。他認為，「電子商務會成為傳統概念，未來會是線下、線上、物流結合的新零售模式。」「純電商時代過去了，未來十年是新零售的時代，未來線上線下必須結合起來。」[84] 根據這一思想，近年阿里系加緊線下佈局。2014 年阿里巴巴戰略入股銀泰商業，開啟百貨業線上線下全面融合探索；2015 年 283 億入股蘇寧，探索數碼家電領域深度融合；2016 年入股三江購物，佈局超市領域；2017 年啟動銀泰私有化；2017 年 11 月 20 日，阿里巴巴又大手筆買入高鑫零售總價值 224 億港元的（約

---

83 淘寶家居體驗館匆忙上陣或將面臨 6 大考驗. 搜狐家居. 2011-06-15. Taobao Home-furnishing Store hastily comes into use and will possibly face six challenges. Sohu Home. Jun 15, 2011.

84 2016 年 10 月 13 日馬雲在杭州雲棲大會上的主題發 Jack Ma's keynote speech at Hangzhou Yunqi Conference on March 13, 2016

合 190 億人民幣）的股份，使其直接或間接擁有的高鑫零售份額達到 36.16%
[85]。與此同時，京東集團也在厲兵秣馬，透過入股、參股等資本運作方式加速佈局線下。2015 年 8 月，京東商城與永輝超市達成戰略合作，以每股人民幣 9 元的價格認購永輝超市新發行的普通股，交易總金額達到人民幣 43.1 億元，透過該筆交易，京東持有永輝超市 10% 的股權，並在採購、供應鏈管理等方面取得話語權。[86]

In October 2016, Jack Ma proposed the concept of 「new retail」. He believes that e-commerce will become a traditional concept, and the future will be a new retail model combining offline, online and logistics. The era of pure e-commerce has passed, and the next decade is the era of new retail, the future must be the combination of online and offline.84 According to this idea, in recent years, Ali has stepped up the layout. In 2014, Alibaba strategically invested in Yintai Commercial, which opened up the online and offline comprehensive exploration of the department store industry. In 2015, Alibaba bought share worth ¥28.3 billion of Suning to explore the deep integration of digital home appliances. In 2016, it became a shareholder in Sanjiang Shopping, starting its layout in the field of supermarket. In 2017, it launched the privatization of Yintai. On November 20, 2017, Alibaba also bought share worth total of HK$22.4 billion (approximately ¥19 billion) of Sun Art Retail Group, which resulted in obtaining a direct or indirect

---

85 阿里巴巴 190 億元入股大潤發母公司 . 新浪科技 . 2017-11-20 Alibaba Bought a Share Worth ¥19 Billion of the RT-Mart's Parent Company. Sina Technology. Nov 20, 2017

86 京東 43 億人民幣入股永輝超市 . 新浪科技 . 2015-08-07 Jingdong Bought a Share Worth ¥4.3 Billion of Yonghui Supermarket. Sina Technology. Aug 7, 2015

share of 36.16% of Sun Art Retail Group.[85] At the same time, the Jingdong Group is also rushing to the top of the line, accelerating the layout of the capital through capital operations such as shareholdings and equity participation. In August 2015, Jingdong Mall reached a strategic cooperation with Yonghui Superstore to subscribe for the newly issued ordinary shares of Yonghui Superstore at a price of ¥9 per share. The total transaction amount reached ¥4.31 billion. Through this transaction, Jingdong held 10% equity of Yonghui Superstore, and gained the right to speak in procurement and supply chain management.[86]

表 Table 2-6 2016 年阿里巴巴投資線下零售軌跡 Alibaba investment in offline retail in 2016

投資時間 Investment date	投資 / 並購標的 Object of investment/M&A	投資 / 並購金額 Amount of investment/M&A	股權 Stock equity
2017.1.10	銀泰 Yintai	198 億港元 HK $19.8 billion	74%
2016.12.23	華聯超市① Hualian Supermarket ①	2.37 億 ¥237 million	21.17%
2016.11.18	三江購物 Sanjiang Shopping	21.5 億元 ¥2.15 billion	32%
2016.11.17	如涵電商 Ruhnn holdings	4.3 億人民幣 ¥430 million	12.90%
2016.8.8	閃電購 52Shangou	2.67 億元 ¥267 million	領投 C 輪
2016.6.2	蘇寧雲商 Suning Commerce Group	282.33 億元 ¥28.233 billion	19.99%
2016.3.28	易果生鮮 Yiguo	未透露① Not yet disclosed ①	未透露② Not yet disclosed ②
2016.3	盒馬鮮生 Freshhema	1.5 億美元 $150 million	領投 A 輪 A round of lead investment
備註：Remarks: ① 易果生鮮直接投資華聯超市，阿里透過易果生鮮間接持有華聯超市。Yiguo invested directly in Hualian Supermarket, and Ali held Hualian Supermarket indirectly through Yiguo. ② 阿里巴巴從 2013 年開始，連續投資了易果生鮮的 A、B、C 三輪融資。Since 2013, Alibaba has continuously invested in the three rounds of financing of A, B and C.			

實體零售企業「觸網」相對滯後，但近年實體零售商已大步邁進，2013年，連鎖百強企業中已有 67 家開展網路零售業務。正如銀泰公司董事長沈國軍所說的：「不創新肯定是要死的，但創新也有風險，但不變革風險會更高。」[87] 2013 年 11 月 15 日，銀泰商業與支付寶錢包又正式達成戰略合作。2014 年 3 月 31 日，阿里集團以 53.7 億元港幣對銀泰商業進行戰略投資。5 月，阿里巴巴集團又與銀泰集團聯合複星集團、富春集團、順豐、三通一達（申通、圓通、中通和韻達等民營快遞巨頭），共同成立「菜鳥網路科技有限公司」，將架構能夠支撐日均 300 億網路零售額，在全國任何地區做到 24 小時內送達的物流網路體系。

Though physical retailers relatively lagged behind in online business, in recent years, they have already made great strides. In 2013, 67 of the top 100 chain enterprises have launched online retail business. As Shen Guojun, chairman of Yintai said, 「No innovation definitely means failure while innovation is risky. However the risk of the former is higher.」[87] On November 15, 2013, Yintai and Alipay Wallet formally reached a strategic cooperation. On March 31, 2014, Ali Group made a strategic investment in Yintai with HK$5.37 billion. In May, Alibaba Group and Intime Group jointly established Cainiao Network Technology Co., Ltd. with Fosun Group, Fuchun Group, SF Express, Santong Yida (Private express giants including Shentong, Yuantong, Zhongtong and Yunda). The architecture will be able to support a daily average of 30 billion online retail sales, and establish a

---

87　諸振家. 銀泰聯手支付寶錢包. 聯商網. 2013-11-16. Zhu Zhenjia. Yintai Teamed Up with Alipay Wallet. Linkshop.com. Nov 16, 2013.

logistics network system that can assure timely delivery within 24 hours in any part of the country.

我們正在看到更多、更快的實體商業與網路商業的互相滲透和融合。實體商業與網路商業的界限正在逐漸消失。應該指出，實體商業與網路商業的融合不僅僅是實體商業開設網店，而是兩者從供應鏈到服務鏈的深度一體化。網路商業和實體店鋪的深度融合，使過去的不可能成為可能。只要消費者有什麼樣的需求，就有什麼樣的發展機會和消費空間。

We are seeing more and faster penetration and integration of physical and online businesses. The boundaries between physical business and online business are gradually disappearing. It should be pointed out that the integration of physical business and online business not only indicates the establishment of online stores by physical business, but also the deep integration of the two from supply chain to service chain. The deep integration of online business and physical stores makes the impossible of the past possible. Development opportunities and consumption space always caters to consumers' needs.

「雲消費」時代透過網路和實體商業的結合，50 平方米的社區便利店完全可以提供比家樂福、沃爾瑪大賣場更低廉的價格、更豐富的商品、更全面的服務。山西太原唐久便利店與京東商城的深度融合，已經讓我們看到了希望。

In the era of cloud consumption, through the combination of online and physical business, a 50-square-meter community convenience store can provide richer products with cheaper prices and more comprehensive services than

Carrefour and Wal-Mart. The deep integration of Tangjiu Convenience Store in Taiyuan, Shanxi province and Jingdong has already made us see the signs of hope.

唐久與京東雙方系統的深度對接，是雙方資源的互補整合。唐久的 800 多家便利店全部被整合到網上大賣場中，門店庫存可以在網上銷售，同時建設總倉。如果用戶購買的商品在門店有庫存，則系統根據 LBS 定位將送貨交給距離用戶最近的門店，實現 1 小時達（甚至可以實現付費的 15 分鐘送達）；如果門店無庫存，則由總倉發貨，實現次日達。唐久擁有「人氣組合」、「店長推薦」、「猜你喜歡」、「最近瀏覽」等特色推薦方式；在結算頁，唐久網上大賣場依託京東物流系統的強大支援，可以支援 4 級位址，可細化到街道。在支付環節，唐久支持貨到付款和貨到刷卡兩種方式。除此之外還支持銀行轉帳、郵局匯款，並有京東信貸——京東白條支持。[88]

The deep docking between Tangjiu and the Jingdong is the complementary integration of the resources of both parties. Tangjiu' s more than 800 convenience stores have all been integrated into online hypermarkets, and store inventory can be sold online while the general warehouse is built. If the goods purchased by the consumer are in stock at the store, the system will deliver the goods to the nearest store according to the LBS positioning, achieving the one-hour delivery (or even 15 minutes if paid extra). If the store has no inventory, then the general warehouse will deliver the goods to achieve the overnight delivery. Tangjiu also

---

88 O2O 模式 PK：京東 + 唐久便利店 vs1 號店 + 美特好 [J/OL]. 聯商網 O2O Modes PK: Jingdong + Tangjiu Convenience Store vs No.1 Store + MEET ALL [J/OL]. Linkshop. com

has special recommendation methods such as 「popular combination」, 「manager recommendation」, 「guess you like」 and 「recently viewed」. On the settlement page, Tangjiu relies on the strong support of Jingdong logistics system to support fourth-level address which can be detailed to the street. In the payment segment, Tangjiu supports both credit card and cash on delivery. In addition to this, it also supports bank transfer, postal remittance, and Jingdong Baitiao (Jingdong Credit).[88]

表 Table 2-7 太原唐久便利店的 O2O 探索 The O2O exploration of Tangjiu convenience store in Taiyuan

線下門店 Number of offline stores	800 家 800
零售業態 Retail format	便利店 Convenience store
線上平臺 Obline platform	京東 "網上大賣場" Jingdong online hypermarket
上線品類 Online categories	日用百貨 / 食品 / 生鮮 Daily necessities/ Foods/Fresh
上線商品 Online products	超過 12000 種 More than 12,000 kinds
物流體系 Logistics system	全溫層物流 Total temperature
門店配送半徑 Delivery radius of store	500 米 500m
物流時效 Logistics duration	總倉發貨：次日到達 General warehouse:overnight delivery 門店發貨：1 小時到達 Store:one-hour delivery

資料來源：O2O 模式 PK：京東 + 唐久便利店 vs 1 號店 + 美特好 . 聯商網
Source: O2O modes PK: Jingdong + Tangjiu convenience store vs No.1 Store + MEET ALL. Linkshop.com

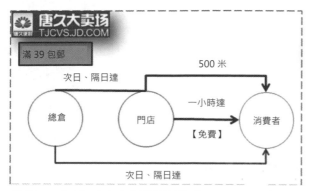

圖 Figure 2-10　唐久大賣場物流體系 Tangjiu's logistics system
Free delivery for 39 yuan

（三）產品開發者與銷售者的距離逐步消失
The distance between product developers and sellers is gradually disappearing

　　「雲消費」環境下，產品開發者和消費者的距離將消失，將實現零通路、零管道的新型模式。產品開發者可以直接面對終端消費者，進一步縮短產品進入市場的管道，管道障礙瀕臨消失，而零售商也會基於對消費者的熟悉，將產業鏈向研發設計延伸，維繫其存在的價值。

In the environment of cloud consumption, the distance between product developers and consumers will disappear, and a new mode of zero-access and zero-channel will be realized. Product developers can directly face the end consumers, further shorten the channels for products to enter the market Channel barriers are likely to disappear, and retailers will extend the industry chain to R&D design based on their familiarity with consumers, maintaining the value of its existence.

小米採用網上售賣的方式，直接面對最終消費者，不透過傳統流通鏈條，不僅從物流到庫存節約巨大的成本，更重要的是縮短了產品進入市場的時間，增強消費者體驗。

Xiaomi adopts online sales to directly face the end consumers. It does not pass the traditional circulation chain, which not only saves huge cost generated from logistics to inventory, but also more importantly shortens the time for products to enter the market and enhances the consumer experience.

國際眾多奢侈品牌也採取直營模式，紛紛收回代理權。傑尼亞 (Ermenegildo Zegna)、萬寶龍 (Montblanc)、勞夫羅倫 (Ralph Lauren) 等多個國際一線品牌相繼收回在中國市場上的代理權，轉向直營模式。2008 年瑞士的高檔鋼筆手錶品牌 —— 萬寶龍中斷了和其上海的合作夥伴 —— 上海國瑞信 (集團) 有限公司代理關係，法國箱包製造商克洛伊結束了和香港資訊科技集團合作代理合同。2010 年 7 月 17 日英國最大的奢侈品零售商巴寶莉 (Burberry) 集團公司以 7000 萬英鎊收購其特許經營夥伴 Kwok Hang Holdings 位於中國大陸地區的巴寶莉特許經營店，從而直接控制其品牌在大陸的業務。[89]

Many international luxury brands have also adopted the direct mode and regained their agency rights. Several international first-line brands such as Ermenegildo Zegna, Montblanc and Ralph Lauren have successively regained their agency rights in the Chinese market and turned to the direct mode. In 2008, Switzerland's high-end pen watch brand-Montblanc discontinued its agency relationship with its Shanghai partner, Shanghai Guoruixin (Group) Co., Ltd. Chloe, a French luggage manufacturer ended its cooperation contract with Hong

Kong Information Technology Group. On July 17, 2010, Burberry Group, the UK's largest luxury retailer, acquired its franchise partner Kwok Hang Holdings' Burberry franchise stores in mainland China for £70 million to directly control its brand's business there.[89]

---

89 收回中國市場代理權 奢侈品轉向直營時代 [J/OL].21 世紀經濟報導 .2010 年 7 月 21 日 Recovering the Agency Rights in the Chinese Market. Luxury Goods Turned to the Era of Direct Sales [J/OL]. 21st Century China Business. July 21, 2010

「雲消費」時代的主流消費模式

Chapter Three
The Mainstream
Consumption Patterns
in the Era of Cloud
Consumption

第三章
「雲消費」時代的主流消費模式
Chapter Three The Mainstream Consumption Patterns in the Era of Cloud Consumption

　　1994 年，美國麻省理工學院教授尼葛洛龐帝所發表的《數位化生存》一書中，提出「資訊技術的革命將把受制於鍵盤和顯示器的電腦解放出來，使之成為我們能夠與之交談，與之一道旅行，能夠撫摸甚至能夠穿戴的物件。這些發展將變革我們的學習方式、工作方式、娛樂方式——一句話，我們的生活方式。」**90** 誠然，人類社會發展以來，伴隨著技術進步、生產力的發展，也深刻影響著人們的生活，改變了人們的生活方式，而消費行為的變化最直觀地反映了生活方式的改變。

　　In 1994, the book Being Digital published by Negroponte, professor of the Massachusetts Institute of Technology, proposed that "the revolution in information technology will liberate computers subject to keyboards and monitors, making it possible for us to have a conversation with, have a trip with, touch or even wear them. These developments will change the way we learn, the way we work, and the way we entertain – in a word, our lifestyles.」 **90** It is true that since the development of human society, with progress of technology and the development of productivity, information technology has profoundly affected and changed people's lifestyles, and the changes in consumer behavior intuitively reflect lifestyle changes the most.

從消費行為的變化看生活方式的改變
1. Looking at lifestyle changes through changes in consumption behavior

（一）消費模式由「節儉原則」轉向「快樂原則」
Consumption patterns shift from the Principle of Parsimony to the Principle of Pleasure

　　國內生產總值（GDP）是反映一個國家經濟增長、經濟規模、人均經濟發展水準、經濟結構和價格總水準變化的一個基礎性指標，也是衡量消費結構變化的基礎性指標。從經濟發展規律看，當某一國家或地區人均 GDP 突破 1 萬美元時，消費者用於文化、健康、休閒的消費能力大為增強，開始步入體驗消費時期。

　　Gross domestic product (GDP) is a basic indicator reflecting a country's economic growth, economic scale, per capita economic development level, economic structure and general price level. It is also a basic indicator for measuring changes in the consumption structure. From the perspective of economic

---

90 Negroponte.《數位化生存》. 海南出版社 . 1997 Negroponte. Being Digital. Hainan Publishing House. 1997

development, when the per capita GDP of a certain country or region exceeds $10,000, the consumption power of consumers for culture, health and leisure is greatly enhanced, and they begin to enter the period of experience consumption.

隨著中國居民人均收入的不斷提升，北京、上海等城市已經開始向中等發達城市邁進，2008 年上海人均 GDP 突破 1 萬美元，2009 年北京人均 GDP 突破 1 萬美元，2013 年中國內地有 55 個城市人均 GDP 超過 1 萬美元，9 個城市人均 GDP 超過 2 萬美元，按照世界發達國家人均 GDP 標準 1.5 萬美元以上計算，達到發達國家標準的城市共計有 24 個。[91]

As the per capita income of Chinese residents continues to rise, cities such as Beijing and Shanghai have begun to move toward medium-developed cities. In 2008, Shanghai's per capita GDP exceeded US$10,000. In 2009, Beijing's per capita GDP exceeded US$10,000. In 2013, the per capita GDP of 55 cities in mainland China exceeded US$10,000, and 9 cities exceeded US$20,000. According to the per capita GDP standard of the developed countries in the world of more than US$15,000, 24 cities had reached the standard of developed countries.

考察日本、韓國、新加坡、臺灣人均 GDP 從 6000 － 10000 美元期間居民消費結構的變動，可以發現，用於基本生活保障方面的消費大幅度下降，其中，居民食品支出比重下降顯著，韓國降幅達 17 個百分點，日本、新加坡和臺灣地區也下降了約 4 個百分點 (見表 3- 1)。

Investigating the changes in the consumption structure of residents in Japan,

---

91　《2013 年中國城市人均 GDP 排名》. 宜居城市研究室 2013 China's Urban Per Capita GDP Ranking. Livable City Research Office

South Korea, Singapore, and Taiwan, China whose per capita GDP ranges from US$6,000 to US$10,000, we can find that the consumption for basic living security has fallen sharply. Among them, the proportion of food expenditure of residents has dropped significantly, and the decline amplitude of South Korea has reached 17 percentage points while Japan, Singapore and Taiwan also fell by about 4 percentage points (see Table 3-1).

表 Table 3-1 日本、韓國、新加坡和臺灣人均 GDP 6000 — 10000 美元
居民消費支出結構比較 (%)Comparison of consumption structure of residents in Japan, South Korea, Singapore, and Taiwan, China whose GDP ranges from US$6,000 to US$10,000 (%)

國別 Country	年份 Year	人均 GDP The per capita GDP	食品 Food	衣著 Clothes	居住 Housing	設備用品 Equipment supplies	醫療保健 Medical care	交通通訊 Transportation and communication	教育文化 Education and culture	雜項 Miscellaneous
韓國 Korea	1991	6546	35	4	11	6	7	11	11	14
	1996	10543	18	6	16	6	7	15	14	19
	變化 Difference		-17	2	5	0	0	4	3	5
日本 Japanese	1977	6034	28	8	17	6	9	9	9	14
	1984	10452	23	7	19	6	11	10	10	15
	變化 Difference		-5	-1	2	0	2	1	1	1
新加坡 Singapore	1983	6346	25.1	8.6	8.9	8.8	2.9	14	12.3	19.5
	1988	10220	21.9	7.6	9.6	9.3	4.2	13.5	14.6	19.4
	變化 Difference		-3.2	-0.93	0.73	0.52	1.26	-0.49	2.26	-0.13
臺灣 Taiwan	1989	6700	30.5	4.7	17.1	4.8	5.3	13.5	15.1	8.9
	1993	10548	27.4	4.7	19.5	5	7.1	12.5	17.3	6.6
	變化 Difference		-3.1	0	2.4	0.2	1.8	-1	2.2	-2.3

資料來源：根據有關統計資料整理
Source: Organized according to relevant statistics

由表 3-1 可見，在人均 GDP6000 — 10000 美元左右這一歷程中，由於基本生活滿足得以保證，居民消費的型態在發生明顯改變，不僅僅體現在規

模的擴大，更體現出結構性的變化。在人均 GDP 突破 1 萬美元時，居民用於文化、健康、休閒的消費能力大為增強。這時的消費模式已從「節儉原則 (Principle of Parsimony)」轉向「快樂原則 (Principle of Pleasure)」。

It can be seen from Table 3-1 that in the period of per capita GDP ranging from US$6,000 to US$10,000, due to the guarantee of basic living, the pattern of household consumption had changed significantly, not only in the expansion of scale, but also in structural changes. When the per capita GDP exceeded US$10,000, the consumption capacity of residents for culture, health and leisure is greatly enhanced. At this time, the consumption pattern had shifted from the Principle of Parsimony to the Principle of Pleasure.

有很多消費者熱衷於在網上、在朋友圈「曬單」，把自己買來的商品一一在網上展示一番，介紹自己的購物經歷，介紹產品的應用，介紹購物的小竅門等等，不僅自我滿足，也與網友共用購物樂趣。這就是典型的消費的快樂原則。宣揚「標記我的生活」的小紅書 APP 和「55 海淘網」論壇等都是中國網友們分享購物樂趣的天地，網友們的分享圖文並茂，體現了充滿樂趣的生活方式和消費方式。

Many consumers are keen on showing off the shopping list on the Internet and in friend zone with introduction of their shopping experiences, application of products and tips of shopping, etc. They not only receive self-satisfaction, but also share shopping enjoyment with net friends. This is typical Principle of Pleasure in consumption. The Little Red Book APP and the 55 Haitao.com forums that promote 「marking my life」 are all the places where Chinese netizens share their shopping enjoyment. Net friends share pictures and texts, reflecting the fun-filled lifestyle and consumption patterns.

這種變化帶來的不僅僅是消費能力的提升，更主要的是帶來生活方式的變化。我們很多有 20 世紀 6,70 年代記憶的人，還記得買任何東西都要憑票供應，平時總穿哥哥姊姊的衣服，過年才能穿一身體面的新衣服，一家有一輛自行車就是大件，偶爾吃頓肉算「打牙祭」，在這種生活環境下，人們的消費只能是遵循節儉原則，以有限的收入滿足最基本的生活需求。現在，人們出行前考慮的是該穿哪套衣服，什麼衣服配什麼鞋，配哪款包，出門開車，有些家庭有數輛車，不同的車有不同的功用，家庭聚會、朋友小酌要找有情調的飯店，有的「80 後」、「90 後」小夫妻基本不開伙，想去哪吃就去哪，節假日經常策劃一次說走就走的旅遊，看看不一樣的世態風景，看到一盤好菜，一個好景就用手機拍下來在朋友圈中分享一下。這種生活方式，明顯不同於以往，體現了較鮮明的快樂原則。

This change brings not only an increase in spending power, but also a change in lifestyle. Many of us who have memories of the 1960s and 1970s remember that they bought everything by ticket. They always wore hand-me-downs from their older brothers and sisters, and they could only wear new clothes in the Spring Festival. One bicycle was a major possession and meat dishes were rare sumptuous meals at that time. In such living environment, people's consumption could only follow the Principle of Parsimony and meet the most basic needs of life with limited income. Now, people think about what clothes to wear before going out, what shoes to wear, bag to carry to match clothes. When it comes to private cars, some families own several cars with different uses. Family gatherings and friends drink must be held in restaurants of sentimental appeal. Some post-80s-and-90s young couples seldom cook at home and eat wherever they want to go. On holiday, they often plan to go on a leave-at-once trip to enjoy different scenery in the world. They take pictures of a nice dish or landscape with mobile phone and share the pictures in friend zone. This kind of lifestyle is obviously different from the past and reflects the more vivid Principle of Pleasure.

（二）「80 後」消費群及生活方式
The post-80s consumer group and its lifestyle

1‧「80 後」消費群體的消費觀
The consumption view of the post-80s consumer group

當前，社會主流消費群體是生於 20 世紀 60 － 90 年代的消費群體，經濟獨立的「80 後」消費群代表了這個時代的消費潮流。

At present, the mainstream consumer group in society is those born between the 1960s and 1990s. The economically independent post-80s consumer group represents the consumption trend of this era.

「80 後」人群多為獨生子女的一代，他們生活在和平發展、經濟穩定、生活富足的年代，良好的家庭環境和教育體系，造就了「80 後」這一獨立、樂觀、自由個性的年輕消費群體。他們在消費上有一些標識性的特點：

The post-80s is mostly the generation of only children. They live in an era of peaceful development, economic stability, and rich life. Good family environment and education system have created the post-80s, an independent, optimistic, and free-spirited young consumer group. They have some identifying characteristics in consumption:

（1）用這月的錢還上月的債

They use month salary to pay the last month debt

「80 後」普遍喜歡刷卡，經常錢包裡的卡比錢多。很多人都是「月光族」，經常透支，往往剛發了工資，就要用去一半來還上月的錢。據香港《文匯報》報導，香港「80 後」年輕人人均欠信用卡債務 7.7 萬港元。[92]

Generally, the post-80s like to pay by credit card. Often, there are more cards in the wallet than cash. Many of them are the "moonlight clan" who often overdraft their cards. They use half of the month salary to pay the last month debt. According to the Hong Kong Wen Wei Po report, post-80s young people in Hong Kong owe and average credit card debt of HK$ 77,000.

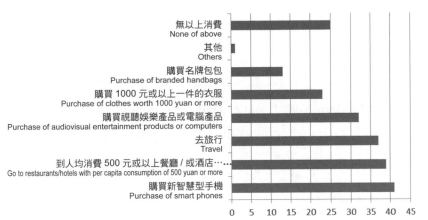

圖 Figure 3-1 「80 後」在欠卡費期間仍進行的消費活動（％）Consumption activities still carried out during the credit card debt of the post-80s(%)
資料來源：香港大學民意研究計畫
Source: Public Opinion Research Program by University of Hong Kong

（2）喜歡淘小店 They like to shop in the thrift shops

「80後」傾向於個性化消費，他們忌諱和周圍人用一樣的東西，穿同款的服裝。對於很多「80後」年輕人，大多有適合自己的特色服裝小店，如果有新貨上架，小店會根據不同消費者，定期發短信告知新貨資訊。

The post-80s tend to consume personally. They abstain from using the same things or wearing the same style of clothing as the people around them. Many of them find special clothing stores catering to their tastes. These stores will inform them of new arrivals by text message according to different tastes.

（3）願意為了漂亮的包裝買單

They are willing to pay for a beautiful package

商品的外包裝是影響「80後」購買與否的重要因素。現代、時尚、特別的包裝，是吸引他們購買的重要原因。他們往往會因為商品的包裝精緻漂亮而衝動性購買。

The outer packaging of goods is an important factor affecting the purchase of the post-80s. Modern, stylish and special packaging is an important reason to attract them to consume. They tend to consume impulsively because the packaging of the goods is exquisite and beautiful.

（4）執著忠誠於品牌

They are loyal and dedicated to the brands that they firmly believe in

「80後」普遍喜歡自己認定的品牌，喜歡耐克就喜歡耐克，喜歡愛迪達就是愛迪達，喜歡喝可口可樂的就不喝百事可樂，喜歡百事可樂的就不喝可哥可樂，喜歡就是喜歡。為此，麥當勞將其主打廣告定義為「我就喜歡」，

可以說，深得「80後」的消費心理。

Generally, the post-80s like the brand that they firmly believe in. Once Nike or Adidas, then always Nike or Adidas. Once Coca-Cola, never Pepsi and vice versa. To this end, McDonald's ad claims that "i' m lovin' it", which can be said that it has deeply won the hearts of the post-80s.

圖 Figure3-2 麥當勞「我就喜歡」廣告 McDonald's ad claims that "i' m lovin' it"

（5）熱衷網路購物 They are keen on shopping online

「80後」購物的首選之地是網路，去 24 小時便利店的時候也要比超市多。他們更樂於接受方便、快捷、隨時隨地的服務。

The best place for the post-80s to shop is on the Internet. Also, they go to a 24-hour convenience store more often than a supermarket. They are more willing to accept convenient and fast services anytime, anywhere.

（6）樂於分享消費經歷 They are happy to share consumption experiences

「80後」有自己的圈子，同學圈、同事圈、閨蜜圈、發小圈等，他們願意找閒暇和朋友共聚，一同購物、泡吧、度假等等，也樂於利用各種聊天工具分享消費經歷。

The post-80s has its own zones like classmate zone, colleague zone, girlfriend zone, childhood zone, etc. They are willing to spend leisure time with friends, shopping, clubbing or going on vacation together, etc., and are happy to use various chat tools to share consumption experiences.

可以說，「80後」消費群在消費觀上更突出表現為體驗性、享受性、獨特性，更重視消費的體驗過程，以及消費接入方式的便捷高效。他們更願意為他們獨特的生活方式買單。

It can be said that consumption concept of the post-80s consumer group is more prominently experiential, enjoyable and unique. They put more emphasis on the experience process of consumption, and the convenience and efficiency of consumer access. They are more willing to pay for their unique lifestyle.

2・「80後」的互聯網消費生活
The post-80s' life of consumption on the Internet

「80後」是與互聯網共同成長的一代，互聯網是他們的主要生活方式之一。據 2012 年中國互聯網路發展狀況統計報告的資料顯示，所有互聯網人群中，「80後」是最主要的人群，線民數量排名第一，他們平均年齡在 20~29 歲，佔線民總人數的比例達到 30.2%；其次是 30~39 歲的線民，佔線民總數的 25.5%；再次是 10~19 歲的線民，佔到線民總數的 25.4%；中老年線民數量相對較少，總數不及線民總數的 1/5，他們年齡在 40－49 歲和 50－59 歲，分別佔線民總數的 12.0% 和 4.3%。

The post-80s is a generation that grows with the Internet, and surfing the Internet is one of their main lifestyles. According to the data of the Statistics Report on 2012 China Internet Development, among the Internet population, the post-80s was the most important group, and the number of Internet users ranks first. The average age was 20-29 years old, accounting for 30.2% of the total number of netizens; followed by netizens aged 30-39, accounting for 25.5% of the total number of netizens; then followed by netizens aged 10-19, accounting for 25.4% of the total number of netizens. The number of middle-aged and elderly netizens was relatively small, less than 20% of the total number of netizens. They were 40-49 and 50-59 years old, accounting for 12.0% and 4.3% of the total number of netizens respectively.

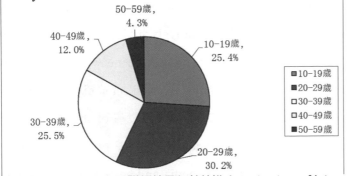

圖 Figure 3-3 2012 年互聯網線民年齡結構 Age structure of Internet users in 2012
資料來源：2012 年中國互聯網路發展狀況統計報告
Source: Statistics Report on 2012 China Internet Development

　　據中國互聯網資訊中心 2009 年下半年公布的各項調查統計資料顯示：中國「80 後」線民的人數規模已經超過 1 億大關，佔據整體線民數量的 1/5 多，相對於整體線民對各類網路應用的使用率，「80 後」線民群體網路應用參與度相對較高，已經成為網路的主力軍。在交流溝通領域，「80 後」線民主要集中在社交網站、即時通信，分別佔用戶數的 52.6% 和 40.2%；在資訊獲取領域，「80 後」線民主要集中在網路媒體和搜尋引擎領域，分別佔用戶數

的 42.6% 和 31.5%；在網路娛樂領域，「80 後」線民主要集中在大型網路遊戲、網路視頻、網路文學應用 3 個方面，分別佔用戶數的 35.5%、28.6% 和 47.0%；在網路商務領域，「80 後」線民主要集中在網路購物、網上支付、團購應用 3 個方面，其中在網路購物應用中的滲透率總共達到 81.7%，80 後線民在網上支付和團購應用中也佔有很大比例。（見表 3-2）

　　According to the survey statistics released by the China Internet Information Center in the second half of 2009, the number of Chinese post-80s netizens had exceeded 100 million, accounting for more than one-fifth of the total number of Internet users. Comparing to the overall usage rate of network applications of netizens, the post-80s network users had relatively high participation in network applications and had become the main force of the network. In the field of communication, the post-80s netizens mainly focused on social networking sites and instant messaging, accounting for 52.6% and 40.2% of the number of users respectively. In the field of information acquisition, the post-80s netizens were mainly concentrated in the field of online media and search engines, accounting for 42.6% and 31.5% of the number of users respectively. In the field of online entertainment, the post-80s netizens were mainly concentrated in three aspects of large-scale online game, online video and network literature applications, accounting for 35.5%, 28.6% and 47.0% respectively. In the field of network business, the post-80s netizens were mainly concentrated in three aspects of online shopping, online payment, and group purchase applications among which the penetration rate of online shopping applications had reached 81.7%. Also, the post-80s netizens accounted for a large proportion in other two aspects. (See Table 3-2)

表 Table 3-2　2012—2013 中國線民對各類網路應用的使用率 2012-2013 Chinese netizens' usage rate of various network applications

應用 Application	2013 年 2013		2012 年 2012		年增長率 (%)Annual rate of growth(%)
	使用者規模（萬）User scale(ten thousand)	線民使用率 (%) Netizens' usage rate(%)	使用者規模（萬）User scale(ten thousand)	線民使用率 (%) Netizens' usage rate(%)	
即時通信 Instant messaging	53215	86.2	46775	82.9	13.8
網路新聞 Netnews	49132	79.6	46092	78.0	6.6
搜尋引擎 Search engines	48966	79.3	45110	80.0	8.5
網路音樂 Network music	45312	73.4	43586	77.3	4.0
博客 / 個人（空）間 Blog/My space	43658	70.7	37299	66.1	17.0
網路視頻 Online video	42820	69.3	37183	65.9	15.2
網路遊戲 Online game	33803	54.7	33569	59.5	0.7
網路購物 Online shopping	30189	48.9	24202	42.9	24.7
微博 Microblog	28078	45.5	30861	54.7	-9.0
社交網站 Social networking sites	27769	45.0	27505	48.8	1.0

資料來源：線民互聯網應用狀況。CNNIC 第 33 次調查報告 Source: Netizens' application of the Internet. The 33rd investigation report by CNNIC

　　圖 3-4 形象地展示了分別在國貿工作和中關村工作的「80 後」白領女性的一日生活，儘管她們的生活態度、生活情趣有所不同，但每一天中都無一例外充斥了資訊互聯、社交分享，互聯網是她們離不開的生活環境。她們是「雲消費」時代典型的消費主流人群。

　　Figure 3-4 vividly shows the day-to-day life of the post-80s white-collar ladies working in international trade and Zhongguancun respectively. Although their attitudes towards life and joy of life are different, their lifes are full of information exchanges and social sharing every day. The Internet is the living environment they cannot live without. They are the typical mainstream consumers in the era of cloud consumption.

Day-to-day life of the lady working in international trade/Zhongguancun

圖 Figure 3-4　北京兩類「80 後」白領女士一天的生活 Day-to-day lifes of two types of post-80s white-collar ladies in Beijing
資料來源：北京國貿女和中關村女一天的生活對比。新華網 Source: Comparison between day-to-day lifes of two types of post-80s white-collar ladies in Beijing. XinhuaNet

據淘寶網發佈的《2012 淘寶十二大網購族群報告》顯示：12 個網購族群裡數量最為龐大的是「夜淘族」，人數高達 2283.2 萬，其中 25－29 歲和 30－34 歲兩個年齡段最多，佔總體比例的 50% 以上，分別是 32% 和 21%，這些人喜歡在晚上 11 點至凌晨 5 點之間逛淘寶，是網購族群中年輕、能「熬」的人群。隨著 80 後逐漸進入中年，更強勢的 90 後甚至 00 後正在崛起，他們是與移動互聯共成長的消費新勢力，他們的消費觀更值得深入研究。

According to the 2012 Report on Taobao Twelve Online Shopping Group released by Taobao.com, the largest number of online shopping group in the 12 online shopping groups was 「night shoppers」 who shopped online at night, with a population of 22.832 million. Among them, the 25-29 year-old and 30~34 year-old were the most, accounting for more than 50% of the total, 32% and 21% respectively. These young people preferred to visit Taobao between 11pm and 5am and were able to stay up late. As the post-80s gradually enter middle ages, the more powerful post-90s and even the post-00s are emerging. They are new consumer forces that grow together with the mobile Internet, and their consumption concept is worthy of further study.

（三）城市消費者傾向選擇「雲消費」相關消費方式
Urban consumers tend to choose cloud consumption-related consumption methods

2012 年、2013 年我們先後在北京市兩個中心城區深入社區居民家庭，開展有關社區生活服務及消費的調研，調研結果也印證了當前城市消費者日益青睞「雲消費」相關的消費方式，如在社區生活服務的選擇上，更傾向於服務整合、希望透過網路或電話等簡單的方式滿足便利服務，在早餐服務的選擇上，對早餐「雲整合」配送表現出較高的接受意願等。資料也顯示，學歷水準、收入水準越高的消費者對「雲消費」相關消費方式的需求度越高，中、年輕消費群體對相關消費方式的接受度高。以下為調研的一些基本分析：

In 2012 and 2013, we successively went deep into community households in two central urban areas of Beijing to conduct research on community life services and consumption. The survey results also confirmed that urban consumers were increasingly favoring cloud consumption-related consumption methods. For example, in the choice of community life services, they were more inclined to service integration and hoped that the convenience services can be accessed in a simple way such as network or telephone. In the choice of breakfast service, they showed a high acceptance to cloud integration distribution. The data also shows that the higher the level of education and income, the higher the demand for cloud consumption-related consumption methods, and the higher acceptance of relevant consumption patterns by middle and young consumer groups. Basic analysis of the survey is as follows:

1．多數居民希望透過網路或電話等便捷方式實現生活服務功能

Most residents wanted to realize life service functions in a convenient and fast way such as network or telephone.

對 2159 戶居民調研顯示，在取得社區服務方面，63.4% 的受訪者對「透過網路或電話等簡單的方式來實現社區生活服務功能」予以了肯定的態度。

According to a survey of 2,159 households, in getting community services, 63.4% of respondents had a positive attitude toward 「realizing life service functions in a convenient and fast way such as network or telephone」.

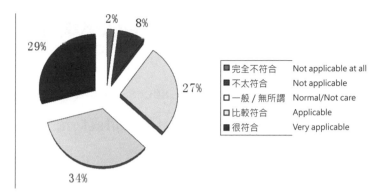

2% 8%

29%

27%

■完全不符合	Not applicable at all
■不太符合	Not applicable
□一般 / 無所謂	Normal/Not care
□比較符合	Applicable
■很符合	Very applicable

34%

圖 Figure 3-5 受訪居民家庭是否希望以網路或電話實現服務圖 Figure of whether the respondents wanted to realize life service functions in a convenient and fast way such as network or telephone

51 歲及以上人群中有 55.1% 的人覺得比較想要透過網路完成，41 － 50 歲有 68.8% 的人覺得比較想要透過網路完成，31 － 40 歲有 66.1% 的人想要透過網路完成，20 － 30 歲有 63.9% 的人想要透過網路完成。

55.1% of people aged 51 or more, 68.8% of people aged 41 to 50, 61.1% of people aged 31 to 40 and 63.9% of people aged 20 to 30 wanted to realize it through the Internet.

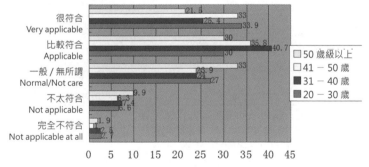

圖 Figure 3-6 按年齡分受訪居民家庭是否希望以網路或電話實現服務 Whether the respondents interviewed by age wanted to realize life service functions in a convenient and fast way such as network or telephone

2. 多數居民希望「透過一個統一的平臺（網站或電話等）滿足各類社區生活服務」

Most residents wanted to「have access to a variety of community life services through a unified platform (website or call, etc.)」

在選擇社區生活服務模式方面，55.8% 的受訪者能夠希望能「透過一個統一的網站或電話就可以滿足各類生活服務」，21.5% 的受訪者認為無所謂，另有 22.8% 的受訪者認為透過撥打不同的電話或瀏覽不同的網站等方式來下單或預定各類社區生活服務比較好。

In selecting the community life service model, 55.8% of the respondents hope to「have access to a variety of community life services through a unified website or call」, 21.5% of the respondents thought it did not matter, and another 22.8% considered that it was better to place orders or book various community life services by dialing different calls or browsing different websites.

表 Table 3-3 受訪居民家庭服務模式選擇意向表
Respondents' intended domestic service models

類型 Type	頻 數 Number	比 例 %Ratio%
透過一個統一的網站或者一個統一的電話等方式就可以滿足各類社區生活服務 To have access to a variety of community life services through a unified website or call	1199	55.8
透過撥打不同的電話或瀏覽不同的網站等方式，來下單或預定各類社區生活服務 To place orders or book various community life services by dialing different calls or browsing different websites	489	22.8
不知道 / 無所謂 No idea/Don't care	461	21.5
總計 Total	2149	100.0

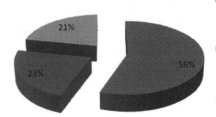

圖 Figure 3-7 受訪居民家庭服務模式選擇意向 Respondents' domestic service models
■ To have access to a variety of community life services through a unified website or call
■ To place orders or book various community life services by dialing different calls or browsing different websites
■ No idea/Don' t care

3· 多數居民願意透過電話預約，通知服務商上門獲得社區生活服務。Most residents were willing to make an appointment by phone to get door-to-door community life services.

　　對於取得生活服務的途徑，受訪者選擇最多的三種方式為「透過電話預約，通知服務商上門服務（72.9%）」、「親自到門店辦理（43.7%）」和「透過網路線上下單，然後有專人上門辦理」（32.2%）。As for the way to get life service, the respondents chose three ways the most, 「reservation on the phone to get door-to-door service (72.9%)」, 「going to the store (43.7%)」 and 「placing orders online to get door-to-door service」 (32.2%).

表 Table 3-4 受訪居民家庭服務模式選擇意向表
Respondents' intended domestic service models

類型 Type	頻數 Number	比例 %Ratio%
透過電話預約，通知服務商上門服務 Reservation on the phone to get door-to-door service	1574	72.9
透過網路線上下單，然後有專人上門辦理 Placing orders online to get door-to-door service	695	32.2
透過向公共號碼發送短信來通知專人上門辦理 Messaging to the public number to get door-to-door service	426	19.7
使用手機用戶端下單，然後有專人上門辦理 Placing orders on the mobile phone client to get door-to-door service	294	13.6
透過數位電視操作下單，然後有專人上門辦理 Placing orders on the digital TV to get door-to-door service	162	7.5
親自到相關門店辦理 Going to the store	943	43.7
總計 Total	2158	189.7

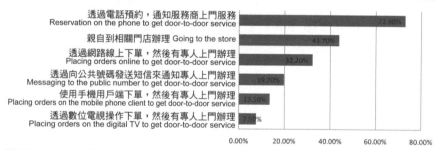

圖 Figure 3-8　受訪居民家庭服務模式選擇意向圖 Respondents' intended domestic service models

4· 居民支付方式多元，選擇最多的三種支付方式為網上支付、現場 POS 機刷卡和公交卡／儲值卡支付 Residents had multiple payment methods, and the three most popular ones were online payment, on-site POS credit card, and public transportation card/ stored-value card payment.

預定社區生活服務時，居民選擇最多的三種支付方式為網上支付（53.3%）、現場 POS 機刷卡（40.7%）和公交卡／儲值卡（18.5%）。另外，選擇手機支付（17.0%）和老年助殘券（12.1%）的比例也較高。

In booking community life services, the three most popular payment methods chosen by residents were online payment (53.3%), on-site POS machine credit card payment (40.7%) and public transportation card/stored-value card payment (18.5%). In addition, the proportion of mobile payment (17.0%) and old-age disability tickets payment (12.1%) was also high.

表 Table 3-5　受訪居民家庭支付方式意向表
Respondents' intended payment methods

支付方式 Payment method	頻數 Number	中選率（%）Ratio%
用網上銀行、支付寶等網上支付 Online payment such as E-bank, Alipay	1149	53.3
手機支付 Mobile payment	367	17.0
現場 POS 機刷卡 On-site POS machine credit card payment	878	40.7
公交卡／儲值卡 Public transportation card/stored-value card payment	399	18.5
老年助殘券 Old-age disability tickets payment	260	12.1
其他 Other payment	18	0.8
不預訂 No reservation	5	0.2
現金 In cash	184	8.5
銀行卡 Bank card	3	0.1
貨到付款 Pay on delivery	2	0.1
	2156	151.4

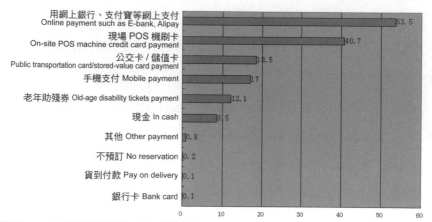

圖 Figure 3-9 受訪居民家庭支付方式意向 Respondents' intended payment methods

5· 多數居民每月有 1-4 次網購消費紀錄 1.3.5 Most residents had online shopping one to four times a month

65.4% 的居民有網購消費，其中，以每月 1 － 2 次（30.9%）和 3 － 4 次（24.3%）的居多；另有 10.2% 的居民每週網購 2 － 3 次以上。幾乎不網購的居民（34.6%）以中老年人為主。

65.4% of the residents have online shopping consumption, of which 1-2 times (30.9%) and 3-4 times (24.3%) are mostly; and 10.2% of residents consume 2-3 times online every week. Almost no online shoppers (34.6%) are mainly middle-aged and elderly.

表 Table 3-6 受訪居民家庭網購頻次情況
Respondents' online shopping frequency

網購次數 Online shopping frequency	頻數 Number	比例 %Ratio%
幾乎不上網買東西 Almost never	746	34.6
每月 1 － 2 次 Once or twice a month	667	30.9
每月 3 － 4 次 Three to four times a month	524	24.3
每週 2 － 3 次 Twice to three times a week	124	5.7
每週 4 － 5 次 Four to five times a week	70	3.2
幾乎每天都買 Almost everyday	27	1.3
總計 Total	2158	100.0

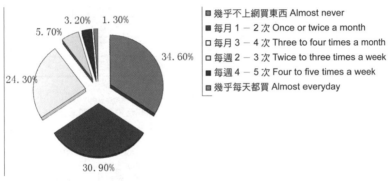

圖 Table 3-10 受訪居民家庭網購情況 Respondents' online shopping frequency

6‧ 多數居民網購月均消費在 500 元以字體統一 Most residents' online shopping spending was less than ￥500 a month

調查結果顯示，網購月均消費 500 元以內的居民比例 72.3%，月均消費 500 － 1000 元的比例達到 18.9%。

According to the survey results, the proportion of residents with an average monthly online shopping consumption within ￥500 was 72.3%, and the proportion of that of ￥500 to ￥1000 was 18.9%.

表 Table 3-7 受訪居民家庭網購花費情況
Respondents' online shopping consumption

費用 Amount	頻數 Number	比例 %Ratio%
100 元及以下 Less than ¥100	783	36.3
101 — 300 元 ¥101 — 300	406	18.8
301 — 500 元 ¥301 — 500	371	17.2
501 — 800 元 ¥501 — 800	255	11.8
801 — 1000 元 ¥801 — 1000	153	7.1
1001 — 2000 元 ¥1001 — 2000	111	5.1
2001 — 3000 元 ¥2001 — 3000	48	2.2
3001 — 4000 元 ¥3001 — 4000	19	0.9
4001 — 5000 元 ¥4001 — 5000	6	0.3
5000 元及以上 More than ¥5000	7	0.3
總計 Total	2159	100.0

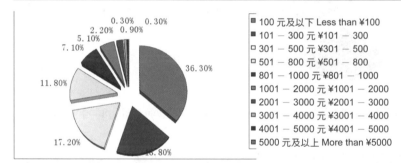

圖 Figure 3-11　受訪居民家庭網購花費 Respondents' online shopping consumption

二
「雲消費」時代的主流消費模式——
體驗化、個人化、社群化、即時化
2. The mainstream consumption pattern in the era of cloud consumption—experiencing, personalization, community and real-timed

（一）關於現代新型消費模式的新理念
New concepts of modern new consumption patterns

隨著互聯網和移動互聯的普及，人們的消費模式隨之產生一系列變化。捕捉到這種變化，近年國外部分商家、學者提出了一系列代表當前新型消費模式的新理念，其中有代表性的是「SoLoMo」理念。

With the popularity of the Internet and mobile internet, people's consumption patterns have undergone a series of changes. Thus in recent years, some foreign merchants and scholars have proposed a series of new concepts that represent the current new consumption patterns, among which the most representative one is the concept of 「SoLoMo」.

2011 年 2 月美國著名 KPCB 風險投資公司 (Kleiner Perkins Caufield &

Byers）的合夥人約翰‧杜爾（John Doerr） 最早提出 SoLoMo 這一概念 [93]。SoLoMo 即 social（社交化）+ local（當地語系化）+ mobile（移動化）。其中社交化（Social）代表各類社交服務網站和應用，如國外的「推特」（Twitter）、「臉書」（Facebook）和國內的人人網、微博；而當地語系化（Local）則代表著以基於地理位置服務（Location Based Service，下文簡稱 LBS）為基礎的各類簽到和定位技術，例如簽到應用「四方」（Foursquare）和「街旁」，社交位置服務「谷歌縱橫」（Google Latitude）；移動化（Mobile）則是由蘋果作業系統（ios）和谷歌公司安卓作業系統（Andriod）帶來的各種移動互聯網應用。在此之後，SoLoMo 概念在國內外產生了巨大的反響，被一致認為代表了互聯網未來發展的趨勢，互聯網企業紛紛增加資源投入相關服務應用的研究和開發。[94]

In February 2011, John Doerr, a partner at the famous KPCB Ventures (Kleiner Perkins Caufield & Byers), first proposed the concept of SoLoMo.93 SoLoMo is the combination of 「social, local and mobile」. 「Social」 refers to various social service websites and applications such as Twitter, Facebook used in foreign countries, and Renren, Weibo in China. 「Local」 represents various sign-in and location technologies based on Location Based Service (LBS), such as sign-in

93 陳雲海 . 移動互聯網 SoLoMo 應用模式分析 . 電信科學 , 2012, 3: 005 Chen Yunhai. Mobile Internet Analysis on SoLoMo Application Patterns. Telecommunications Science, 2012, 3: 005

94 邵文麗 . 基於社交位置的移動零售服務的研究與設計 [D]. 長沙：湖南大學碩士論文 . 2013. Shao Wenli. Research and Design of Mobile Retail Service Based on Social Location [D]. Master's Thesis of Hunan University, Changsha. 2013.

applications "Foursquare" and "Jiepang (streetside)", and social location service "Google Latitude". 「Mobile」 is a variety of mobile Internet applications brought by iOS, Apple's operating system and Android, Google's operating system. Since then, the concept of SoLoMo has generated tremendous repercussions at home and abroad, and is consistently considered to represent the future development trend of the Internet. Internet companies have increased their resources input to invest in the research and development of related service applications.[94]

關於現代新型消費模式的研究見仁見智，我們認為，與當前社會主流消費群的生活方式相適應，主流消費模式日益表現出四大基本屬性：消費的體驗化、個人化、社群化和定位化。

The research on modern new consumption patterns is a matter of different opinions. We believe that, in line with the current lifestyle of the mainstream consumer groups, mainstream consumption patterns are increasingly showing four basic attributes: experiencing, personalization, socialization and positioning of consumption.

（二）消費的體驗化
Experiencing of consumption

1 · 什麼是消費的體驗化 What is the experiencing of consumption

1998 年美國俄亥俄州戰略地平線（Strategic Horizons LLP）顧問公司的共同創辦人 B. 約瑟夫•派恩 （B. Joseph Pine II）與詹姆斯•H. 吉爾摩（James H.

Gilmore）在美國《哈佛商業評論》雙月刊 1998 年 7 － 8 月號發表的〈體驗式經濟時代來臨〉（Welcome to the Experience Economy）一文中首次提出「體驗式經濟」的概念。他們指出，「體驗」是一種創造難忘經驗的活動，其理想特徵是：消費是一種過程，當這一過程結束後，體驗的記憶將永恆存在。而提供體驗的企業及其員工，必須準備一個舞臺，如同表演一般來展示體驗。

In 1998, B. Joseph Pine II and James H. Gilmore, co-founders of Strategic Horizons LLP in Ohio, US, first proposed the concept of "experience economy" in the article Welcome to the Experience Economy published in Harvard Business Review issue July/August 1998. They pointed out that「experience」is an activity that creates unforgettable experiences. Its ideal feature is that consumption is a process, and when this process is over, the memory of experience will last forever. The company and its employees who provide the experience must prepare a stage to show the experience just like a play.

如前所述，隨著社會經濟發展，當某一國家或地區人均 GDP 突破 1 萬美元後，居民用於文化、健康、休閒的消費能力大為增強。人們的消費模式從「節儉原則 (Principle of Parsimony)」轉向「快樂原則 (Principle of Pleasure)」。以快樂原則為主導，人們在消費時，更注重過程的體驗和感受，更加注重透過消費獲得個性的滿足。為此我們認為，「雲消費」時代消費的首要特徵，就是體驗化。消費的體驗化，強調消費過程的個性滿足，鼓勵嘗試與互動，透過情景化的環境，以氛圍、感受、用戶體驗達成消費意向。

As mentioned above, with the development of social economy, when the per capita GDP of a certain country or region exceeds US$10,000, the consumption capacity of residents for culture, health and leisure is greatly enhanced. People's consumption patterns shift from the Principle of Parsimony to the Principle of Pleasure. Taking the latter as the leading factor, when consuming, people pay more attention to the experience and feelings of the process as well as the individual satisfaction. To this end, we believe that the primary feature of consumption in the era of cloud consumption is experience. The experiencing of consumption emphasizes the individual satisfaction of the consumption process, encouraging experimentation and interaction, and achieving consumption through the contextualized environment with the atmosphere, feelings and user experience.

以下是一段關於北京特色商業街區南新倉體驗化消費的精彩描寫：

The following is a wonderful description of the experiencing of consumption in Nanxincang, a characteristic commercial district in Beijing:

「改造後的皇家糧倉，主要是第十七號倉和第十八號倉。第十七號倉裡邊裝修成兩層樓的餐廳，餐桌就坐落在樑柱與磚地之間，自助式的牡丹宴精緻、古色古香，而十八號倉與十七號倉僅有一條紅地毯相連，這就是演出的區域，在這裡，演員謝絕所有的麥克風和揚聲器，全憑原始的唱功，讓人們去欣賞原汁原味的牡丹亭。改建後的演出區總共只有六十幾個座位，票價卻不菲，最低的 580 元，最貴的 3 個包廂達到 1.2 萬元，然而，這樣昂貴的票價，仍然有眾多的人趨之若鶩，上座率在八成以上，楊振寧博士、于丹、陸川等很多名人都曾欣然前往。坐在這樣特定的環境，感受著斑駁的牆磚，雅致的紫檀家具，『裙裾蓮步，暗香迫近眼前』，演員穿著蘇州繡娘手繡的戲服輕歌曼舞，曲笛幽咽婉轉，弦歌如泣如訴。彷彿已穿越過數百年歷史的隧道，

置身於 400 年前的歌寮酒肆，在『牡丹亭』的跌宕起伏的戀情故事中感受古代人的情感。」[95]

The reconstructed royal granary is mainly the No. 17 Warehouse and the No. 18 Warehouse. The No. 17 Warehouse is decorated into a two-storey restaurant. The dining table is located between the pillars and the brick floor. The self-serving Peony Banquet is exquisite and antique. The No. 18 Warehouse is connected to the No. 17 Warehouse with only one red carpet, where the shows are on. Performers refuse all the microphones and speakers and singing is done originally, allowing audience to appreciate the authentic Peony Pavilion. The reconstructed performance area has only a total of over sixty seats. However, the fare is very high, with the cheapest costing ¥580 and the most expensive 3 boxes costing ¥12,000 each. Despite of such expensive fares, still many people flock here and the attendance rate is over 80%. Many famous people such as Dr. Yang Zhenning, Yu Dan, Lu Chuan and so on have been there joyfully. Having a seat in such environment, you can feel the mottled wall tiles and elegant rosewood furniture. 「Beautiful dress and lotus step, the secret fragrance is close to the eyes」. The singing and dancing performers wear costumes made by embroiderers from Suzhou, with the music of flutes resembling whispering and stringed instrument sobbing. It seems that people present have crossed the tunnel for history and come to the romantic places 400 years ago. They feel the feelings of the ancient people in the ups and downs of the Peony Pavilion.[95]

---

95   在皇家糧倉聽「牡丹亭」. 溫州日報 .http://www.wzrb.com.cn/. 2008 年 01 月 11 日
Listen to the "Peony Pavilion" in the Royal Granary. Wenzhou Daily. http://www.wzrb.com.cn/. January 11, 2008

透過這段描寫，我們可以體會到一種皇家古倉與國粹昆曲藝術結合所形成的濃厚的體驗文化的氛圍。在這 600 年皇家古倉文化的烘托下，弘揚著濃郁高雅的時尚品位，傳遞著舒適精緻的人文內涵。人們聽著昆曲悠遠的旋律，感受新的在舊的中，時尚在歷史中的奇特文化體驗。這是一種典型的體驗化消費。

Through this description, we can feel the atmosphere of a rich experiencing culture formed by the combination of the royal ancient warehouse and the Chinese cultural quintessence of Kun Opera. Under the backdrop of the 600-year-old royal ancient warehouse culture, it promotes a rich and elegant fashion taste, conveying the humanistic connotation of comfort and exquisiteness. People listen to the melodies of Kun Opera and enjoy the cultural experience with the integration of the old and the new, fashion and history. This is a typical experience consumption.

2. 如何滿足消費的體驗化——案例觀察

How to satisfy the experiencing of consumption-case studies

（1）蘋果—傳達「偏執創新」的消費體驗

Apple-conveying the consumer experience of "paranoid innovation"

蘋果公司的系列產品契合了當前主流的消費文化，傳達了「偏執創新」的消費體驗。據蘋果公布的 2014 財年 Q1 財報顯示，2014 財年第一季度就賣出了 5100 萬臺 iPhone 產品。

Apple Inc's line of products fits the current mainstream consumer culture and conveys the consumer experience of 「paranoid innovation」. According to Apple' s FY14 Q1 financial report, 51 million iPhones were sold in the first quarter of the year.

■ 產品訴求：與消費者產生情感共鳴

創始人約伯斯認為產品創新跟研發資金沒有必然聯繫，關鍵是對產品的訴求。IBM 在蘋果推出 Mac 時，研發上投入是蘋果的 100 倍以上。蘋果的產品追求是「與消費者產生情感共鳴」，「製造讓顧客難忘的體驗」。

Product appeal: emotional resonance with consumers

Steve Jobs, founder of Apple Inc believes that product innovation is not necessarily related to R&D funds, the key is the demand for products. When Apple launched Mac, IBM' s investment in R&D was more than 100 times that of Apple. Apple's pursuit for product is "to resonate with consumers" and "to create an unforgettable experience for customers."

■ 只生產千錘百鍊的精品

約伯斯堅持：超一流的產品會帶來超一流的利潤。蘋果的產品品種非常少，但每一種都是經過千錘百鍊的精品，每一種都讓消費者追隨，讓擁有者自豪。

Only produce quintessential boutiques

Steve Jobs insists that super-class products will bring super-class profits. Despite of very few product varieties, each Appl' e product is a quintessential boutique, which makes consumers to follow and the owners proud.

■ 設計超酷體驗

蘋果設計極其簡約，堅持「酷」的特色。在品牌塑造上，蘋果不採用傳

統的硬性行銷手法，而是製造酷的體驗，成為一種個性化的標誌。

Experience super cool design

Apple's design is extremely simple, adhering to the "cool" features. In terms of branding, Apple does not use traditional hard marketing methods, but creates cool experience and makes it become the personalized symbol.

■ 形成品牌俱樂部

蘋果讓全球眾多的粉絲加入品牌俱樂部，透過開設自己的網站、出版自己的雜誌，參與到品牌的行銷促銷活動中。在中國，蘋果用戶被稱為「果粉」，甚至擁有自己的「果粉網」，有自己的社交系統。

Form the brand club

Apple has brought many fans around the world to the brand club. By opening website and publishing magazine of its own, it participates in the marketing promotions of the brand. In China, Apple users are called 「iFans」 and even have their own websites and social systems.

■ 蘋果旗艦店——360 度全方位體驗的消費空間

彙聚全球發燒友的蘋果體驗店就是典型的體驗化消費場所。我們看到，紐約第五大道蘋果旗艦店外觀由 90 塊玻璃改成 15 塊超大玻璃，每級造價 5000 美元的玻璃臺階蜿蜒而下，上千平方米的開敞銷售空間各類最新產品任人使用，還有「The Genius Bar」（天才吧），「one to one」（私人培訓服務）、青少年活動等特色服務，創造了 360 度全方位體驗的消費空間。

Apple flagship store - a 360-degree experience of consumption space

The Apple Experience Store, which brings together global enthusiasts, is a typical place of experience consumption. We can see that the appearance of Apple's flagship store on the Fifth Avenue in New York has changed from 90 pieces of glass to 15 pieces of super-large glass. The winding-down glass steps cost US$5,000 each. In the open sales space of thousands of square meters, the latest products are available. People can also use the Genius Bar, One to One (aprivate training service), youth activities and other special services, creating a 360-degree experience of consumption space.

圖 Figure 3-12　紐約第五大道蘋果旗艦店 Apple's flagship store on the Fifth Avenue in New York

（2）美國女孩 —— 商業與娛樂結合的體驗
American Girl-the combination of business and entertainment

全球知名玩具品牌「美國女孩」，利用美國女孩文化的大眾性特點，主要針對 8 歲以上的美國小女孩顧客，緊密圍繞美國小女孩的生活，提供洋娃娃、服裝、書及時尚配飾等目錄產品線，將娛樂和零售有機結合起來，讓她們參與、互動、分享，與自己的「美國女孩」玩具一起成長，在生活體驗中融合情感，傳遞關愛。這個僅誕生 20 餘年的年輕品牌，已成功佔據世界兒童

奢侈玩具的頭號寶座，公司網站每年訪問量高達 2300 萬人次。[96]

American Girl, the world-renowned toy brand, uses the popular characteristics of American girl culture to mainly aim to American little girls over 8 years old, closely surrounding the life of them and providing products such as dolls, clothing, books and fashion accessories. The brand combines entertainment and retail, letting the girls participate in, interact with, share with and grow with their own American Girl toys and integrate emotions in life experience and convey love. Although it has only been created for more than 20 years, the brand has successfully become the No. 1 of luxury toy brands in the world. The company's website has a daily visit of 23 million times.[96]

■ 每個女孩都能找到與自己一樣的一款「美國女孩」

在美國女孩專賣店中，每一款「美國女孩」都是獨一無二的，每一個女孩都可以在店鋪裡根據自己的眼睛、頭髮和膚色等個性特徵選購到一款玩具，使來店鋪購物的每一個女孩實現自己的夢想和渴望。

Every girl can find the same 「American Girl」 as herself.

In the American Girl franchised store, every piece of doll is unique. Girls can buy dolls in the store according to their own characteristics such as eyes, hair and skin color. They realize their dreams and desires through buying.

■ 美國女孩和「美國女孩」擁有自己的生活空間

在「美國女孩」專業劇院，定期上演女孩們喜歡的戲劇或時裝秀。劇碼

96　朱翊敏、李蔚 . 美國女孩：讓夢想照進現實 . 新行銷 . 2009-06-04 Zhu Samin, Li Wei. American Girl: Let the Dream Shine into Reality. New Marketing. Jun 4, 2009

大都圍繞友情、親情，讓女孩們在娛樂的同時更加珍惜朋友、父母與家庭。

Both American girls and American Girl dolls have their own living space

At the professional theaters of American Girl, girls can watch their favorite dramas or fashion shows regularly. Most of the plays revolve around friendship and affection between members so that girls can cherish friends, parents and families while entertaining.

在以黑白為基本色調的咖啡館裡，女孩和她們的娃娃可以在正式的用餐氛圍中盡情享受餐飲服務，娃娃也有一個特別的座位。精美的亞麻布、閃亮的銀器、美食誘人的香味、女孩們歡快的笑聲，都創造著美好的就餐體驗。正式用餐前，孩子們可以與桌子上的留言機進行有趣的遊戲問答。餐牌上沒有碳酸飲料，但是有專門為女孩們準備的粉紅檸檬水。

In cafes with black and white as the basic color, girls and their dolls can enjoy the dining service in a formal dining atmosphere, and the dolls also have their special seats. Exquisite linen, shiny silverware, tempting scent of the food, and cheerful laughter of the girls create a wonderful dining experience. Before the formal dining, children can have fun game quiz with the answering machine on the table. There are no carbonated drinks on the menu but pink lemonades for girls.

女孩們還有和玩偶共同的攝影室、美髮沙龍。笑容美麗的照片會登上《美國女孩》雜誌封面上，讓女孩們體驗做明星的感受。美髮沙龍不僅為小主人，也為娃娃打造和主人一樣的髮型。

Girls also share studios and hair salons with their dolls. Pictures of the girls with beautiful smile will be posted on the cover of the American Girl magazine, allowing them to experience the feeling of being a star. The hair salon makes a hair style not only for the little master but also for the her doll.

■「美國女孩」有自己的專賣產品

我們生活中擁有的所有產品,「美國女孩」也都有,包括鞋帽、錢包、項鍊、腰帶、手鐲,甚至還有玩物小狗、小洋娃娃和小書本等等。在美國大街上經常可以看到小女孩抱著與她穿著同樣衣服、戴著同樣帽子、背著同樣包包的「美國女孩」玩偶。

「American Girl」 has its own dedicated products

American Girl dolls have all the products we human beings have in our lives, including shoes and hats, wallets, necklaces, belts, bracelets, and even pet dogs, small dolls and small books. In America, you can see little girls holding the American Girl dolls wearing the same clothes and hats, and carrying the same bags as they do quite often on the street.

■美國女孩和「美國女孩」有自己的交友天地

美國女孩公司經常舉辦各類聚會和活動,比如美容聚會(愛護娃娃,為它做護膚美容)、手工聚會(學習製作娃娃用的小枕頭和睡袋)、時裝設計聚會(為娃娃設計時裝)、美食聚會(學習製作小蛋糕,宴請小客人)、生日聚會、美國女孩廣場一日遊等等活動。還透過雜誌、網站傳播美國女孩文化。

American girls and American Girl dolls both have their own friends.

The American Girl Company often hold various parties and events, such as beauty parties where girls do skin care and hairdressing for their dolls; hand-making parties where girls learn to make small pillows and sleeping bags; fashion design parties where girls design the fashionable dresses for dolls; food parties where they learn to make small cakes and fete little guests; birthday parties and one-day trips to the American Girl's Square, etc. The company also spreads American Girl culture through magazines and websites.

圖 Figure 3-13　美國女孩博物館式專賣店 American Girl' s museum-style store

圖 Figure 3-14　「美國女孩」享受美髮服務 American Girl dolls are enjoying hairdressing service

（三）消費的個人化
Personalization of consumption

1· 什麼是消費的個人化
What is personalization of consumption

顧名思義，「個人化」代表了獨特性和專屬性。一般意義上，擁有專屬的產品或服務，通常是高階層人士享有的某些或單一的特權，顯示了一種很高的進入門檻。如價格不菲的高爾夫俱樂部會員資格、個人化訂製某種限量版奢侈品等。

As the name implies, 「personalization」 represents uniqueness and specificity. In general, having dedicated products or service is usually certain or single privilege enjoyed by the high-level group, showing a high entry threshold, such as the high-priced golf club membership and personalization of a limited edition luxury goods, etc.

我們認為，在「雲消費」的時代，消費的個人化又有新一重含意：即商業智慧和雲資料的發展使個人化訂製不再是少數人的特權，每個消費者都能享受獨一無二的商品和服務，享受消費的尊崇感、自豪感。

We believe that in the era of cloud consumption, the personalization of consumption has a new meaning: the development of business intelligence and cloud data makes personalization no longer a privilege of a few people, and each consumer can enjoy unique commodity and service as well as the sense of respect and pride through consumption.

2. 如何實現消費的個人化——案例觀察 How to realize the personalization of consumption-case studies

（1）尚品宅配—最大限度滿足消費者個人化家居服務需求

Shangpin Home Delivery-Satisfy consumers' needs for personalized home services to a maximum

尚品宅配打破了傳統的家具生產、經營模式，深度洞悉消費者的需求，然後結合自身優勢、利用資訊技術無縫滿足消費者需求，為消費者訂製個人化的家具產品，使家具訂製成為一種生活方式，僅用 10 年時間由軟體研發公司跨界發展而成的全屋家具訂製服務商，成就了「C2B 商業模式的中國樣本」，被讚為「傳統產業轉型升級的典範」。[97]

Shangpin Home Delivery breaks the traditional furniture production and operation models. It deeply understands the needs of consumers, and then combines its own advantages and uses information technology to seamlessly meet consumer needs, customizing personalized furniture products for them and making furniture customization become a lifestyle. The software R&D company developed into a custom service provider of full-house furniture in 10 years and has become the 「sample of C2B business model in China」. It is praised as 「the paragon of the transformation and upgrading of traditional industries」.[97]

■ 把握需求，研究消費者生活行為

為準確把握消費者需求，尚品宅配研究了不同消費者在不同生活空間的

---

97 宗禾 . 尚品宅配打造 C2B 模式中國樣本 . 中國商網 . 2014-05-08 Zong He. Shangpin Home Delivery Creates a Sample of C2B model in China sample. China Business Website. May 8, 2014

生活行為。例如，在臥室，處於育兒期的人要給小孩餵奶、換尿布，其生活行為就和處於新婚期的人有很大差別。此外，尚品宅配還研究了不同消費者的審美需求——不同年齡、不同性別、不同教育背景、不同職業的人，審美觀也不同。不同的需求，需要不同的產品，滿足不同的定位。

Grasp the demand and study life behavior of consumers

In order to accurately grasp the needs of consumers, Shangpin Home Delivery has studied the behaviors of different consumers in different living spaces. For example, in the bedroom, people in childcare period need to feed and change diapers for infants, thus their behavior is very different from the newlyweds. In addition, Shangpin Home Delivery also studies the aesthetic needs of consumers of different ages, genders, educational backgrounds and different professions, who have different aesthetics from each other. Consumers with different needs require different products to meet different positioning.

■ 確認需求，打造個人化家具成品

在捕捉了更多需求後，更大的困難在於，面對消費者近乎漫無方向、難以捉摸的個性化需求，如何幫助消費者確認真正的需求？尚品宅配採用「店網一體化經營」模式，透過推動雲設計的發展來解決這一問題。店網一體化經營的「網」是在網上賣東西，而是先在地面上建數百家實體店，然後再配合公司專有的「新居網」，店、網結合，互相配合、支持。尚品宅配在全國有幾千名家居設計師，均可以透過尚品宅配下屬新居網的產品庫和房型庫，根據消費者的不同需求，利用網路雲計算服務，設計可以匹配不同房型、不同風格的家居空間解決方案。

Confirm the demand and create personalized furniture

After capturing more demand, greater difficulty arises: how to help consumers confirm the real needs among their individual needs that are almost undirected and elusive? Shangpin Home Delivery adopts the 「integrated operation of both online and physical stores」 model to solve this problem by promoting the development of cloud design, that is, it builds hundreds of physical stores first and then combines it with the company's proprietary "Xinjuwang (www.homekoo.com)" to operate. There are thousands of home designers in the country, all of which can refer to the product and room-type libraries of 「Xinjuwang」 to work out home space solutions that match different room types and styles according to the needs of different consumers, with the help of the online cloud computing service.

目前，尚品宅配已透過基於互聯網的即時交易和互動設計系統建立的「新居網」線上服務平臺，採集了全國數千個樓盤的數萬種房型資料，建立了「房型庫」。同時，也採集數百家家居企業及數千名協力廠商設計師的素材建立了「產品庫」，透過「雲計算和大數據」技術對不同人群在不同生活空間的行為和功能需求進行深入研究，研發出數百萬個海量的「空間整體解決方案」的「方案庫」。再加上全國 600 多間地面實體體驗店以及佛山工廠的「大規模訂製」系統無縫連接和全流程資訊化，實現了真正的 C2B 和 O2O 商業模式。

At present, Shangpin Home Deliery has collected tens of thousands of room-type data of thousands of real estates across the country and established the room-type library through the 「Xinjuwang」 online service platform established by the Internet-based real-time transaction and interactive design system. At the same time, it has also collected the materials of hundreds of home furnishing companies and thousands of third-party designers to establish the product library. Moreover, it has developed the scheme library of millions of integrated solutions of space by conducting in-depth research on the behavior and functional needs of different people in different living spaces through cloud computing and big data technology. Together with more than 600 physical experience stores across the country and seamless connection and informatization of whole process of 「Mass Customization」 system of factories in Foshan, real C2B and O2O business models are realized.

■ 最大限度滿足需求，加速升級傳統產業

尚品宅配採用個性化訂製的生產技術，並資訊科技緊密結合，讓消費者主動參與到產品的設計、製造中，一方面更大程度上滿足了消費者個性化需求；另一方面透過改革創新技術和商業模式，改變了傳統家具生產經營方式庫存量大、資金週轉慢等生產方式和商業模式的弊端。

Satisfy consumer demand to a maximum and accelerate the upgrading of traditional industries

Shangpin Home Delivery adopts customized production technology closely integrated with information technology, allowing consumers to actively participate in the design and manufacture of products. On the one hand, the individualized needs of consumers can be met to a greater extent. On the other hand, through reform and innovation of technology and business model, drawbacks of production methods and business models such as large inventory and slow capital turnover in traditional furniture production and management methods are changed.

（2）特斯拉私人訂製 Private Customization of Tesla

特斯拉由斯坦福大學的碩士輟學生馬斯克與碩士畢業生 J.B.Straubel 於 2003 年創立，專門生產純電動車。特斯拉採用網路訂製模式，包括車體顏色、輪轂尺寸、車頂天窗及內飾配置，以及可選充電配套裝置等，都需要客戶在預約時加以明確，然後按單生產。

Tesla, founded in 2003 by Elon Musk, dropout graduate student of Stanford University, and J.B. Straubel, graduate student of Stanford University, specializes in the production of pure electric cars. Tesla adopts the model of online customization which allows consumers to make a clear reservation on body color, wheel size, sunroof and interior configuration, as well as optional charging kits of the car so that the order can be put into production.

Tesla 為瑞士名表廠 TAG Heuer 豪雅 150 週年紀念特別打造一款頂級的限量版跑車 Tesla Roadster，由 Tesla 的首席設計師 Franz von Holzhausen 操刀設計，在原版 Tesla Roadster 基礎上，以沉穩的銀灰色調，搭配碩大的 TAG Heuer 的白色 LOGO 標識，車內還設計了一個特殊的中控臺，以便安置一隻 1/5 秒的機械式豪雅碼錶。這款豪雅版 Tesla Roadster 和普通版的 Tesla Roadster 配置一樣，做為首款純電動的超級跑車，其百公里加速僅為 3.9 秒，完全可以媲美燃油動力的跑車，一次充電可以行駛超過 320 公里以上，而據說最高紀錄為 500 公里。

Tesla created a top-of-the-range limited-edition sports car Tesla Roadster for the 150th anniversary of the Swiss watchmaker TAG Heuer, designed by Tesla's chief designer Franz von Holzhausen. Based on the original Tesla Roadster, the new Tesla Roadster was built in a calm silver-gray tone and attached a large white logo of TAG Heuer. A special center console was also designed in the car to accommodate a mechanical TAG Heuer stopwatch of 1/5 second. The TAG Heuer-editoned Tesla Roadster has the same configuration as the regular ones. As the first pure electric supercar, its 100-kilometer acceleration costs only 3.9 seconds, which is comparable to fuel-powered sports cars. It can travel more than 320 kilometers on a single charge and the highest record is said to be 500 kilometers.

圖 Figure 3-15 特斯拉用 17 吋觸控式螢幕代替傳統儀表盤 Tesla replaces traditional dashboard with 17-inch touch screen

圖 Figure 3-16 訂製版特斯拉 Customized Tesla sports car

（3）愛定客 IDX—全球首家線上個性化訂製潮鞋品牌

IDX-The first brand of online customized shoes in the world

愛定客成立於 2012 年 6 月 18 日，是一家為消費者提供個性化訂製鞋服務的電子商務公司，較之於 Nike、Adidas、Converse 動輒數千元的訂製鞋，在愛定客上訂製一款鞋的價格區間約為 200 － 500 元，價格較為「親民」，符合年輕時尚潮人的消費定位。目前，愛定客平臺從休閒運動鞋品類擴充到衛衣、T 恤、牛仔褲、手機殼、3C 產品、箱包、家居，文具等全品類商品。愛定客所有商品都為個人訂製，真正實現及落實「零庫存」的環保概念，從真正接到訂單之後才投入生產，保證了所有商品都是最新鮮的並且沒有庫存，擁有多項專利技術，行銷管道遍及全網，並擁有 O2O 線下體驗店。

Founded on June 18th, 2012, IDX Holding Company is an e-commerce company that provides customized shoes for consumers. Compared with Nike, Adidas and Converse whose customized shoes easily cost thousands RMB, the price range of a pair of customized shoes is between ¥ 200 － 500 on IDX, which is relatively wallet-friendly and in line with the consumption orientation of fashion young people. At present, the IDX platform has expanded from casual sports shoes to sweaters, T-shirts, jeans, mobile phone cases, 3C products, luggage, home furnishing, stationery and other products. All products are made customized for the purpose of realizing and implementing the environmental protection concept of 「zero inventory」. Products are put into production only after being ordered, ensuring that all products are the freshest and are no inventory. The platform have many patented technologies and marketing channels that are throughout the network, as well as O2O offline experience stores.

在愛定客的店鋪中可以看到張小盒、刀刀等設計師的身影，設計師通常都有自己的圈子，有自己的粉絲，透過與愛定客的合作，一方面使設計師拓

寬了品類，有助於其品牌影響力的延伸，因為實物產品可以加強設計師和粉絲之間的互動；另一方面豐富了愛定客消費者的選擇，同時對塑造品牌形象也有一定幫助。

In the stores of IDX, you may find Zhang Xiaohe, Daodao and other designers, who usually have their own circles and fans. By cooperating with IDX, the designers can broaden work categories, helping them to extend the influence of themselves for physical products can enhance the interaction between designers and fans. On the other hand, such cooperation enriches the choice of consumers, and also helps to shape the brand image.

愛定客整合設計、原料採購、生產、訂製、銷售、物流、售後等諸多環節，網路後端直接與工廠對接，消費者的訂單生產流水線的工人能夠直接透過 IT 系統看到個性化訂單的尺碼、面料和顏色等設計資料及要求，3 分鐘後就可以啟動生產，實現了全產業鏈運作模式，解決了傳統服裝類廠商採取 OEM 形式尋找代工產既無法掌握生產流程，又損耗時間的「瓶頸」。

IDX integrates design, raw material procurement, production, customization, sales, logistics, after-sales and many other steps. Its network back-end is directly docked with the factory, which allows the workers of the production line behind consumers' customized orders to look directly at the design data and requirements of size, fabric and color through the IT system. The order can be put into production in 3 minutes, realizing the operation mode of the whole industry chain and solving the bottleneck that traditional clothing manufacturers adopting OEM form can not master the production process and wastes time.

圖 Figure 3-17 微博用戶分享自己的訂製鞋作品 Weibo users share customized shoes of their own

圖 Figure 3-18 愛定客的個性化訂製網頁 Customization webpage of IDX

（4）統帥現象：家電企業的訂製模式 Tongshuai (Leader) phenomenon: customization mode of home appliances enterprises

統帥電器是海爾集團旗下的家電品牌。做為互聯網時代的定製品牌，統帥堅持「你設計，我製造」的產品企劃和生產模式，透過搭建開放的互聯網用戶參與平臺，廣泛收集用戶個性化需求，提升用戶訂製體驗，形成了「為需要的功能買單，不需要的功能免單」的設計研發理念。

Tongshuai is a home appliance brand under the Haier Group. As a customization brand in the Internet era, Tongshuai insists on the product planning and production mode of「you design, I manufacture」. By setting up an open platform for Internet user participation, it widely collects the individualized needs of users and enhances the user customization experience, forming a design and development concept of 「paying for the needed functions and others are for free」.

透過聯手天貓聚划算平臺推出「大屏時代的幸福生活」活動，統帥電器吸引了大量網友的關注，2012年2月14日支付訂金階段伊始，僅僅6個小時，海爾及統帥訂製家電意向訂購金超過5000萬元，創造了一個銷售奇蹟，更體現出消費者對於訂製電器產品旺盛的需求。

Tongshaui, together with Tmall, launched the "Happy Life in the Big Screen Era" campaign, having attracted a lot of netizens' attention. On February 14, 2012, at the beginning of the deposit payment phase, Haier and Tongshuai customized home appliance received orders worth more than ¥50 million in only 6 hours, which has created a sales miracle and reflected the strong demand of consumers for customized electrical products.

（四）消費的社群化
Community of consumption

1‧ 什麼是消費的社群化 What is community of consumption

（1）社群與消費密不可分 Community and consumption are inseparable

社群本屬於社會學概念。社會學所指的社群（Community）通常指在某些邊界線、地區或領域內發生作用的一切社會關係。它可以指實際的地理區

域或是在某區域內發生的社會關係，亦指存在於較抽象的、思想上的關係。

Community is a sociological concept. It usually refers to all social relationships that occur within certain boundaries, regions or domains including social relationship that takes place in an actual geographical area or a certain area, and also relationship that is more abstract and ideological.

與消費者行為相關，近年出現的熱點概念「品牌社群」，最早是 Muniz 和 O. Guinn 在 1995 年的消費者研究協會年會上提出的。他們在 2001 年的研究中，將其定義為基於品牌崇拜者的一系列社會關係的非地域性專業化社群。他們認為品牌社群具有類似於傳統社群的三個基本特徵：共同意識、共同的儀式慣例以及基於倫理的責任感。[98] Mcalexander 等認為從消費者體驗的角度來看，品牌社群是一個以消費者為中心的關係網絡，這些重要的關係包括消費者與品牌的關係、消費者和企業的關係、消費者和產品的關係以及消費者之間的關係。[99]

Related to consumer behavior, the hot concept of 「brand community」 that emerged in recent years was first proposed by Muniz and O. Guinn at the Consumer Research Association Annual Meeting in 1995. According to their study in 2001, they defined concept as a non-regional specialized community based on a series of social relationships among brand admirers. They believe that the brand community has three basic characteristics similar to the traditional community: common sense,

---

98　Muniz Jr. A. M.‧O　Guinn T. C.. Brand Community [J] . Journal of Consumer Research‧2001‧27（4）：412-432.

99　Mcalexander J. H.‧Schouten J. W.‧Koenig H. F..Building Brand Community [J] . Journal of Marketing‧2002‧66（1）：38-54.

common rituals and conventions, and ethic-based responsibility.[98] Mcalexander and other scholars believe that from the perspective of consumer experience, the brand community is a consumer-centric relationship network. These important relationships include the relationship between consumers and brands, consumers and enterprises, consumers and products, and among consumers.[99]

2012 年美國智慧手機在 15 歲以上人群中的覆蓋率達到了 51%，平板電腦的覆蓋率達到 25%。人們每天平均 9.6 分鐘看一次手機，每天看手機次數達 150 次。據 Business Insider 的資料顯示，社交媒體已成為美國消費者消費決策的重要因素。[100] 而根據互聯網資料研究資訊公司 We Are Social 的研究，中國大陸線民的上網時間中，有 41% 是用在社會化網站上。

In 2012, the coverage of smartphones and tablets among people over the age of 15 reached 51% and 25% respectively. People check their smartphones once an average of 9.6 minutes a day, and 150 times in total. According to data released by Business Insider, social media has become an important factor in consumer spending decisions in the United States.[100] According to a study made by We Are Social, an Internet data research consulting company, Internet users in mainland China spend 41% of time online on community websites.

這裡我們分享部分美國專業機構關於關於社群化消費的權威統計：

The following is some of the authoritative statistics on community

---

100 美兩國移動互聯用戶的消費行為調查. 中國廣告協會互動網路分會（IIACC）聯合美國互動網路廣告署 Survey on Behavior of Mobile Internet Users in the United States. Interactive Internet Advertising Committee China (IIACC) and US Interactive Network Advertising Agency

consumption released by some professional organizations in the United States:

All Facebook：有 3/4 的美國消費者的購買決策會先參考臉書上的評論，且有一半的受訪對象會因為社會化媒體上的推薦而嘗試新品牌。

All Facebook: Three-quarters of consumers in the United States will first refer to Facebook's comments when making purchasing decisions, and half of the respondents will try new brands recommended on social media.

Social Times：41.5% 的 18 － 43 歲消費者認為社會化媒體上的內容會影響他們的購買決策，女性消費者比男性所受的影響比例更高。

Social Times: 41.5% of consumers aged 18 to 43 believe that content on social media will influence their purchasing decisions, and female consumers are more likely to be affected than men.

iMedia：當消費者從他們的朋友那聽聞到某個品牌後，會驅動他們比平常人 2 倍的意願想與該品牌接觸，4 倍的意願想去購買該品牌。

iMedia: When consumers hear about a brand from their friends, they will be willing to try the products twice as much as those don' t, and even more, have 4 times the willingness to buy the products.

Search Engine Land：52% 的消費者認為網路上的正面評論 (Reviews) 會促使他們更願意去當地的企業消費。

Search Engine Land: 52% of consumers believe that positive reviews on the Internet will make them more willing to consume in local businesses.

ApEngines：80% 的人在第 1 次光顧餐館前會上網查詢餐館的訊息，

88% 的人會根據網路上對餐館的評價來決定到底去哪一家。

ApEngines: 80% of people will check the restaurant's information on the Internet before visiting the restaurant for the first time and 88% will decide which one to go to according to the evaluation of the restaurant on the Internet.

Mashable：44% 的汽車購買者，會先在相關論壇上做研究。

Mashable: 44% of car buyers will do research on relevant online forums first.

Nielsen：消費者對品牌訊息來源的信任度調查，92% 的人最相信所認識的人的推薦，70% 為線上的消費者意見。

Nielsen: In regard of trust in brand informatrion sources, 92% of consumers most believe in the recommendations made by people they know, and 70% believe in consumer opinions on the Internet.

empathica：2012 年年中調查 6500 名美國消費者近期至零售商店或餐廳，主要受何種社會化平臺影響？答案是 73% 受 Facebook 影響，38% 受 Google 上搜尋到的評論影響。

empathica: In the middle of 2012, a survey of 6,500 consumers in the United States was conducted to tell what kinds of social platforms had affected them in retail stores or restaurants recently. The findings came to the conclusion that 73% were affected by Facebook and 38% were affected by comments searched on Google.

All Facebook：2012 年年中針對超過 6500 名的美國消費者所做的調查顯示，在做出購買決策時，有 3/4 的消費者會將臉書上的評論做為重要的購買參考，並且在選擇購買何種品牌時，有一半的受訪對象會參考社會化媒體上

的推薦意見。

All Facebook: In the middle of 2012, a survey of 6,500 consumers in the United States showed that three-quarters of consumers took comments on Facebook for an important reference when making a purchase decision. When choosing which brand to buy, half of the respondents will refer to the recommendations on social media.

Accenture Interactive： 93% 的美國消費者傾向購買有經營社會化媒體的品牌。

Accenture Interactive: 93% of consumers in the United States tend to buy brands that operate social media of their own.

ComScore：接受粉絲團資訊的消費者購買星巴克咖啡的比例，比起沒有接受粉絲團資訊的消費者高出 38%。[101]

ComScore: Consumers, who have received fan group information, buy Starbucks coffee at a 38% higher rate than those who haven't.[101]

（2）消費與社交一體化 Integration of consumption and social interaction

毋庸置疑，我們正處在一個高度資訊化的社會，每個人都有固定的交際圈，同時每個人都處於一個又一個、一環套一環的資訊輻射圈中，驢友圈、社區鄰里圈、家長圈、同事圈、親友圈、粉絲圈、同學圈、老鄉圈等等，每個人都可能被他人影響，每個人又都可能影響他人，隨著 QQ、微博、微信

---

101 臺灣牛 . 50 個調查資料：社會化媒體如何影響消費者的購物決策 . SocialBeta. 2012-12-21 Taiwanese cattle. 50 Peices of Survey Data: How Social Media Influences Consumers' Decisions. SocialBeta. Dec 21, 2012

等網路平臺的擴散式傳播，網路大 V、意見領袖、明星、論壇達人，身邊的時尚達人等等都以他們的消費愛好、偏好，自然地帶動消費，引領時尚。按照小米 CEO 雷軍的話：「我的朋友買手機的時候都會問我，因為他們沒有我瞭解手機。」一般消費者在購買專業化產品時，往往會詢問發燒友朋友的意見，這些意見很可能就是消費者的最終選擇。而更多的消費者會透過網路社區或線上評論選擇產品實施消費。

Undoubtedly, we are in a highly information-based society. Everyone has a fixed social circle. At the same time, everyone is in the information circles that are linked with one another, such as the circles of tour pals, neighborhood, parents, colleagues, relatives，fans, schoolmates and fellow villagers, etc. Everyone may be affected by others and in the same time, affect others. With the spread of QQ, Weibo, WeChat and other network platforms, big Vs(the verified people who have many followers on Sina Weibo), opinion leaders, celebrities, forum leaders, fashionistas take advantage of their hobbies and preferences to naturally drive consumption and lead fashion. Lei Jun, the CEO of Xiaomi, once said "My friends ask me for advice when buying a mobile phone, because they don't know about mobile phones better than me." When buying specialized products, consumers often ask for the opinions of friends who are enthusiasts, which will be very likely to become the ultimate choice. And more consumers will choose and purchase the products according to online communities or reviews.

據普華永道對全球消費者的研究，積極地使用社交媒體的中國消費者中，57%的受訪消費者在社交媒體上關注喜愛的品牌或零售商的最新消息（在全球受訪者中，這一比例為 38%）。同時，更多的中國網購者使用社交媒體

與品牌互動，提出對企業及產品的評價，以及尋找新品牌。[102]

According to PricewaterhouseCoopers' research on global consumers, 57% of Chinese consumers who actively use social media are following the latest news on their favorite brands or retailers on social media (among global respondents, the ratio is 38%). At the same time, more Chinese online shoppers use social media to interact with brands, propose evaluations of companies and products, and seek new brands.

根據艾瑞諮詢 2011 年初的調研資料顯示，SNS 網站、專業旅遊點評網站和博客是旅遊用戶分享出遊經歷的主要途徑。半數以上的女性表示自己喜歡微博平臺推薦的旅遊資訊，她們非常關注旅遊行銷帳號，並願意與朋友分享出境遊資訊。

According to research data from iResearch in early 2011, SNS websites, professional travel review websites and blogs are the main ways for travel users to share their travel experiences. More than half of the women said that they liked the travel information recommended by the Weibo (the microblog). They were very concerned about the marketing accounts of travelling and were willing to share the outbound travel information with friends.

---

102 《揭祕網購者：多管道零售的 10 個迷思》. 普華永道研究報告 . 2013 Unveiling Online Shoppers: 10 Myths of Multi-Channel Retailing. PricewaterhouseCoopers Research Report. 2013

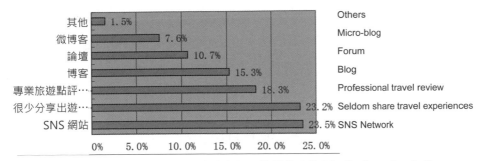

圖 Figure 3-19　2010 年中國用戶分享出遊經歷的網站分佈 Distribution of websites where Chinese users shared travel experiences in 2010
資料來源：iResearchinc.2011 年 6 月
Source: iResearchinc. June 2011

　　為了更好地適應社群化消費，近年來越來越多的國外商家開始採用真人秀的形式，邀請意見領袖（也是典型消費者）代言產品，把一款鞋、一件衣服的設計到製作、試穿全過程網上直播，讓消費者充滿期待，體驗化參與，使品牌植入人心。用當前流行的術語，網路大 V、意見領袖、明星、論壇達人，乃至身邊的時尚達人，都是創造消費、引領社群化消費的「自媒體」。

　　In order to better adapt to community consumption, more and more foreign merchants have adopted the form of reality show in recent years, inviting opinion leaders (also typical consumers) to endorse products. They are asked to put the whole process of designing, making and fiitting of a pair of shoes or a piece of clothing in a live webcast, so that consumers will be full of expectations. With experience participation, the brand will go deep into the hearts of the people. Current popular terminologies like big Vs(the verified people who have many followers on Sina Weibo), opinion leaders, celebrities, forum leaders, fashionistas around us are all 「we-media」 that creates consumption and leads community consumption.

我們認為，所謂「社群化」的消費模式，即在「雲消費」時代，消費與社交一體化，消費場所即為社交場所。「消費的社群化」強調消費的社群認同，消費意見與消費結果透過意見領袖或「群友」推薦而透過 QQ、微信等網路平臺擴散式傳播達成消費意向，引領消費潮流。

We believe that the so-called community consumption pattern, in other words, is the integration of consumption and social interaction in the era of cloud consumption. Consuming places are where social interaction occurs. The community of consumption emphasizes social identity of consumption. Consumer opinions and consumption results are recommended by opinion leaders or 「group friends」 and spread through the online platforms such as QQ and WeChat to achieve consumer intentions and lead the consumption trend.

2‧ 如何利用消費的社群化——案例觀察
How to use the coummunity of consumption-case studies

（1）小米論壇—「米粉」的消費社群
Xiaomi Forum-Consumer Community of "miboys" and 「migirls」

成立於 2010 年 4 月的小米做為移動互聯行業第一家米粉參與開發，全靠口碑傳播的企業，最大可能地集聚了「米粉」的需求，2013 年銷售份額已佔據中國第一，世界第六，該公司 2013 全年共銷售手機 1870 萬支，增長了160%；銷售額達到 316 億元，增長 150%。

Founded in April 2010, Xiaomi is the first enterprise in the mobile internet industry which its fans (called 「miboys」 and 「migirls」) participate in the development of products. Its branding relies on word-of-mouth communication and the enterprise has met as much demand made by its fans as possible. In 2013, the sales share ranked the first in China and the sixth in the world. The company sold a total of 18.7 million mobile phones in 2013, an increase of 160%. Its sales reached ¥31.6 billion, an increase of 150%.

小米科技的官方論壇小米社區是「米粉」的聚集地，也是小米的重要品牌社群。小米的品牌社群化主要體現在共同的儀式和傳統、共同意識、責任感三個方面。

Xiaomi Community, the official forum of Xiaomi Technology is the gathering place of 「miboys」 and 「migirls」, and also an important brand community of Xiaomi. The community of Xiaomi is mainly reflected in the three aspects, they are common ritual and tradition, common sense and responsibility.

■ 共同的儀式和傳統 Common rituals and traditions

品牌社群的共同儀式和傳統是指在品牌社群的形成與發展的過程中積澱下來的一些共同的習慣以及一些不成文的規定。在小米社區，消費者必須購買小米手機才能成為小米的認證用戶，探索小米手機的功能並接受老用戶的一些經驗。米粉們上手後，他們開始發帖分享經驗、上傳照片，展示才藝，還參加小米舉辦的「隨時拍」、「兩週年祝福徵集」、「小米故事」等活動，這一系列過程中，米粉在社區中感受集體的認同，有傳教體驗、娛樂體驗、沉浸體驗、審美體驗、創造和認同體驗等等。共同儀式和傳統讓小米發燒友稱其為「米粉」。

The common rituals and traditions of the brand community refer to some common habits and unwritten rules that have accumulated during the formation and development of the brand community. In the Xiaomi community, consumers must purchase Xiaomi mobile phones to become certified users of Xiaomi, thus explore the functions of Xiaomi mobile phones and accept some experience of old users. After having got hand of the phones, the 「miboys」 and 「migirls」 begin to post experiences and photos, show talents in the community, and participated in activities such as 「Photographing Anytime」, 「The Second Anniversary Blessings」 and 「Xiaomi Story」 organized by Xiaomi. In these activities, 「miboys」 and 「migirls」 can feel the collective identity like missionary experience, entertainment experience, immersive experience, aesthetic experience, creation and identity experience, etc. Common rituals and traditions make Xiaomi enthusiasts a "miboys" and 「migirls」.

■ 共同意識 Common sense

品牌社群的共同意識不僅僅是指態度分享或者簡單的接收，它更是一種歸屬感的分享。小米社群的共同品牌意識是指其成員自覺認為自己是「米粉」，是小米社區大家庭中的一員。他們並不是簡單地接收或者分享小米「因為米粉，所以小米」的品牌承諾，而是從內心覺得，我是一名「米粉」，「為發燒友而生」的品牌承諾就是米粉的品牌主張。

The common sense of the brand community is not only about attitude sharing or simple reception, but also sharing a sense of belonging. The co-brand awareness of the Xiaomi community means that its members consciously consider themselves to be 「miboys」 and 「migirls」, and are members of the Xiaomi community.

They are not simply accepting or sharing the brand promise of "Xiaomi for miboys and migirls", but from the heart believe that the brand promise of "born for the enthusiasts" is the brand claim of them.

■ 責任感 Responsibility

品牌社群的責任感指的是對品牌社群及其他成員的責任感。米粉具有共同意識後，開始對小米社區產生歸屬感，開始以主人的態度對待小米社區，積極主動地維護小米，並漸漸地對其他競爭品牌產生排斥。例如，米粉積極地回答新手的提問、米粉積極為小米「品質門」尋找原因、米粉與魅族用戶的口水戰、回應求救帖、資訊交流、參加活動等都體現了米粉的責任感。[103]

The sense of responsibility of the brand community refers to the sense of responsibility of the brand community itself and other members. After having the common sense, 「miboys」 and 「migirls」 begin to have a sense of belonging to the Xiaomi community, and treat the community with the owner's attitude, actively maintaining the community and gradually becoming rejective to other competing brands. For example, 「miboys」 and 「migirls」 actively answer questions from novices, look for the causes of Xiaomi's 「quality-gate」 scandal, wars of words with Meizu users, response to call-for-help posts, exchange information and participate in activities, all of which reflect the sense of responsibility of Xiaomi fans.

---

[103] 曾郭鈴，黎小林 .. 基於網路品牌社群的行銷戰略——以北京小米科技有限責任公司為例 [J]. 企業活力 ,2012(12). Zeng Guoling, Li Xiaolin. Marketing Strategy Based on Network Brand Community-Taking Beijing Xiaomi Technology Co., Ltd. as an Example[J]. Enterprise Vitality, Dec 2012.

據不完全統計，小米論壇現有將近 1000 萬的用戶，空間用戶過 1000 萬，微博粉絲為 300 多萬，微信粉絲約 280 萬，透過這樣的社交矩陣，粉絲們源源不斷為小米提供各種產品、服務建議，並自發進行口碑傳播。為了保持溫度感，創業之初雷軍、黎萬強等幾個聯合創始人要保持每天在論壇上 1 個小時，企業做大後每天再忙也要保持在論壇上十幾分鐘，而所有工程師也被鼓勵透過論壇、微博和 QQ 等管道和用戶直接取得聯繫，要讓「這些宅男工程師覺得他寫程式不是為了小米公司寫的，是為了他的粉絲在做一件工作」。

According to incomplete statistics, Xiaomi Forum has nearly 10 million users, over 10 million space users, more than 3 million Weibo fans, and 2.8 million WeChat fans. Through such a social matrix, fans continue to provide various kinds of products and service recommendations for Xiaomi, and communicate the brand spontaneously in a word-of-mouth manner. In order to maintain the sense of temperature, at the beginning of the business, Lei Jun, Li Wanqiang and other co-founders must ensure their time online of one hour every day. After the company becomes bigger, they are still online for ten more minutes every day even if they are busy. Plus, all engineers are encouraged to directly get in touch with users through channels like forum, Weibo and QQ so that 「these nerdy engineers would think that they write the programs not for the company, but work for the fans."

《全球商業經典》雜誌曾經梳理出一個小米使用者扭曲立場的金字塔結構：塔尖是可以參與決策的發燒友，比如小米論壇的神祕組織「榮組兒」，以賦予粉絲特權的方式鼓勵其參與決策；塔中間是米粉群體，他們信賴和追隨小米的價值主張，購買小米產品的意願強烈；塔基則是普通的大眾用戶，

他們能夠從微博、微信、事件行銷以及米粉的自發傳播中接觸小米，繼而轉化為產品購買者或晉級為米粉。[104]

Global Business has sorted out a pyramid structure in which Xiaomi users have distorted their positions. The spire is composed of enthusiasts who can participate in decision-making, such as the mysterious organization "Rong Zu' er" (monophonous with the name of a Hongkong singer) on Xiaomi forum, where fans are admitted and given priviledges to participate in decision-making. The middle is the group of fans who trust and follow the value propositions of Xiaomi, and have strong willingness to buy Xiaomi products. The foot consists of the ordinary users who know Xiaomi from Weibo, WeChat, event marketing and the spontaneous communication of Xiaomi fans. They are likely to be converted into consumers or qualified as fans.

（2）Shoes of Prey 與意見領袖合作引導消費

Shoes of Prey collaborate with opinion leaders to guide consumption

Shoes of Prey 是澳大利亞的一家網路鞋店，消費者可以在這家網店裡自由選擇鞋子的尺寸、樣式和顏色，自主設計鞋跟的款式，還能自選面料。DIY 設計好自己的鞋子之後，網上下單訂做，兩週後送貨上門。不同款式顏色和形狀可以有 4 萬種鞋子的組合。該網站給消費者提供了一種自主設計師的自豪感，並能充分展示自我價值。在客服稱呼上面，鞋店也突破了以往的常規，進行客服實名制，消費者透過和客服建立良好的信任感，從而打消跟

---

104 揭祕小米崛起背後的文化戰略 . 福布斯中文網 Demystifying the Cultural Strategy Behind the Rise of Xiaomi. Forbes China

鞋店的疏遠感。鞋店還透過尋找意見領袖和有一定影響力的微博名人合作，展示製鞋的服務過程，贏得了很高的訪問量。

Shoes of Prey is an online shoe store in Australia where consumers can decide the size, style and color of their shoes, design heel styles and choose fabrics for themselves. After finishing designing their shoes in a DIY manner, they place orders online and the products will be delivered to their doors two weeks later. With different styles of colors and shapes, there can be 40,000 combinations of shoes. The online store provides consumers with a sense of pride of being designers for themselves and fully demonstrate their self-worth. In regard of calling service reps, the shoe store has also broken through the past routines and adopted the real name system, which establishs a good sense of trust between consumers and service reps, thus eliminating the sense of alienation with the shoe store. The store also cooperated with opinion leaders and influential Weibo celebrities to showcase the service process of shoemaking, winning a high volume of visits.

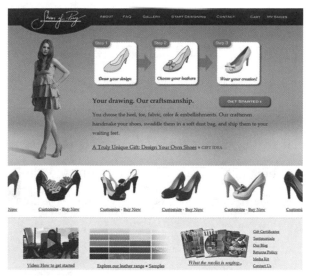

圖 Figure 3-20 消費者可以在 Shoes of Prey 網頁上 DIY 自己獨一無二的鞋子
Consumers can DIY their unique shoes on the Shoes of Prey website.

JONAYE: The new Red with White Spots Cotton Blend just makes my heart swoon! Mixed with a fabulous clashing tone of Pink Soft Leather and a 4.5 inch Square Heel, this is absolutely my dream shoe of the moment!

LUCY: I'm really feeling the nautical theme at the moment so I absolutely adore our new Navy and White Striped Cotton Blend material, with a touch of Midnight Blue Patent on the stiletto and platform to add a dressy edge to the look!

圖 Figure 3-21　意見領袖、微博名人推薦心愛的鞋子 Opinion leaders, Weibo celebrities recommend their beloved shoes

JONAYE：這雙新紅底白點棉混鞋簡直讓我癡迷！混搭著粉色軟皮革絕妙的對比色調和 4.5 英寸的方跟，絕對是我現在的夢幻之鞋！

JONAYE: The new pair of red and white cotton mixed shoes makes me obsessed! With the mix-and-match of superb contrasting tones of pink soft leather and the 4.5-inch square heel, it is definitely my dream shoes!

LUCY：我目前對航海樣式特有感覺，所以我絕對喜歡我們新的航海樣式藏青和白色條紋棉混鞋，鞋跟和鞋底具有午夜藍模式的感覺，在外觀上還增加了裝飾性邊緣！

LUCY: I have a special feeling about the nautical style at present, so I absolutely like our new navy blue and white striped cotton-mixed shoes. The heel and sole have a midnight blue pattern and a decorative edge is added to the appearance!

（五）消費的實時化 Real-time consumption

1‧ 什麼是消費的實時化 What is the real-time consumption

消費的即時化是指消費者透過智慧終端機鎖定消費目標，以參與互動、移動搜索等滿足消費者隨時隨地消費需求，特別強調消費的專屬性和移動性。

The real-time consumption means that consumers lock their consumption targets through smart terminals, meet their demand for consumption anytime and anywhere by interaction and mobile search, etc., with particular emphasis on the specificity and mobility of consumption.

大量精準的網路地圖搜索服務的出現，為消費的即時化提供了基礎。在谷歌地圖、蘋果地圖之後，中國消費者已經能夠很熟練地應用百度地圖、搜狗地圖、高德地圖等地圖搜索服務。以百度地圖為例，其 4.0 版本主打免費語音瀏覽、室內定位、即時公交、生活搜索四大功能。如消費者需要找尋一

家附近的目標餐廳，只要在百度地圖的輸入框中進行搜索可以很方便地搜到目標餐廳，或者透過「附近」按鈕，可以查看所在地周邊的所有類型的餐廳，遊客透過點擊餐廳就能實現點餐、訂座、結帳、點評等。

The emergence of a large number of accurate web map search services provides the basis for real-time consumption. Besides Google Maps and Apple Maps, Chinese consumers have now been able to use other map search services such as Baidu Maps, Sogou Maps, and Gaode Maps. Taking Baidu map as an example, its 4.0 version features four functions: free voice navigation, indoor positioning, real-time bus, and live search. If the consumer needs to find a nearby target restaurant, he or she can easily locate the target restaurant by typing the name of the restaurant in the input box, or view all types of restaurants nearby by touching the 「near」 button. People can place the order, book the seats, check out, review and more by touching the restaurant on the devices.

隨著移動終端的普及及大量服務應用的開發，越來越多的消費者已經適應了移動「即時化」消費，購物、訂餐、訂座、買票、預定酒店等生活需求，均可以「即時化」滿足。如在旅遊領域，遊客可以透過移動媒體隨時隨地查詢、分享旅遊資訊，預定旅遊產品。2012 年 Travelzoo 旅遊族和 CNNGo 調查發現，中國內地年輕一代的海外旅行者 75% 表示願意透過手機端瞭解甚至預定旅行產品，70% 手機使用者表示會在旅行中打開手機應用。更多旅遊消費者把美食與手機關聯，就地尋找一些特色、優惠等餐廳用餐 [105]。

---

105 海外遊趨向多維度競爭 體驗性旅遊受寵 . 解放日報 .2012 年 11 月 4 日 Overseas Travel Tends to Multi-dimensional Competition Experience Travel is Favored. Liberation Daily. November 4, 2012

With the popularity of mobile terminals and the development of a large number of service applications, more and more consumers have adapted to the real-time mobile consumption, where shopping, ordering, booking, buying tickets, booking hotels and other life needs can all be done. For example, in the field of tourism, visitors can use the mobile media to inquire, share travel information and book tourism products anytime, anywhere. In 2012, the survey made by Travelzoo and CNNGo found that 75% of the younger generation of overseas travellers in mainland China expressed their willingness to learn about or even book the tourism products through mobile phones, and 70% of mobile phone users said they would launch mobile applications during their travels. More travel consumers associate food with mobile phones and look for some restaurant with specialties and special offers.[105]

2· 如何實現消費的即時化——案例觀察

How to realize the real-time consumption-case studies

（1）移動餐飲——即時化的餐飲服務享受

Mobile catering-real-time catering service enjoyment

■ 淘寶「淘點點」的移動餐飲服務平臺

2013 年 12 月，淘寶正式推出其移動餐飲服務平臺——「淘點點」，希望重新定義「吃」，希望將餐飲行業做成「淘寶＋天貓」的模式，即每個菜品都是一個 SKU(庫存商品)，一些熱銷的菜品，相當於淘寶中的熱銷款。將餐飲服務變成商品，讓買賣雙方直接交易。「淘點點」從點菜切入生活服務，

使用者透過移動用戶端，就可以對各個餐館的出品和價格一目瞭然。

「Taodiandian」 of Taobao, the mobile catering service platform

In December 2013, Taobao officially launched 「Taodiandian」, a mobile catering service platform, hoping to redefine the concept of 「dining」 and make the catering industry into a 「Taobao+Tmall」 model, that is, each dish is considered as a SKU (Stock Keeping Unit). Some hot dishes are equivalent to the best sellers in Taobao. The model turns catering services into commodities and let buyers and sellers trade with each other directly. 「Taodiandian」 cuts into the life service from catering where users can see the output and price of each restaurant at a glance through the mobile client.

■ 騰訊微信訂餐服務

2014 年 2 月，騰訊宣布與大眾點評達成戰略合作，在微信「我的銀行卡」裡添加了「今日美食」欄目，大眾點評的商戶資訊、消費點評、團購、餐廳線上預定等本地生活服務未來都將與 QQ、微信等騰訊產品合作。騰訊新推出的微信訂餐服務名為「半小時微信訂餐」，做為連接微信用戶與速食外賣店的橋樑，將線下服務與互聯網結合在一起，透過線上攬客，線下服務，達成交易。該服務透過公眾號 banxiaoshi086 或二維碼，讓微信用戶隨手訂閱；透過用戶分組和地域控制，實現精準的消息推送，直指目標使用者；借助個人主頁和朋友圈，透過用戶口碑傳播推廣。

Tencent WeChat catering service

In February 2014, Tencent announced a strategic cooperation with Dianping. com, and added the 「Food」 section to WeChat 「My Bank Card」 menu.

Dianping will cooperate with QQ, WeChat and other Tencent products in local live services like merchant information, consumer reviews, group purchases, online restaurant reservations, etc. Tencent's new WeChat catering service is called 「WeChat Half-hour catering」, which serves as a bridge between WeChat users and fast-food take-out stores. It combines offline services with the Internet to complete orders through attracting customers online and providing services offline. WeChat users can subscribe the catering service through the official account 「banxiaoshi086」 or QR code right at their fingertips. Also, messages can be accurately pushed to the target users through the user grouping and geographical control. Moreover, the service can be popularised through the personal homepage and circle of friends by word-of-mouth communication.

圖 Figure 3-22 半小時微信訂餐 WeChat Half-hour catering

（2）「無線（限）1號店」隨時隨地購物

Shop anytime, anywhere in the 「Limitless No. 1 Store」

2012 年 10 月 15 日，1 號店正式發佈「無限 1 號店」專案，目的是線上下建造出千家虛擬 1 號店——「無限 1 號店」。首先，「無限 1 號店」在其手機應用中預置了虛擬「無限 1 號店」的整體模型，以及 1000 家線下虛擬店的 POI(Point of Interest，即興趣點); 其次，用戶在任意預設的 POI 點附近，打開該應用。用戶可以在其手機攝像頭透過 AR（增強現實）技術，使虛擬立體商店的全貌呈現在使用者的智慧手機螢幕中。在使用者真實移動步伐的同時，螢幕中場景會根據使用者的移動距離隨之變化，接近後商品會逐一陳列在該虛擬商店的牆上，以體驗現實購物的感覺。

On October 15th, 2012, No. 1 Store officially released the 「Limitless No. 1 Store」 project, aiming to build a thousand offline virtual No. 1 Stores, that is, the Limitless No. 1 Store. First of all, the Limited No. 1 Store presets the overall model of the virtual store in its mobile phone app, and the POI (Point of Interest) of 1000 offline virtual stores. Secondly, users open the app near any preset POI. They can use the AR (Augmented Reality) technology together with the phone cameras to present the full view of the virtual 3D store on the screen. At the same time as users are actually moving around, the scene in the screen will change according to the moving distance. The products will be displayed one by one on the wall of the virtual store to reproduce the feeling of real shopping.

「無限 1 號店」專案整合了地鐵戶外廣告、手機屏、互聯網等多種管道，使用戶只需在手機上下載「掌上 1 號店」應用，對著線下車站貼出的相關廣告上的產品二維碼進行掃描，就能透過手機完成購買，讓消費者可以利用等車的「碎片時間」隨時、隨地購物。

The 「Limitless No. 1 Store」 project integrates subway outdoor advertising, mobile phone, the Internet and other channels, so that users only need to download the 「No. 1 Store」 app on the mobile phone and scan the product QR codes on related ad posted in the offline stations. By scanning the product QR codes, they can complete the purchase on their mobile phones, so that consumers can take advantage of the fragmented time of standing on the station to shop anytime, anywhere.

圖 Figure 3-23 「無線 (限)1 號店」專案 TThe 「Limitless No. 1 Store」 project

「雲消費」時代交易模式的根本性變化

Chapter Four
Fundamental Changes in the Transaction Patterns in the Era of Cloud Consumption

第四章
「雲消費」時代交易模式的根本性變化

Chapter Four Fundamental Changes in the Transaction Patterns in the Era of Cloud Consumption

一

不同經濟發展階段交易模式的變化
1. Changes in transaction patterns in different stages of economic development

（一）交易成本的概念
The concept of transaction costs

交易成本是新制度經濟學的核心概念。本書首先從交易成本理論入手，探討不同社會經濟發展階段交易方式的變化。

Transaction Costs is the core concept of neo-institutional economics. The book begins with the transaction costs theory and explores the changes in transaction patterns in different stages of economic development.

交易成本（Transaction Costs）又稱交易費用，最早由諾貝爾經濟學獎得主科斯 (Coase, R.H., 1937) 所提出的。[106] 制度經濟學的第二代掌門人，諾貝爾經濟學獎得主奧列佛·威廉姆森 (Oliver Eaton Williamson) 將交易成本分為六大類：搜尋成本、資訊成本、議價成本、決策成本、監督交易進行的成本和違約成本（1975 年）。1985 年威廉姆森（Williamson）進一步將交易成本加以整理區分為事前的交易成本、事後的交易成本、討價還價的成本、建構及營運的成本、為解決雙方的糾紛與爭執而必須設置的相關成本、約束成本。在這一系列成本中，核心為資訊的搜尋、獲得及契約保障的成本[107]。1979年達爾曼 (C.J.Dahlman) 則將交易活動的內容加以類別化處理，認為交易成本包含搜尋資訊的成本、協商與決策成本、契約成本、監督成本、執行成本與轉換成本。[108] 張五常在《論新制度經濟學》中將交易成本歸納為資訊成本、

---

106 羅奈爾得科斯（Ronald H. Coase）.《企業的性質》（The Nature of the Firm（1937））
Ronald H. Coase. The Nature of the Firm (1937)
107 威廉姆森（Williamson）.《交易成本經濟學》Williamson. Transaction Cost Economics
108 C.J.Da hlma n The Problem of Externality Journa l of Legal Studies，1979（22）：141-162

談判成本、擬定和實施契約的成本、界定和控制產權的成本、監督管理的成本和制度結構變化的成本等。

The concept of Transaction Costs was first proposed by Coase, R.H., the Nobel Prize winner in 1937.[106] Oliver Eaton Williamson, the second-generation head of institutional economics and the Nobel Prize winner, divided transaction costs into six categories in 1975: search costs, information costs, bargaining costs, decision-making costs, supervision costs and default costs. In 1985, Williamson further divided the transaction costs into costs before the transaction, costs after the transaction, bargaining costs, construction and operation costs, related costs that must be set to resolve disputes between the two parties and committed costs. Among these costs, the core is the costs of information search and acquisition and contract guarantee.[107] In 1979, C.J. Dahlman classifies the content of transaction activities and believes that transaction costs include the costs of information search, negotiation and decision, contract costs, supervision costs, execution costs, and conversion costs.[108] In On Neo-Institutional Economics, Steven N.S. Cheung summarized transaction costs into information costs, negotiation costs, costs of drafting and implementing contracts, costs of defining and controlling the property rights, costs of supervision and management, and costs of institutional changes.

由以上概念可知，交易必然存在成本，交易成本表現在各個層面，由於交易雙方資訊不對稱而產生的資訊成本是交易成本的核心內容，掌握了更多資訊優勢的企業，在交易中就必然佔據了更多主動，改變資訊獲得方式，暢通資訊獲得管道，就可以降低資訊成本從而降低交易成本。

It can be seen from the above concept that the transaction must have

cost, there are always transaction costs, which are manifested at all levels. The information cost generated by the information asymmetry of the two parties is the core content of the transaction costs, thus the enterprises that have more information advantages will certainly take more initiative in transaction. Changing the way to obtain information and unblocking the access to information can reduce the information costs, thus reducing transaction costs.

（二）交易發展的三個階段
Three stages of transaction development

應用交易成本理論，我們認為現代企業的經濟活動主要經歷了三個階段：內部交易階段、外部交易階段和「雲消費」時代階段（「雲交易」）。

We believe that application of the transaction costs theory in the economic activities of modern enterprises mainly went through three stages: the internal transaction stage, the external transaction stage and the cloud consumption era stage (cloud transaction).

內部交易是指由母公司與其所有子公司組成的企業集團範圍內，母公司與子公司、子公司相互之間發生的交易。內部交易成本就是透過企業內部管理的方式協調供需雙方的矛盾而發生的成本，主要包括企業內部管理成本、員工及相關福利成本等。

Internal transactions refer to transactions between the parent company and its subsidiaries or different subsidiaries within the group composed of the parent company and all its subsidiaries. The internal transaction costs are the costs incurred by the internal management of the company to coordinate the contradiction between the supply and demand sides, mainly including the internal management costs, employees and related welfare costs.

外部交易 (成本) 是指透過市場交易的方式來協調供需雙方的矛盾而發生的成本，主要包括廣告費用、企業稅負、籌資手續費、發行費、公關費等。

External transaction costs refer to the costs incurred by the market transactions to coordinate the contradiction between the supply and demand sides, mainly including advertising costs, corporate taxes, financing fees, issuance fees, public relations fees.

「雲交易」(成本) 是指「雲消費」時代，企業透過自由開放的市場和資訊資源平臺而產生的成本。資訊成本逐漸成為交易成本中比重最大的成本。

Cloud transaction refers to the cost generated by free and open markets and information resource platforms of enterprises in the era of cloud consumption. Information costs are gradually becoming the largest part of transaction costs.

1・ 內部交易階段企業內部交易效率高於外部交易 In the internal transaction stage, efficiency of internal transactions is higher than external transactions

科斯在〈企業的性質〉一文中，在交易成本的框架下很好地回答了內部交易時期企業內部交易優於外部交易的基本問題。 [109] 這一理論可以解釋在

---

[109] 羅奈爾得科斯 (Ronald H. Coase)．《企業的性質》(The Nature of the Firm (1937))

中國電子商務發展的初級階段，由於產業鏈相應配套服務不完善，電商企業為了更快、更經濟地發展，紛紛採用大而全的產業擴張模式，通常採取自建物流、自主開發全套應用軟體、自建機房以及網路安全系統，來支援電子商務交易的順利完成。如當當網就在全國 11 個城市建設了 20 個物流中心，全國庫房面積超過 50 萬平方米。卓越網在北京大興建立了 4.5 萬平方米的配送中心，還成立了自有配送公司（世紀卓越快遞）。

In The Nature of the Firm, Coase gave a good answer to why internal transactions is superior to external transactions within the framework of transaction costs.109 This theory can explain that in the initial stage of the development of e-commerce in China, due to the imperfect corresponding supporting services fo the industrial chain, e-commerce companies have adopted large and comprehensive industrial expansion models for faster and more economic development. They usually adopt self-built logistics, independent development of a full set of application softwares, self-built computer rooms and network security system to support the successful completion of e-commerce transactions. For example, Dangdang.com has built 20 logistics centers in 11 cities across the country, and the national warehouse space is over 500,000 square meters. Joyo.com has established a 45,000-square-meter distribution center in Daxing, Beijing, and has established its own distribution company, the Century Express.

2・外部交易階段大企業與大企業交易的優勢更為明顯 The advantages of external transactions among large companies is more obvious in the external transaction stage.

隨著社會化大生產的發展，連鎖等新型流通方式的興起，企業規模不斷擴大，企業向外部尋求發展成為可能，企業經濟活動進入外部交易階段。根據威廉姆森的研究，交易成本產生的重要原因之一在於資訊不對稱（Information Asymmetric）。即由於環境的不確定性和自利行為產生的機會主義，而交易雙方握有不同程度的資訊，使得市場的先佔者（First Mover）擁有較多的有利資訊而獲益，並形成少數交易。

With the development of socialized large-scale production and the emergence of new types of circulation such as chain management, the scale of enterprises has continued to expand, and it has become possible for enterprises to seek development from the outside, and the economic activities of enterprises have entered the stage of external transactions. According to Williamson's research, one of the important reasons for the transaction costs is Information Asymmetric. That is to say, due to the uncertainty of the environment and the opportunism generated by self-interested behavior, the two sides of the transaction hold different levels of information, which makes the First Mover in the market benefit from more beneficial information they have, thus causing small-member transactions.

美國經濟學家約瑟夫·斯蒂格利茨 (Joseph Eugene Stiglitz)、喬治·阿克爾洛夫 (George A. Akerlof) 和邁克爾·斯彭斯（A. Michael Spence）等認為：市場經濟活動中，個人及企業及搜尋和收集資訊都要付出相應的成本，然而各類人員對有關資訊的瞭解存在著差異，掌握資訊比較充分的人員，往往處於比較有利的地位，而資訊貧乏的人員，則處於比較不利的地位。在商品交換時，資訊獲取者要想獲得準確的資訊，就應該為這些資訊產權的轉移或轉讓或使

用付費。如果沒有獲得準確的資訊或資訊掌握不充分,而導致交易行為失敗,實際上就付出了更大的交易成本。在某些領域,表面看來降低了某些環節的交易成本,但又勢必增加一些其他的成本,整體看來,經濟效率並沒有提高。(由於在「對充滿不對稱資訊市場進行分析」領域所做出的重要貢獻,以上三位經濟學家分享了 2001 年諾貝爾經濟學獎)。

American economists Joseph Eugene Stiglitz, George A. Akerlof, and A. Michael Spence believe that in market economic activities, to search and gather information, both individuals and enterprises have to pay the corresponding costs. However, there is a difference in the understanding of relevant information among all types of personnel. Persons with adequate information are often in a favorable position, while those with poor information are at a disadvantage. When exchanging goods, in order to acquire the accurate information, acquirers should pay for the moving, transfer or the use of information property rights. More cost would be paid because of the failure of transactions caused by inaccurate or inadequate information. In some areas, although it appears that the transaction costs of certain aspects have been reduced, others are bound to increase. Economic efficiency has not improved on the whole.

PS. The above three economists shared the 2001 Nobel Prize in Economics for their important contributions in the field of analysis of markets full of asymmetric information.

根據以上理論,我們就可以理解為什麼「農超對接」長期叫好不叫座。所謂農超對接是指農產品與超市直接對接,市場需要什麼,農民就生產什麼。在理想狀態下,透過農超對接既可避免生產的盲目性,穩定農產品銷售管道

和價格，還可以減少流通環節，降低流通成本，給消費者帶來實惠。有專家測算，透過農產品產地直採，可以降低 20% － 30% 流通成本。[110] 從表面上看，農超對接少了一次交易成本，但同時又增加了組織成本，增加了組織管理者的工作量。若以超市為主體對接分散農戶，企業將不可避免地會增加資訊搜尋、商品議價、商品決策、專業人員組織甚至自建物流、建設生產基地等成本，一般超市企業缺乏規模經濟，採購量有限、配送效率不高，要實現農超對接近乎奢侈。由此可見，在外部交易階段，大企業的發展優勢更為明顯，大企業與大企業交易的優勢更為明顯。大企業有條件掌控更多有價值的資訊，掌握大量資源，大企業間的交易流向更為通暢、便捷、成本最低廉。為此，企業間交易模式主要為大企業對接大企業、大零售商對接大批發商、大批發商對接大生產商。

According to the above theory, we can understand why the "agricultural&super docking" has long been a castle in the air. The so-called agricultural &super docking refers to the direct docking between agricultural products and supermarkets, that is, farmers produce whatever the market needs. Under ideal conditions, through agricultural&super docking, the blindness of production can be avoided, the sales channels and prices of agricultural products can be stabilized, the circulation links can be simplified, the circulation costs can be reduced, all of which can bring benefits to consumers. Some experts have

---

110 陳奇 . 閩台農產品電子商務發展研究 [D]. 福建農林大學碩士論文 . 2013 年 4 月 1 日
Chen Qi. Research on the Development of Agricultural Products E-commerce in Fujian and Taiwan [D]. Master Thesis Fujian Agricultural and Forestry University. April 1, 2013

estimated that direct production of agricultural products can reduce the circulation cost by 20%~30%.[110] On the surface, agricultural&super docking has reduced transaction costs, but at the same time increased organizational costs and workload of organizational managers. If supermarkets are the mainstay for docking scattered farmers, enterprises will inevitably increase the cost of information search, commodity bargaining, commodity decision-making, professional organization, or even self-built logistics, and construction of production bases. Generally, supermarket enterprises lack economies of scale, large purchases volumes and high distribution efficiency, so it is close to luxury to achieve the docking. It can be seen that in the external transaction stage, the advantages of large enterprises development and external transactions among large companies are more obvious. Large enterprises have the conditions to control more valuable information, master a large number of resources, and the transaction flow amomg large enterprises is more smooth and convenient with the lowest cost. To this end, the inter-enterprise transaction model is mainly for large enterprises to dock large enterprises, large retailers to dock large wholesalers, and large wholesalers to dock large producers.

3· 「雲消費」時代交易信息獲得方式的變化及交易成本的降低導致交易模式的根本性變化 Changes in the way of acquiring transaction information in the era of cloud consumption and fundamental changes in transaction patterns by transaction costs reduction

「雲消費」時代,資訊成本逐漸成為交易成本中比重最大的成本。正如

麥肯錫在《大數據：下一個創新、競爭和生產力的前沿》專題研究報告中認為，對企業而言，「海量資料的運用將成為未來競爭和增長的基礎」。一方面，大數據帶來海量資訊，互聯網提供了關於消費的全球性的海量資訊，既有消費者的資訊，又有大量廠商的商品資訊、技術、資本、人才等生產要素的資訊。同時，資訊獲得更為透明，資訊的評價更為透明（如淘寶賣家信用等級評價）；另一方面，資訊獲得的方式更趨虛擬化、快捷化。網路交易中交易雙方從洽談、簽約到訂貨、支付等，均可透過互聯網完成，交易過程完全虛擬化；衛星、光纜等先進傳輸手段的運用，則使交易活動突破時間、空間，使交易方式更為便捷化。這種變化也為交易雙方帶來更多的發展機會，更大的發展可能。因此，「雲消費」時代自由開放的市場和資訊資源平臺能夠為越來越多的交易提供更加低廉的產品或服務，這就進一步解放了企業生產力，促進企業更多依賴社會分工、細分化服務，企業間的交互性進一步增強，商業體系的整體運行也變得越來越有效率，發展速度倍增。商業做為市場經濟運行體系中的重要組成部分，同樣將受益於網路經濟環境下自由開放的市場和多元共用的資訊資源平臺，交易資訊獲得方式、交易成本的降低、交易模式的根本性變化等，驅動了現代零售革命。

In the era of cloud consumption, information costs have gradually become the largest part of total transaction costs. As McKinsey&Company said in its research report Big Data: The Next Frontier for Innovation, Competition and Productivity, for enterprises, 「the application of massive data will become the basis for future competition and growth.」 On the one hand, big data brings a lot of information, and the Internet provides a huge amount of global information about consumption, including information about consumers, information about the products of a large number of manufacturers and technology, as well as information about capital, and talents. At the same time, information acquisition and information evaluation (e.g. credit rating of Taobao sellers) are more transparent. On the other hand, information acquisition is more virtualized and faster. In the online transaction, the transaction process is completely virtualized for both parties, from negotiation and signing to ordering and payment, etc., all can be completed through the Internet. The use of advanced transmission means such as satellite and optical cable makes the transaction break through the boundary of time and space and more convenient, which will also bring more development opportunities and greater development possibilities to both parties of the transaction. Therefore, the free and open market and information resource platform in the era of cloud consumption can provide more low-cost products or services for more and more transactions, which further liberates the productivity of enterprises and promotes enterprises to rely more on social division of labor and segmented services. The inter-enterprise interaction is further enhanced, and the overall operation of the business system becomes more efficient and the development speed is doubled. As an important part of the market economy operation system, business will also benefit from the free and open market and the multi-shared information resource platforms in the network economy environment. The changes in the way to obtain transaction information, the reduction of transaction costs, and the fundamental changes in the transaction models have driven the modern retail revolution.

二

「雲消費」時代的 P2P 模式
2. The P2P mode in the era of cloud consumption

（一）什麼是 P2P What is P2P

電子商務按照交易對象不同可以分為企業對企業的電子商務（B2B），企業對消費者的電子商務（B2C），企業對政府的電子商務（B2G），消費者對政府的電子商務（C2G），消費者對消費者的電子商務（C2C），企業、消費者、代理商三者相互轉化的電子商務（ABC），以消費者為中心的全新商業模式（C2B2S），以供需方為目標的新型電子商務（P2D）。

The types of E-commerce, according to different transaction objects, can be divided into business-to-business e-commerce (B2B), business-to-consumer e-commerce (B2C), business-to-government e-commerce (B2G), and consumer-to-government e-commerce (C2G), consumer-to-consumer e-commerce (C2C), agents-business-consumer e-commerce(ABC), a new consumer-centric business model (C2B2S) and another new business model targeting the provider and the demander (P2D).

其中 B2B（Business to Business）是商家（泛指企業）對商家的電子商務，B2C（Business to Customer）是企業對消費者的電子商務，當前中國 B2C 電

子商務網站非常多，比較大型的有天貓商城、京東商城等。C2C（Consumer to Consumer）則是消費者對消費者的電子商務。

Currently, there are many B2C e-commerce websites in China, among which the relatively large ones include Tmall, Jingdong, etc.

近年來源於美國的 O2O 概念盛行。O2O（Online To Offline）是指將線下的商務機會與互聯網結合，讓互聯網成為線下交易的前臺。這樣線下服務就可以用線上來攬客，消費者可以用線上來篩選服務，成交可以線上結算，很快達到規模。有觀點認為，一家企業能兼備網上商城及線下實體店兩者，並且網上商城與線下實體店全品類價格相同，即可稱為 O2O。也有觀點認為，O2O 是 B2C 的一種特殊形式。[111]

In recent years, the concept of O2O (Online To Offline), which was originated in the United States, has prevailed. It refers to the combination of offline business opportunities and the Internet, making the Internet a front-end for offline transactions. In this way, consumers can be collected online to consume the offline services. Also, consumers can filter the offline services online. Transactions can be settled online and quickly reach certain scale. There is a view that a company can have an online store and an offline store at the same time, where the same goods have the same price. Then it is O2O. There are some other views that O2O is a special form of B2C.[111]

---

111 O2O 商城如何打開 O2O 市場？除非打造 O2O 商場．當地 O2O. 2013-11-14 How Does the O2O Mall open the O2O market? Unless You Build an O2O Mall First. Local O2O. Nov 14, 2013

我們認為,「雲消費」時代電商雲服務平臺的發展使中小企業獲得電子商務發展環境下與大企業同等的資訊獲得機會,其信譽評價更為直接、透明,從而獲得更好的成長機會,也勢必將改變電子商務產業發展格局,使原有產業鏈上大企業對大企業、大企業集聚小企業的經營模式變為小企業對其他企業、企業對消費者,甚至生產企業直接對應消費者、消費者對消費者的模式,我們可稱其為 P2P 模式。

We believe that the development of e-commerce cloud service platform in the era of cloud consumption enables SMEs to obtain the same information acquisition opportunities as large enterprises in the e-commerce development environment, and their reputation evaluation can be made more directly and transparently, thus obtaining better growth opportunities. Such develoment is also bound to change the development pattern of the e-commerce industry, so that the business models among the large enterprises or large enterprises gathering small ones in the original industrial chain will be changed into business models among small enterprises and other ones, and between enterprises and consumers, or even between manufacturers to consumers, or consumers and consumers, which we can call the P2P model.

P2P（Peer-to-Peer）中的「Peer」在英語裡有「（地位、能力等）同等者」、「同事」和「夥伴」等意義,即交易雙方互為買賣雙方,傳統交易結構中買方、賣方、合作夥伴、供應鏈上下游企業等概念都面臨新的詮釋。[112] P2P 一是直接將人們聯繫起來,讓人們透過互聯網直接交互;二是使得網路上的溝通變

---

112 陳雲紅. 基於 P2P 的資訊交流與資源分享模式探究 Chen Yunhong. Research on Information Exchange and Resource Sharing Models Based on P2P

得更容易、更直接、更加共用和交互，真正地消除中間商；三是改變互聯網現在的以大網站為中心的狀態，重返「非中心化」，並把權利交還給用戶。

「Peer」 in P2P (Peer-to-Peer) suggest equivalence in (status, ability, etc.), colleagues and partners. The concept indicates that both parties in the transactions can be buyers and sellers. Traditional concepts in transaction structure such as buyers, sellers, partners, and upstream and downstream companies in the supply chain are all facing new interpretations.[112] Firstly, P2P connects people and allows them to interact with each other directly through the Internet. Secondly, it makes communication on the network easier, more direct, more shared and interactive, and truly cuts out the middlemen. Thirdly, it decentralizes the current large-scale website-centered Internet and returns power to the users.

（二）P2P 模式典型案例
Case study of the P2P model

電子商務市場的開放和門檻的降低，使大量個人賣家在網上找到一席之地。2011 年中國個人網店峰值時，其數量一度達到 1620 萬家。儘管此後兩年理性回落，2013 年為 1122 萬家，但這相對於天貓約 7 萬家的規模，仍顯示了強大的個人化消費的力量。[113]

The opening and the lowering of the threshold of the e-commerce market have

---

113 劉曉凱 . 個人網店數量下降 . 中國產經新聞報 . 2014-03-23 Liu Xiaokai. The Number of Individual Online Stores Dropped. China Industry&Economy News. Mar 23, 2014

enabled a large number of individual sellers to find a place on the Internet. The number of individual online stores peaked in 2011, reaching 16.2 million. Despite the number rationally declined in the next two years (11.22 million in 2013), it still shows the strong power of personalized consumption compared to about 70,000 online stores of Tmall.[113]

在「雲消費」的環境下，隨著消費者分享購物心得和交易資訊，使透過海外管道代購奶粉的市場逐漸興起，降低了傳統奶粉貿易管道流通環節中檢測檢疫、關稅等成本。據不完全統計，僅淘寶網上從事奶粉代購的商家就有1.76 萬家左右。P2P 的交易模式使奶粉代購成為一個產業，也造就了大量中小代理商的事業。

In the context of cloud consumption, as consumers share their shopping experience and transaction information, the market for purchasing milk powder through overseas channels has gradually emerged, which saves the costs of testing and quarantine, and customs duties in the circulation of traditional milk powder trade channels. According to incomplete statistics, there are about 17,600 merchants engaged in the purchase of milk powder on Taobao.com alone. The P2P model has made the purchase of milk powder an industry and also created a large number of small and medium-sized agents.

1. 趕集網 P2P 交易平臺 Ganji.com, a P2P transaction platform

為眾多普通消費者熟悉的趕集網是 P2P 交易的典型平臺。2005 年成立的趕集網是一家分類資訊門戶網站。任何有相關需求的個人都可以及時、有效地在該網站發佈個人分類廣告，讓所有人知道您需要或是可以提供什麼樣的

商品、服務和說明。任何人都可以借助這個平臺將自己的二手物品、租售房產、教育培訓、交友等供需資訊發佈在這一平臺上，找到 P2P 的對應者。截至 2013 年 2 月，趕集網日均新增發帖 235 萬條，每日訪客超過 2169 萬人次，頁面日訪問量超過 25366 萬人次。[114]

Ganji.com, which is familiar to many ordinary consumers, is a typical P2P transaction platform. Established in 2005, Ganji.com is a classified information portal. Anyone with relevant needs can post a personal classified advertisement on the website in a timely and effective manner, so that everyone knows what kind of goods, services and help you need or can provide. Anyone can use the platform to post their second-hand items, real estate rental, education and training, dating and other information of supply and demand on the platform to do the P2P matching. As of February 2013, the number of daily new posts reached 2.35 million, with more than 21.69 million visitors and more than 253.66 million page views everyday.[114]

2 · 小豬短租網以 P2P 模式堅守短租市場 Xiaozhu.com adheres to the short-term rental market in P2P model

小豬短租網也是 P2P 交易的典型案例。小豬短租網選擇用 P2P 模式發展短租市場，透過種子房東的方式逐步擴大和培育市場，打造安全保障體系，收取交易佣金。小豬短租網為房東和房客搭建了一個誠信、有保障的線上溝通和交易平臺，並透過財產、人身安全保障方案及身分識別等機制，建立綠

---

114 趕集網相關統計資料 Statistics from Ganji.com

色平臺生態系統。有效地將房東的閒置房產資源透過網路進行分享，充分利用並發揮最大價值，同時加強房東和房客間的社交關係及交互互動，為房客提供有別於傳統酒店且更具人情味的住宿體驗。這家成立於 2012 年的企業，到 2018 年，全球房源已經突破 35 萬，覆蓋國內 395 個城市和海外 225 個目的地，在全國超過 20 座城市設有運營中心。

Xiaozhu.com, a short-term rental website, is also a typical case of P2P transactions. It chooses to use the P2P model to develop the short-term rental market, gradually expand and cultivate the market through the seeded landlords, create a security system and earn the commissions. Xiaozhu.com has established a credible and guaranteed online communication and transaction platform for landlords and tenants, and also created a green ecosystem through mechanisms such as property, personal safety protection and identification, thus effectively shares the landlord's idle property resources through the network and makes full use of the maximum value of the resources, strengthening the social relationship and interaction between the landlords and the tenants and providing the tenants with a more humane accommodation experience that is different from the traditional hotel. The company, founded in 2012, has exceeded 350,000 global housing listings by 2018, covering 395 cities in China and 225 destinations overseas, with operations centers in more than 20 cities across the country.

3 · 「人人車」P2P 二手車交易模式

Renrenche.com′s P2P transaction model of second-hand cars

「人人車」是一個為個人用戶提供優質二手車交易保障的網上交易平

臺。「人人車」的模式可以概括為「上門質檢」+「無店鋪交易」，實為一個買家和賣家直接交易的 P2P 二手車市場。其交易的步驟是：第一步，車主線上提交二手車出售需求，由「人人車」平臺派出專業評估師，攜帶設備上門檢驗車況並給出檢驗結果和建議售價。第二步，完成質檢、估價、車況資訊採集後，在「人人車」平臺上掛牌出售。第三步，買家看到中意的二手車，可與車主直接溝通，預約線下驗車並成交，全程由平臺方人員提供線上和線下支援。第四步，平臺一方透過交易擔保和售後追蹤等方式保障交易成功。由於免去了場租和車輛採購成本，「人人車」僅向車主收取售價 3% 的服務費，這對傳統經銷商 12% － 20% 的差價體系造成了衝擊。

Renrenche.com is an online transaction platform that provides transaction security of high-quality secondhand car for individual users. The model can be summarized as 「on-site quality inspection」 and 「non-store Transaction」, which is actually a P2P second-hand car market where buyers directly trade with sellers. The steps of the transaction are as follows:

Firstly, the car owner submits the demand for selling online. Then Renrenche dispatches professional appraisers with equipments to check the condition of the car, give the result and the suggested selling price. Secondly, after all the aforementioned steps, the selling demand is posted on the website. Thirdly, once the car takes a certain buyer' s fancy, he or she can communicate directly with and make an appointment with the owner to finish the inspection and transaction offline. Renrenche provides online and offline support during the whole process. Fourthly, Renrenche guarantees the success of the transaction through transaction guarantee and after-sale tracking. With no rental and purchase cost, Renrenche only charges the owner a service fee of 3% of the transaction price, which has a great impact on the traditional dealer's price difference system of 12% to 20%.

在市場分層上，「人人車」只做 6 年 10 萬公里以內、品質相對較高的車輛，同時確保所有車源均為個人車主，杜絕二手車販的翻新車流入。由於隔絕了劣質車源，用戶才會在整體上信任這個市場的交易環境，同時也防止市場內生的劣幣逐良幣現象。

In terms of market stratification, Renrenche only sells cars with relatively high quality, which are less than 6 years old with a travel distance less than 100,000 kilometers. At the same time, it ensures that all cars are privately owned wipes out refurbished cars from secondhand car dealers. By getting rid of the inferior cars, users start to trust the market in general. Besides, this prevent bad cars from driving out good ones.

在中立訂價上，「人人車」與中國品質認證中心合作，依靠專業評估師進行約 249 項現場檢測，盡量還原真實車況並給出訂價建議 (當然買賣雙方還可以自由議價)。傳統二手車買賣中，經銷商的差價模式通常基於單筆交易的資訊不透明，而且不透明越深，獲利空間越大。因此經銷商必須在一定程度上使買賣雙方資訊隔絕，甚至會採用調里程表、私下翻新等不正當手段。而在平臺模式下，平臺方收入主要體現為服務費 + 交易費，收入規模與市場整體交易規模相匹配。從長期來看，只有整個市場做大，平臺方的蛋糕才能分得更多，這導致其在短視行為上更加謹慎。

In terms of neutral pricing, Renrenche cooperates with the China Quality Certification Center and relies on professional appraisers to conduct about 249 on-site inspections to restore real-life conditions and give pricing advice (of course, buyers and sellers can also negotiate freely at their will). In traditional secondhand car sales, the differential pricing model used by dealers is usually based on the opaque pricing of a single transaction, and the deeper the opacity, the greater the profit margin. Therefore, dealers must cut off the information between buyers and sellers to a certain extent, and even use improper means such as tampering with the odometer or refurbishing the car in secret. In the platform model, revenue of the platform side is mainly composed of service fee and transaction fee, the scale of which matches the overall transaction scale of the market. In the long run, only the 「pie」 of the entire market is bigger will the platform side get more share. That sis the reason why it is more cautious in short-sighted behavior.

在去仲介化上，「人人車」最大限度壓縮了中間環節，讓買家和賣家距離更近，同時交易成本更低。專業性和信任度始終是決定二手車交易鏈條長短的關鍵因素。由於買賣雙方不懂行，這就不可避免需要一個具備專業能力的仲介存在。但如果仲介過分強大，就又變成了一個資訊不對稱的製造者，從而損害交易者對市場的信任。「人人車」採取了一個折衷的方法，僅向交易雙方輸出專業服務，而不參與交易本身，讓買賣雙方自由撮合。

In terms of de-intermediation, Renrenche minimizes the intermediate links, making buyers and sellers closer and transaction cost lower. Professionalism and trust are always key factors in determining the length of transaction chain of secondhand car. Since buyers and sellers are no expert, an intermediary with professional competence is inevitably in need. But if the intermediary is too strong, it then becomes a maker of information asymmetry, thus damaging the both parties' trust in the market. In light of this, Renrenche adopts a compromise method, that is, it only outputs professional services to both parties of the transaction and does not participate in the transaction itself, allowing buyers and sellers to freely match.

4・「人人快遞網」P2P 模式
P2P model of Rrkd.cn (everyone courier)

「人人快遞網」隸屬於四川創物科技有限公司,是中國第一個眾包平臺,旨在將全社會公眾都發展成為自由快遞人。截至 2014 年,其全國註冊用戶數已逾 600 萬人,其中透過實名認證和培訓考核的自由快遞人數達 180 萬人,日交易單量突破 8 萬,交易額超過 200 萬人民幣。

Rrkd.cn is affiliated to Chuangwu Technology Co., Ltd. in sichuan province. It is the first crowdsourcing platform in China, aiming to turn the public into freelance couriers. As of 2014, the number of registered users nationwide has exceeded 6 million. Among them, the number of couriers who passed the real-name certification and training evaluation reached 1.8 million. The number of daily transaction orders exceeded 80,000 and the transaction volume exceeded 2 million RMB.

「人人快遞網」的網路平臺實際上是移動互聯網思維與 P2P 思維對快遞行業的滲入。「人人快遞網」透過用戶、自由快遞人、商家形成資源集結,運用交易體系、支付體系、信譽體系為主的三大體系實現資源利益眾包共用,是全民均可參與的一種業務型態。做為協力廠商電子商務網路資訊服務平臺,「人人快遞網」不收取任何平臺使用者支付的平臺服務使用費,只需下載一

款帶有社交功能的名為「人人快遞」的 P2P 應用軟體 ，完成帳號註冊、身分資訊審核 、信用卡綁定後就可成為自由快遞人。該應用可即時查詢身邊的寄件資訊，願意接件的使用者可根據自身的行程安排，攜帶「順路」的快件，並將其送到目的地。透過軟體搶單，自由快遞人可以順路將快件送到目的地，並獲得快遞費的 80%。透過該軟體，市民可以快件的方式寄送鮮花、蛋糕、檔、票據、鑰匙等。

Rrkd.cn is actually the penetration of mobile Internet thinking and P2P thinking into the express industry. It forms a resource assembly through users, freelance couriers and merchants, and realizes the sharing of resource benefits by using the three major systems of transaction system, payment system and credit system. It is a business form that all citizens can participate in. As a third-party e-commerce network information service platform, Rrkd.cn does not charge any platform service fee paid by any platform user. To become a freelance courier, sser just needs to download a P2P app called 「Renrenkuaidi」 which has the social function and complete the account registration, identity information review and credit card binding. By using the app, one can query the delivery demand around in real time and pick up the parcel if he or she would like to stop by according to his or her own schedule and send it to the destination to finally earn 80% of the courier fee. Through the app, citizens can send flowers, cakes, documents, bills, keys, etc. by express.

三
「雲消費」時代小微企業成長空間增強
3. The growth of small and micro enterprises in the era of cloud consumption

（一）小微企業發展面臨兩大「瓶頸」
Development small and micro enterprise faces two major "bottlenecks"

　　據統計，中國中小微企業佔到企業總數的 99.7%，其中，小微企業佔到企業總數的 97.3%，而其實現資訊化的比例不足 10%。[115] 相較大型企業的成熟完善，小微企業的各種資源有限，經營成本偏高。一般意義上，小微企業發展主要面臨兩大「瓶頸」：一是資訊獲得的成本。小微企業資訊化預算投入有限，資訊獲得量有限。二是信譽評價的成本。小微企業品牌價值低，社會公眾認識度低，資訊評價不透明。信用度不高會造成小微企業融資難，銀行不願意向小微企業放貸，進而推高了小微企業的發展成本。與之相較的，大企業控制著產業鏈，擁有巨大的資訊資源優勢，市場壁壘高，能夠獲得更

---

115 做低成本的「小微」信息化. 中國資訊產業網 Informationization of Low-cost Small and Micro Enterprises. CNII

多發展的機會。因此越是資訊不透明，則大企業強者越強，在市場上佔據絕對強勢地位。

According to statistics, China's small and medium-sized enterprises account for 99.7% of all enterprises in China, of which small and micro enterprises account for 97.3%, and its proportion of informationization is less than 10%.115 Compared with the maturity and perfection of large enterprises, small and micro enterprises have limited resources and high operating costs. In the general sense, the development of small and micro enterprises mainly faces two major「bottlenecks」:

a. the cost of obtaining information. Small and micro enterprises have limited investment in informationization budgets and limited access to information.

b. The cost of reputation evaluation. Small and micro enterprises have low brand value, low public awareness, and opaque information evaluation.

Low credit rating will make it difficult for small and micro enterprises to raise funds. Banks are not willing to lend money to them, which in turn inflates the development costs of small and micro enterprises. In contrast, large enterprises control the industrial chain. They have huge information resources advantages and high market barriers, thus enabling them to obtain more opportunities for development. Therefore, the more opaque the information is, the stronger the big companies are and the more absolute strong dominance they show in the market.

（二）「雲消費」環境破解小微企業發展「瓶頸」
Cloud consumption cracks the development "bottlenecks" of small and micro enterprises

1·「雲消費」環境下信息獲得成本更低 Lower information acquisition costs in the cloud consumption environment

在「雲消費」時代，基於大數據的資料開放和共用以及應用是推動商業企業發展的重要動力。資料開放利用程度越高，資訊知識做為生產要素的程度就會越高。在「雲消費」時代的「雲消費」資源環境下，由於資訊的互聯互通、資訊獲得管道的多元化，使企業交易資訊獲得的難度下降，獲得廣域資訊的可能性增加，這讓更多企業，尤其是小微企業有與大企業同樣的資訊獲得機會，讓更多小微企業能夠在利用其他企業資源的基礎上獲得更多的發展機會。

In the era of cloud consumption, data opening, sharing and application based on big data is an important driving force for the development of commerce and enterprises. The higher the degree of openness of data, the higher the level of information knowledge as a factor of production. In the resourcing setting of cloud consumption, thanks to the interconnection of information and the diversification of information access channels, the difficulty of obtaining enterprise transaction information is reduced and the possibility of obtaining wide-area information is increased, which makes more enterprises, especially small and micro enterprises, have the same information acquisition opportunities as large enterprises, so that more small and micro enterprises can obtain more development opportunities on the basis of using resources of other enterprises.

美國聯邦政府認為,「資料是一項有價值的國家資本,應對公眾開放,而不是把其禁錮在政府體制內。」自 20 世紀以來,美國國會、政府先後出臺了一系列法規,對資料的收集、發佈、使用和管理等諸環節都做出了具體的規定。經過數十年修改完善,形成較為成熟的法律框架和道德規範。2010 年,美國國會更新法案,進一步提高了資料獲取精準度和上報頻度,使得美國資料獲取和彙聚體系更加成熟。[116]

The US federal government believes that "data is a valuable national capital, it should be opened to the public rather than being imprisoned within the government system." Since the 20th century, the US Congress and the government have issued a series of laws and regulations on the collection, distribution, use and management of data. After several decades of revision and improvement, a relatively mature legal framework and ethics have been formed. In 2010, the US Congress updated the bill to further improve the accuracy of data collection and reporting frequency, making the data collection and convergence system in the US more mature.[116]

2009 年歐巴馬推出 Data.gov,是其上臺後推行透明政府計畫的關鍵舉措。為方便公眾使用和分析,Data.gov 平臺還加入了資料的分級評定、高級搜索、用戶交流以及和社交網站互動等新功能。比如在 Data.gov 上提供的白

---

116 美國數位政府戰略提升資訊服務效率. 中國電子報、電子資訊產業網. 2013-07-12 US Digital Government Strategy Improves Information Service Efficiency. CENA. July 12, 2013

宮訪客搜索工具，不僅能夠搜索到訪客資訊，而且可以將白宮訪客與其他微博、社交網站等進行關聯，進一步增加了訪客的透明度。

In 2009, Obama launched Data.gov, a key move to implement the plan of transparent government after he took office. For public use and analysis, Data.gov also includes new features such as data rating, advanced search, user communication and interaction with social network. For example, by using the White House Visitor Search Tool available on Data.gov, one can not only search for visitor information, but also link White House visitors to Weibo and other social network, further increasing the transparency of visitors.

為了更方便民眾使用開放資料，方便應用領域的開發者利用政府開放資料開發多種應用， Data.gov 彙集了 1264 個應用程式和軟體工具、103 個手機應用外掛程式。另外，Data.gov 還發佈了政府 API 索引，使得這些資源可以更易被找到和使用。透過開放 API 介面，Data.gov 讓政府的資訊和服務交付更加便捷，也讓公眾和企業家在構建更佳政府、提升服務的過程中成為合作夥伴。[117]

To make it easier for people to use open data, and for developers in the application domain to use the government's open data to develop multiple applications, Data.gov brings together 1,264 applications and software tools, and 103 mobile application plug-ins. In addition, Data.gov has released a government API index that makes it easier to find and use these resources. Through an open

---

[117] 美國：大數據國家戰略，中國電子政務網 The United States: National Strategy of Big Data, www.e-gov.org.cn

API interface, Data.gov makes information and service delivery of government easier and allows the public and entrepreneurs to become partners in building better government and improving services.[117]

據瞭解，美國政府資料開放已創造巨大的市場和商機。典型事例是 1983 年美國政府將用於軍事的衛星定位系統 GPS 向公眾開放使用，由此帶動了一系列與此相關的生產和生活服務創新，包括汽車導航、精準農業、通信等技術創新及應用，同時創造了大量的就業崗位，僅美國國內就有約 300 萬的就業崗位依賴於 GPS。近年，美國的資料開放為其帶來了更大的價值。[118] 據統計，透過資料開放，2013 年美國在政府管理、醫療服務、零售業、製造業、位置服務、社交網路、電子商務 7 個重點領域產生的潛在價值已經達到了 2 萬億美元。[119]

It is understood that the opening of US government data has created huge markets and business opportunities. A typical example is the open use of the GPS satellite system used by the US government in the military in 1983, which has led to a series of related production and life service innovations, including automotive navigation, precision agriculture, communications and other technological innovations and applications. At the same time, a large number of jobs had been created. There were about 3 million jobs that relied on GPS in the United States alone. In recent years, the openness of data in the United States has brought greater value to the country.[118] According to statistics, through data opening, the potential

---

118 徐繼華, 馮啟娜, 陳貞汝. 智慧政府大數據治國時代的來臨 [M]. 北京 : 中信出版社, 2014. Xu Jihua, Feng Qina, Chen Zhenru. Smart government: The coming of the Era of Big Data Governance [M]. Beijing: CITIC Publishing House, 2014.
119 馬建勳. 資料開放——大數據發展的基礎. 國家資訊中心網站. Ma Jianxun. Data Openness-The Foundation for Big Data Development. www.sic.gov.cn

value generated by seven key areas as government management, medical services, retail, manufacturing, location services, social networking, and e-commerce in 2013 had reached $2 trillion.[119]

近年來，中國政府在開放資料方面已做出切實努力。國家統計局繼中國統計、資料中國兩大手機用戶端上線運行後，2013年5月統計政務微信的「統計微訊」正式開通，在當年9月舉辦的第四屆「中國統計開放日」上，國家統計局新版資料庫「國家資料」正式上線運行，並向普通公眾開放。新版「國家資料」資料庫存儲600多萬筆資料，比以往增加了3倍。公眾完成網上註冊後就可以在「國家資料」方便地查詢全國及31個省市的月度、季度、年度資料。[120]

In recent years, the Chinese government has made practical efforts in open data. Following the launch of China Statistics and ChinaIDC, two major mobile phone clients, the National Bureau of Statistics officially opened the Statistical Information, the officail Wechat account of government affairs in May 2013. On the fourth China Statistical Open Day in September, National Data, the new database of the National Bureau of Statistics was officially launched and opened to the general public. It stored more than 6 million cases of data, a three-fold increase over the last version. After completing the online registration, the public can conveniently query the monthly, quarterly and annual data of the country and 31 provinces and cities in the National Data.[120]

---

120 新版「國家資料向公眾開放」[N]. 北京日報．2013-09-16. The New National Data is Open to the Public [N]. Beijing Daily, Sept 16, 2013.

中國很多電子商務企業也透過開放平臺實現了資料的共用，進一步降低了資訊獲得成本，個人和企業能第一時間獲得電子商務市場資料，從而使個人、商家和企業及時做出正確合理的決策，避免電子商務投資和競爭的盲目性。如百度、eBay、亞馬遜、樂天、淘寶等電子商務企業均競相開放資料。

Many e-commerce companies in China have also realized data sharing through open platforms, which further reduces the cost of information acquisition. Individuals and enterprises can obtain e-commerce market data in the first time so that they can make correct and reasonable decisions in time and avoid the blindness of e-commerce investment and competition. E-commerce companies such as Baidu, eBay, Amazon, Lotte, Taobao are competing for open data.

基於雲計算創新技術和開放運營理念，2014 年百度建立了資料開放平臺。將包括「雲」、「資料工廠」、「百度大腦」三大元件在內的核心大數據對外開放，使得與百度合作的機構和傳統企業都能夠線上使用百度的大數據架構，處理自身積累的大數據或融合百度大數據，並以此來改造和優化傳統行業的企業管理、產品服務、商業模式等。[121] 由此為中國小微企業帶來巨大的資料資源。

Based on cloud computing innovation technology and open operation concept, Baidu established a open data platform in 2014. The core big data including the three components of the cloud, data factory and Baidu brain was opened to the

---

121 陳中. 百度正式開放大數據引擎 開展對外合作 [N]. 證券時報，2014-04-25. Chen Zhong. Baidu Officially Opened the Big Data Engine to Carry Out External Cooperation [N]. Securities Times, Apr 25, 2014.

outside world, enabling institutions and traditional enterprises that cooperated with Baidu to use Baidu's big data architecture online to handle their own accumulation of big data or integrate big data of Baidu to transform and optimize enterprise management, product services, business models, etc. in traditional industries.[121] This brought huge data resources to small and micro enterprises in China.

阿里巴巴也建立了「資料超市」，需求者可透過開放的 API 獲取淘寶消費者的資訊和銷售商品的名稱、類目、型號、介紹、索引、分類明細及上傳、編輯、修改資訊，以及店鋪資訊、交易明細資訊、淘寶商品管理（淘寶商品的等介面）等資訊，並建立相應的電子商務應用。由此需求者可以透過更暢通的管道、更低廉的價格、更直接的方式獲得海量資訊，這一方面節約了小微企業的資訊搜尋成本，另一方面也解決了小微企業在交易過程中由於資信問題等資訊不對稱造成的發展「瓶頸」。[122]

Alibaba has also launched the data supermarket. Through the open API, the demander can obtain the information of the Taobao consumer and the name, category, model, introduction, index, classification details of the goods, and also upload, edit, and modify store information and transaction details. In the same time, the demander can manage goods on Taobao and establish corresponding e-commerce applications. Therefore, the demander can obtain massive information through more smooth channels, cheaper prices, and more direct ways, which saves the information search cost of small and micro enterprises, and also solves the 「bottlenecks」 of development of small and micro enterprises caused by

---

122 淘寶首度開放數據：宏觀資料免費商業資料將收費. 新浪科技 Taobao's First Open Data: Macro Data and Free Commercial Data will Start Charging. Sina Technology

information asymmetry in the transaction process.[122]

隨著產業分工的不斷細化，以及雲計算、資料採擷等資訊技術的不斷發展，形成了一批協力廠商甚至第四方的獨立物流、軟體發展、客戶管理、客戶關係管理、精準媒體行銷、系統維護、客戶資料分析挖掘、資訊評價、網路安全等電商綜合服務企業，而且這些電商服務產業鏈得以重構並不斷專業化、分工化、小型化，形成基於電商產業的大量微細胞、微環節。由此，形成了現代電子商務的新的生態發展環境，即電子商務「雲」服務的平臺。在電商「雲」服務平臺中，各種資源（軟體、硬體、管理、服務、資訊諮詢、資料檢索等）就像水、電等基礎設施一樣，價格便宜而又無處不在，並且隨著加盟企業的增加和產業的不斷細分其基礎不斷擴大。從基本的計算、存儲資源到具體的電子商務應用服務，所有電子商務的後臺服務都可以放在「雲」服務平臺的資源池中。在「雲」服務平臺中，任何規模、任何經營基礎的企業都可以根據自己的需求訂製自己所需要的各種服務和設施。企業在降低成本的同時，使資源得到了充分的共用。「雲」服務平臺為現代化的企業，特別是中小企業提供了很大的發展機會和平臺。

With the continuous refinement of industrial division and the continuous development of information technology such as cloud computing and data mining, a number of third-party and even fourth-party e-commerce integrated service enterprises of independent logistics, software development, customer management, customer relationship management, precision media marketing, system maintenance, customer data analysis and mining, information evaluation, and network security have been formed. These e-commerce service industry chains have been reconstructed and continuously specialized, divided, and miniaturized to form a large number of micro cells or links based on the e-commerce industry. As a result, a new ecological development environment of modern e-commerce, namely the cloud service of e-commerce, has been formed. On such platform, just like water, electricity and other essential supplies, various resources such as software, hardware, management, services, information consultation, data retrieval, etc. are cheap and ubiquitous. With the increase in franchisees and the continuous segmentation of the industry, such resources have continued to expand. From basic computing and storage resources to specific e-commerce application services, all e-commerce back-end services can be placed in the resource pool of the cloud service platform. Enterprises of any size and any business base on the platform can customize the services and facilities they need according to their own needs. Enterprises are fully sharing resources with one another while the costs are reduced. The cloud service provides great development opportunities and platforms for modern enterprises, especially small and medium-sized enterprises.

在「雲消費」時代，電商「雲」服務平臺成功實現了消費者的個性化資料分析，破解了商業個性化服務的難題。如一款在北京開通的購物應用 APP「趣逛」，意在為零售商提供更精準有效的移動互聯網行銷服務，並為零售商提供專業的消費行為資料採擷服務。「趣逛」能夠全面地記錄消費者到達某個商場或超市的次數、購物路線、停留的時間等，並針對每個消費者的資料進行分析，使每個消費者擁有獨特的資料資訊。「趣逛」可以基於資料資訊進行分析，從而實現精準行銷、促銷。

In the era of cloud consumption, the e-commerce cloud service platform successfully realized the personalized data analysis of consumers and solved the problem of commercial personalized service. For example, a shopping app called "Quguang" (Happy Shopping) in Beijing is intended to provide retailers with more accurate and effective mobile Internet marketing services, and to provide retailers with professional consumer behavior data mining services. Quguang can comprehensively record the number of times a consumer arrives at a certain mall or supermarket as well as the shopping route, the time of stay, etc. Unique data can be produced by analyzing the data of each consumer. The app can achieve precise marketing and promotion based on such data.

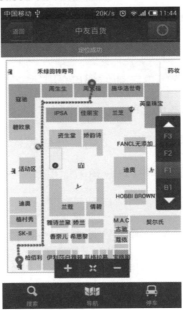

圖 Figure 4-1 「趣逛」地圖手機頁面 Map page of Quguang on the phone

2· 「雲消費」環境下企業信譽評價更為直接、透明 Enterprise reputation evaluation under the cloud consumption environment is more direct and transparent

從電子交易評價系統的發展看，當前企業、消費者個體都有幸可以享受更多、更系統、更全面的網路資訊評價服務。其中既有如大眾點評網、口碑網這樣的公共消費評價平臺，也有如微博、微信、易信那樣的個性化資訊交流平臺。2013年6月，大眾點評網移動用戶端累計獨立用戶數已超過7500萬，電子會員卡發卡量突破1000萬，月綜合瀏覽量（網站及移動設備）超20億。2011年1月21日推出的騰訊微信，截至2013年4月，經過短短的2年時間，註冊用戶量已經突破4億，成為亞洲地區最大使用者群體的移動即時通信軟體。淘寶推出以信譽等級衡量店鋪信譽度的指標體系，凡是在淘寶上購買商品的消費者都可以對店鋪的信譽度進行評價。當消費者對店鋪的好評超過1萬筆，該店鋪就成為「皇冠店鋪」；如果好評超過5萬筆就可以成為「金冠店鋪」。螞蟻金服旗下獨立的協力廠商徵信機構——芝麻信用，透過雲計算、機器學習等技術客觀呈現個人的信用狀況，已經在信用卡、消費金融、融資租賃、酒店、租房、出行、婚戀、分類資訊、學生服務、公共事業服務等上百個場景為用戶、商戶提供信用服務。2015年1月5日，中國人民銀行發佈了允許騰訊徵信、芝麻信用等8家機構進行個人徵信業務準備工作的通知，被視為中國個人徵信體系向商業機構開閘的信號。

From the development of the electronic transaction evaluation system, currently, enterprises and consumers are fortunate to enjoy more systematic and comprehensive network information evaluation services. Among them, there are public consumption evaluation platforms such as Dianping.com and Koubei.com, as well as personalized information exchange platforms such as Weibo, WeChat and Yixin. In June 2013, the number of independent users of Dianping.com has exceeded 75 million, the number of electronic membership cards has exceeded 10 million, and the monthly pageviews on websites and mobile devices have exceeded 2 billion. As of April 2013, the number of registered users of Tencent WeChat, launched on January 21, 2011, has exceeded 400 million in just two years, becoming the mobile instant messenger with the largest user group in Asia. Taobao launched an index system that measures the creditworthiness of stores with credit ratings. Anyone who buys goods on Taobao can evaluate the credibility of the store. When consumers have made decent comment for the store for more than 10,000 times, the store becomes an 「imperial crown shop」; and 50,000 times for a 「golden crown shop」. Credit Sesame, an independent third-party credit reporting agency affiliated with Ant Financial, which objectively presents personal credit status through cloud computing, machine learning and other technologies, has already been providing credit services to users and merchants in hundreds of scenes such as in credit card, consumer finance, financial leasing, hotels, rental, travel, marriage, classification information, student services, and public services. On January 5, 2015, the People's Bank of China issued a notice to allow eight institutions, including Tencent Credit and Credit Sesame, to prepare for the personal credit business. This is regarded as a signal that the Chinese personal credit system has opened its gate to commercial institutions.

當前，更多消費者傾向於透過網路上的評論判斷產品品質並選擇購買，來自同等消費者的正面評價對他們的消費判讀具有積極意義。據美國有關機

構統計：52% 的消費者認為網路上的正面評論會促使他們更願意去當地的企業消費，有 72% 的消費者相信線上評論如同親友們的親自推薦，58% 的消費者會因為網路上的正面評論而對該企業／品牌有信心，76% 的消費者會先看線上評論來決定去哪家本地企業消費，97% 根據線上評論而購物的消費者發現線上評論是正確的。在公開的網路環境下，直接、透明的評價對小微企業和大企業是同樣公平的。

At present, more consumers tend to judge the quality of products and make the purchasing choices according to the reviews on the Internet. Positive reviews from the other consumers of the same level have great significance for them. According to statistics from relevant agencies in the US, 52% of consumers believe that positive comments on the Internet will make them more willing to go to local businesses, 72% believe that online reviews are just like personal recommendations from friends and relatives, 58% become condident in a certain enterprise or brand because of the positive reviews on the Internet, 76% will first check the online reviews to decide which local business to go to and 97% who shop according to the online reviews find that such reviews are correct. In an open network environment, direct and transparent reviews are equally fair for both small and micro enterprises and large enterprises.

由此，在「雲消費」時代，資訊獲得成本的降低、信譽評價機制的變化，賦予小微企業更平等的競爭環境和發展基礎，使其獲得更高、更快的成長空間，使企業跨越式發展成為可能。在某些領域新成長起來的小微企業甚至有可能超越機構臃腫、效率低下的大企業，實現「新魚吃舊魚」、「小魚吃大魚」。

This means that, in the era of cloud consumption, the reduction of information acquisition costs and the change of reputation evaluation mechanism have given small and micro enterprises a more equal competitive environment and development foundation, so that they can obtain higher and faster growth space and develop by strides. Small and micro enterprises that have grown up in some areas may even surpass the bloated and inefficient large enterprises to achieve 「new-born fish eats old one」 or 「small fish eats big one」.

（三）小微企業的成功實踐 —— 案例觀察 Successful practice of small and micro enterprises——case studies

1‧「淘品牌」成為大品牌 Taobrand becomes a big brand

「淘品牌」最早出現在服裝領域，是淘寶商城推出的基於互聯網電子商務的全新的「淘寶商城和消費者共同推薦的網路原創品牌」。「淘品牌」大多來自於設計師的個人愛好的原創設計。如「裂帛」品牌就是 2006 年年底北京兩姊妹拋棄流行服飾的元素，強調「參照本心、無拘無束」的品牌理念，運用誇張的設計詮釋了設計師的生活態度，成為網上知名品牌。此後玩具、化妝品等領域的「淘品牌」也利用網路崛起。

Taobrand first appeared in the field of clothing. It was the brand-new "Internet original brand recommended by Tmall and consumers" launched by Tmall based on Internet e-commerce. Most of the Taobrand products originated from designers' personal hobby. For example, the brand Liebo was started by two sisters in Beijing who abandoned elements of fashion clothing at the end of 2006. The brand emphasized the brand concept of 「following your heart and being unrestrained」 and used exaggerated design to interpret the designer's attitude towards life, and became a well-known online brand. Since then, other Taobrands in the fields of toys and cosmetics have also appeared by using the Internet.

利用網路平臺，很多「淘品牌」在短短幾年間就成為擁有良好口碑、千萬忠誠粉絲的大品牌。例如，「裂帛」創業僅 3 年會員人數就超過 3 萬。2008 年 4 月入駐淘寶商城的「歐莎」女裝，不到 4 個月就上升到品牌女裝類銷量第一的位置。發端於小分子原料 OEM 的化妝品「姿美堂」在 2009 年 12 月 22 日開設了淘寶商城旗艦店，從 2010 年 8 月開始連續 18 個月創造淘寶銷量第一的傳奇紀錄。

With the help of the Internet, many Taobrands have become big brands with good reputation and millions of loyal fans in just a few years. For example, Liebo attracted more than 30,000 members in only 3 years since its start. OSA, a female attire brand that entered the Tmall in April 2008, rose to the number one position in the category in less than four months. The cosmetics brand Simeitol, which originated from the OEM of small-molecular raw material, opened its flagship store in Tmall on December 22, 2009. Since August 2010, it has created a legendary record of top-selling in Taobao for 18 consecutive months.

據淘寶資料監測工具資料魔方獲得的資料顯示，2013 年天貓「雙十一」女裝銷售前十位依次為茵曼、韓都衣舍、阿卡、裂帛、歐時力、波司登、ONLY、歌莉婭、Vero Moda、初語，其中茵曼、韓都衣舍、阿卡、裂帛、初

語 5 家均為「淘品牌」，茵曼、韓都衣舍、阿卡的銷售額都突破了 1 億元。2017 年「雙十一」，「淘品牌」三隻松鼠成交額 5.22 億，蟬聯零食堅果特產類榜首；開場 3 分鐘銷售即破億元的「淘品牌」林氏木業與天貓簽訂 2018 年戰略協定，目標銷售 100 億元。

According to data obtained by Data Cube, a data monitoring tool of Taobao, sales top ten in female attire of Tmall "Double Eleven" in 2013 are Inman, Hstyle, Artka, Liebo, Ochirly, Bosideng, ONLY, Gloria, Vero Moda and Toyouth, in which Inman, Hstyle, Artka, Liebo and Toyouth are all Taobrands. Among them, sales volume of Inman, Hstyle and Artka exceeded ¥100 million. At the Double Eleven in 2017, sales volume of Three Squirrels, a Taobrand, reached ¥522 million, ranking the top of the category of snacks and nuts specialty. Sales volume of Lshmy, also a Taobrand, exceeded ¥100 million yuan in only three minutes after opening. Lshmy signed the 2018 Strategy Agreement with Tmall, with the objective of reaching sales volume of ¥10 billion.

2·「來伊份」實現彎道超車
Laiyifen achieved the overtaking around the corner

成立於 1999 年的「來伊份」公司，主營業務為休閒食品經營，產品覆蓋炒貨、蜜餞、肉製品等九大系列，達到 700 多種。主要消費群體定位為白領和較高收入的家庭。不同於一般的傳統商超，「來伊份」店鋪只銷售自有品牌的休閒食品。在經營理念上採取合作共贏的理念，不依靠收取場地費、廣告費盈利，而是完全依靠產品銷售和服務獲利，與供應商之間實現共同成長。

Founded in 1999, Laiyifen is mainly engaged in the business of snacks. The products cover nine series including roasted seeds and nuts, candied fruit and meat products, reaching more than 700 kinds. The main consumer groups are positioned as white-collar workers and higher-income families. Different from the traditional supermarkets, Laiyifen only sells snacks of its own brand. As for the operation theory, Laiyifen adopts concept of cooperation and win-win. It does not profit from venue fees or advertising fees, but from sales and service, and achieves common growth with suppliers.

「來伊份」在連鎖直營店直接面對消費者的基礎上，採用團購、網上訂購以及電話訂購的經營方式。除了與各大購物網站合作外，還開通了官方網站商城和手機 APP 平臺，並在業界率先推出「線上下單，門店預約取貨」和「門店預定線上貨，當場取貨或送貨上門」等 O2O 的服務，更讓消費者盡享便捷。在顧客層面方面，「來伊份」的品類較多，線下門店空間無法陳列全部的商品，線上延伸了虛擬陳列。顧客可以選擇線上下單，送貨或闇店自提，且線上線下積分通兌。在管理層面方面，利用互聯網和移動設備，實施扁平化管理，建立移動門店管理系統，隨時瞭解門店細節，實現門店督導功能和銷售員互動培訓。

On the basis of facing consumers directly, Laiyifen uses group purchase, online ordering and telephone ordering as operation patterns. In addition to cooperation with major shopping websites, it also opened the official website mall and mobile APP, and was the first in the industry to launch O2O services like 「ordering online and picking up goods by reservations」 and 「ordering online goods and picking up or delivering goods on the spot", which offered convenience to consumers. At the customer level, Laiyifen has so many categories of products that the offline stores usually have no enough space to display all of them, then the online virtual shop expanded the storage. Customers can place orders online and choose either home delivery or store pick-up at will. Both online and offline shopping points can be interchanged. At the management level, Laiyifen uses the Internet and mobile devices to implement flat management and establish the mobile store management system in order to keep abreast of the details of the store and achieve store supervision and interactive training for the sales staff.

截至 2013 年，「來伊份」線上線下擁有會員近 700 萬人，擁有連鎖直營專賣店超過 2500 多家，在淘寶、京東 1 號店、易迅等平臺有旗艦店，而且擁有電商官網和手機 APP 平臺。正如「來伊份」公司 CIO 王戈軍在 2014 年中國連鎖業 O2O 大會上的發言所說，在當前市場環境下，小企業完全有機會實現彎道超車，超越大企業。這家小企業已於 2016 年 10 月 12 日在上交所掛牌上市，開盤大漲 44%，股價攀升至 16.8 元，市值 40 億元。

As of 2013, Laiyifen had nearly 7 million members online and offline and more than 2,500 chain direct-selling stores. It owned flagship stores in Taobao, Jingdong No. 1 Shop and Yixun, as well as the official website and mobile APP. As Wang Gejun, CIO of the company, said on the 2014 O2O Conference for China's Chain Industry, in the current market environment, small enterprises have the opportunity to overtake big ones. The small company was listed on the Shanghai Stock Exchange on October 12, 2016. It rose by 44% on the opening day and its share price climbed to ¥16.8, with a market capitalization of ¥4 billion.

3· 絕味鴨脖——小產品創造大市場
Juewei Duck Necks-small product creates a big market

絕味鴨脖是長沙絕味軒公司旗下的品牌產品，該產品行銷打破了傳統的門店銷售，利用網路新媒介的行銷方式，加強與消費者的深度溝通，搶佔年輕消費者，用最便捷、最習慣的方式與品牌進行互動，增強消費者的體驗和黏性。

Juewei Duck Necks is a brand of Xuweixuan Company in Changsha. It breaks the traditional store sales by adopting the marketing method of network new media to strengthen the deep communication with consumers and seize the young consumers. It enables consumers to interact with the brand in the most convenient and custom way, thus enhancing the consumer experience and stickiness.

2013 年 11 月，絕味鴨脖把鴨脖子搬上手機，在淘寶網發佈絕味鴨脖 15.6 元抵 20 元的代金券，可在北京、上海、深圳等 9 大城市 400 餘家門店自由兌換。並利用淘寶微淘、微信、新浪微博以及天涯論壇等線上網路管道，配合門店宣傳，在花費少量宣傳成本的情況下將活動廣泛傳播開來。活動期間，絕味鴨脖活動的瀏覽量總計逾億次，並且引發了絕味鴨脖粉絲的一次線上大集合；銷售代金券的淘寶網因為這次活動，僅 120 小時就吸引到了全國各地的絕味鴨脖粉絲 50 萬人。2015 年絕味鴨脖全國門店已突破 5000 家，每天賣出鴨脖 100 萬份，成為鴨脖連鎖領域的領導品牌。

In November 2013, Juewei Duck Necks put their products on the mobile phone and released vouchers valued at ¥20 for ¥15.6 in Taobao, which can be freely used in more than 400 stores in 9 cities including Beijing, Shanghai and Shenzhen.

It also used other online network channels such as Weitao of Taobao, WeChat, Sina Weibo and Tianya Forum in conjunction with store promotion to spread the activities widely with a small publicity cost. During the event, the total number of visits totaled more than 100 million times, and this triggered a big online gathering of fans of the brand. By selling the vouchers, Taobao attracted 500,000 fans all over the country in just 120 hours. In 2015, the number of outlets of the brand exceeded 5,000 and daily sales reached one million sets. Juewei Duck Necks became the leading brand in the duck neck chain.

「雲消費」時代傳統零售業的變革

Chapter Five The Change of Traditional Retail Industry in the Era of Cloud Consumption

第五章

「雲消費」時代傳統零售業的變革

Chapter Five The Change of Traditional Retail Industry in the Era of Cloud Consumption

一

危機與挑戰 1. Crises and challenges

（一）危機始終伴隨挑戰

Crisis is always accompanied by challenges

　　零售業是高度競爭的行業。從一定意義上說，自零售業誕生那天起，就不斷面臨危機，也伴生著挑戰。每一次零售革命的發生，更是危機與挑戰博弈的結果。據王成榮教授的研究，零售革命往往在出現以下三種情況的時候

發生：一是經濟危機，二是跨界競爭，三是技術革命。在經濟危機時，購買力下降、消費低迷，往往會催生新的業態，尤其是「價格殺手」。歷史上歷次零售革命的出現都符合這一規律，本輪零售革命更是如此。最近一輪經濟危機已持續數年，同時網店遍地開花，毫無疑問網店是最為典型的「價格殺手」。[123] 網店這種新型業態無疑是零售業應對危機、發展創新的標誌，同時網上零售的迅猛發展也影響了傳統零售業的發展生態，加深了傳統零售業的危機。

The retail industry is a highly competitive industry. In a certain sense, since the day of the birth of the industry, it has been facing crisis, associated with challenges. The occurrence of every retail revolution is the result of a game of crisis and challenge. According to Professor Wang Chengrong's research, the retail revolution often occurs in the following three situations: economic crisis, cross-border competition and the technological revolution. When economic crisis happens, the decline in purchasing power and low consumption often lead to new formats of business, especially the 「price killer」. The emergence of every retail revolution in history has been consistent with this law, and this round is no exception. The latest round of economic crisis has lasted for several years, with online stores appearing everywhere. There is no doubt that online stores are the most typical "price killers."[123] Online store, as a new format of the business, is undoubtedly a sign of the retail industry's response to the crisis, development

---

123 全管道零售勢不可擋 回歸零售本質是關鍵. 中國商報 Full-channel Retail is Unstoppable. Returning to the Essence of Retail is the Key. China Business News

and innovation. At the same time, the rapid development of online retailing has also affected the development of and deepened the crisis of the traditional retail industry.

（二）傳統零售商面臨的危機
Crisis faced by traditional retailers

1. 網路零售發展與實體商業發展形成反差 Contrast between online retail development and physical business development

我們來看一組數字，2009—2013 年 5 年間中國社會消費品零售總額由 132678 億元增長至 234380 億元，增幅達 76.7%。其中網路零售市場規模由 2009 年的 2.1%（2786.238 億元）增長至 2013 年的 7.8%（18281.64 億元），增幅達 271%，超過社會消費品零售總額增幅 194 個百分點。[124]

Here is a set of figures. In the five years from 2009 to 2013, the total retail sales of consumer goods in China increased from ¥1,3267.8 billion to ¥23438 billion, an increase of 76.7%. The online retail market grew from 2.1% (¥278.6238 billion) in 2009 to 7.8% (1828.164 billion) in 2013, an increase of 271%, exceeding the growth of total retail sales of consumer goods by 194 percentage points.[124]

2013 年，國家商務部監測的 3000 家重點零售企業中，網購銷售增長達

---

124 中國電子商務研究中心發佈《2012 年度中國網路零售市場資料監測報告》Report on 2012 China Online Retail Market Data Monitoring released by China E-Commerce Research Center

31.9%，增速與傳統零售業的百貨店、超市和專業店相比，分別高出 21.6%、23.6% 和 24.4%。

In 2013, among the 3,000 key retail enterprises monitored by the Ministry of Commerce, online shopping sales increased by 31.9%, which was 21.6%, 23.6% and 24.4% higher than that of traditional retail stores, supermarkets and specialty stores respectively.

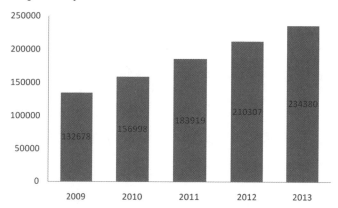

圖 Figure 5-1　2009—2013 年中國社會消費品零售額 2009-2013 Social consumer goods retail sales in China
資料來源：中國統計年鑑 Source: China Statistical Yearbook

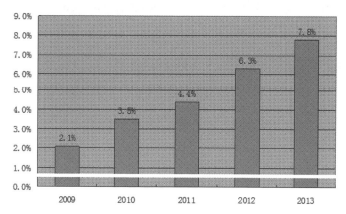

圖 Figure 5-2　2009—2013 年網購規模佔社會消費品零售總額的比例 2009-2013 Online shopping scale' s portion in total retail sales of consumer goods
資料來源：中國電子商務研究中心 Source: China E-commerce Research Center

近年中國大型實體零售商業業績普遍不理想。據國家統計局公布的資料，2013 年 1—11 月，限額以上企業消費品零售額增長為 11.4%，而商務部監測的 3000 家重點零售企業銷售額增長僅為 8.9%，增速明顯低於平均水準。

In recent years, performance of China's large-scale physical retail business has been generally unsatisfactory. According to the data released by the National Bureau of Statistics, from January to November in 2013, the retail sales of consumer goods of enterprises above designated size increased by 11.4%, while the sales of 3,000 key retail enterprises monitored by the Ministry of Commerce increased by only 8.9%, the growth rate of which was significantly lower than the average.

2013 年上半年統計年報顯示，中國約一半超市的利潤增幅出現下滑。其中永旺和蒟峰蓮花分別虧損達 2634 萬港元和 4674 萬元。華潤萬家、京客隆、寧波三江、京客隆等企業利潤增幅下滑，其中華潤萬家利潤下滑幅度達 63.7%。

The annual statistical report for the first half of 2013 showed that the profit growth of about half of China's supermarkets declined. Among them, AEON and CP Lotus lost HK$26.34 million and ¥46.74 million respectively. The profit growth of China Resources Vanguard, Jingkelong and Sanjiang Shopping in Ningbo declined, among which China Resources Vanguard's profit fell by 63.7%.

表 Table 5-1　　2013 年度零售上市公司營收排名（部分）2013 ranking of revenue of listed retail enterprises（partial）

業態 Format	排名 Ranking	企業 Name of enterprise	營收（億元）Revenue (100 million yuan)	增幅（%）Rate of increase	淨利（億元）Net profit	增幅（%）Rate of increase	門店數 Number of outlets
市場 Super-market	1	高鑫零售（大潤發一歐尚）Sun Art Retail(RT-Mart-Auchan)	839.58	10.60	29.420	16.10	323
	2	華潤萬家 China Resources Vanguard	763.00	14.00	8.018	-65.20	4600
	3	永輝超市 Yonghui Superstores	305.42	23.73	7.144	46.07	288
	4	華聯超市 Hualian supermarket	303.80	4.80	0.500	84.40	4530
	5	物美商業 Wumart	188.86	9.00	4.590	23.70	549
	6	人人樂 RenRenLe	127.10	-1.53	0.230	126.00	—
	7	步步高 BBK	113.89	13.82	4.190	22.19	179
	8	京客隆 Jingkelong	104.04	5.82	0.570	45.72	279
	9	蔔蜂蓮花 CP Lotus	108.81	1.90	-0.970	75.00	58
	10	新華 New Hua Du	73.80	10.70	-2.300	-244.35	—
	11	三江購物 Sanjiang Shopping	46.90	8.47	1.530	-12.95	—
	12	紅旗連鎖 Hongqi Chain	44.37	13.62	1.570	-9.61	—

百 貨 Depart- ment store	1	重慶百貨 Chongqing Department Store	302.46	7.84	7.840	13.25	—
		百貨業態 Department store format	268.67	10.29	—	—	—
		超市業態 Supermarket format	98.49	5.64	—	—	—
		電器業態 Electric appliance format	42.09	0.33	—	—	—
	2	豫園商城 Yuyuan Tourist Mart	225.23	10.96	9.810	1.34	—
	3	王府井百貨 Wangfujing Department Store	197.9	8.35	6.940	3.09	48
	4	百盛百貨 Parkson	174.81	4.30	3.540	-58.40	58
	5	金鷹商貿 Golden Eagle	168.3	3.10	12.300	1.40	26
	6	天虹商場 Rainbow Department Store	160.32	11.51	6.150	4.73	62
	7	銀泰商業 Yintai	156.92	12.60	10.650	9.50	30
	8	萬達百貨 Wanda	154.9	39.00	1.120	-7.00	75
	9	銀座股份 Ginza	142.18	5.04	2.660	-23.10	135
		商業營收 Business revenue	139.36	10.81	—	—	—
	10	首商集團 Beijing Capital Retailing Group	120.47	-1.30	3.270	-17.28	—

資料來源：聯商網 Source: Linkshop.com

2‧ 連鎖百強企業經營效益明顯下滑 The operation performance of the top 100 chain enterprises has dropped significantly

據中國連鎖經營協會發佈的「2012 中國連鎖百強」及相關榜單顯示，2012 年連鎖百強銷售規模 1.87 萬億元，同比增長 10.8%，為歷年來增長最慢的一年，也低於社會消費品零售總額 14.2% 的增幅。連鎖百強企業銷售額佔社會消費品零售總額的 9.3%，低於 2011 年 11% 的水準。資料顯示，2012 年連鎖百強企業新開店速度達 8%，為 10 年來最低水準，而 2006—2010 年的該項資料分別為 26%、17%、24%、19% 和 9.8%。[125]

According to the 2012 China Chain Top 100 and related lists released by China Chain Store & Franchise Association, the sales volume of the top 100 chain enterprises in 2012 was ¥1.87 trillion, a year-on-year increase of 10.8%, which was the lowest in history and lower than the total retail sales of consumer goods' increase rate of 14.2%. The sales of the top 100 chain enterprises accounted for 9.3% of the total retail sales of consumer goods, which was also lower than 11% in 2011. The data shows that in 2012, the growth rate of new stores of the top 100 chain stores reached 8%, the lowest in the past decade, while the figures for 2006-2010 were 26%, 17%, 24%, 19% and 9.8% respectively.[125]

---

125 2013 年度中國零售業十大事件：傳統零售增長放緩 . 贏商網 Top Ten Events in China's Retail Industry in 2013: Traditional Retail Growth Slowed down. Winshang.com

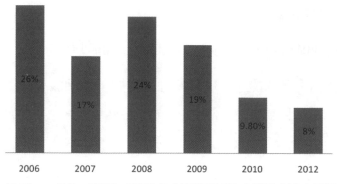

圖 Figure 5-3　2006—2012 年中國連鎖百強企業開店速度情況 2006-2012 growth rate of new stores of top 100 chain enterprises in China

2012 年中國連鎖百強中近半數企業利潤增幅在 0% － 5%，利潤出現負增長或虧損的企業明顯多於 2011 年。百貨、大型超市、標準超市和便利店各主要指標大多同比下滑。以每平方米效率來看，便利店業態從 2011 年的 4.27 萬元 / 平方米年下降到 2012 年的 4.2 萬元 / 平方米年，大型超市業態從 2011 年的 2.94 萬元 / 平方米 / 年下降到 2012 年的 2.6 萬元 / 平方米年，降幅最大的當屬百貨業態，從 2011 年的 3.13 萬元 / 平方米·年下滑到 2012 年的 2.8 萬元 / 平方米年。此外，超市業態客單價從 2011 年的 62 元下降到 2012 年的 58 元，百貨業客 (態) 單價則從 2011 年的 385 元大幅下滑到 2012 年的 337 元。便利店業態毛利率從 2011 年的 19.1% 下滑到 2012 年的 17.2%。[126]

In 2012, nearly half of the top 100 chain enterprises in China saw profits increase by 0% to 5%, among which those with negative growth or loss of profits were significantly more than in 2011. Most of the main indicators of department stores, large supermarkets, standard supermarkets and convenience stores fell year-on-year. In terms of efficiency per square meter, the convenience store format has dropped from yearly ¥42,700 per square meter in 2011 to ¥42,000 yuan in 2012, and the large supermarket form ¥29,400 to ¥26,000. The largest decline occured in the department store format, from ¥31,300 to ¥2,800. In addition, the unit price of supermarket format dropped from ¥62 in 2011 to ¥58 in 2012, and department stores from ¥385 to ¥337 yuan. Gross margin of onvenience store format fell from 19.1% to 17.2%.[126]

2013 年第三季度零售業便利店和超市銷售額增長為 13.3% 和 13%，百貨店和專賣店增長分別為 7.4% 和 6.5%。從毛利率情況看，2013 年第三季度各業態分別比上年同期下降了 0.6% － 1.7%。[127]

In the third quarter of 2013, sales of retail convenience store format and supermarket format increased by 13.3% and 13% respectively, and department store format and specialty store format by 7.4% and 6.5% respectively. In terms of gross profit margin, the formats in the third quarter of 2013 decreased by 0.6% to 1.7% compared with the same period of the previous year.[127]

3‧ 商業地產租金、人工成本等壓力趨大 High pressures form ommercial real estate rent and labor costs, etc

與此同時，零售企業還面臨商業地產租金的居高不下甚至不斷增長。一些主要城市核心黃金商圈的物業租金年均增幅達到兩位數。

---

126 銷售增速放緩 中國連鎖百強座次重排．第一財經日報 Sales Growth Slowed Down: Reranking of China Chain Top 100. China Business News
127 2013 年度中國零售業十大事件．聯商網、《中國商報‧超市週刊》聯合推出 Top Ten Events in China's Retail Industry in 2013. Linshop.com, China Business Daily‧ Supermarket Weekly

At the same time, retail enterprises are still facing high or even growing commercial real estate rents. The average annual increase in property rents in the core gold business districts of some major cities has reached double digits.

2010—2011 年僅廣州地區，商鋪整體租金年增幅在 5% — 10%，熱門區域租金的上漲幅度更是達 15% 以上。[128] 租金成本已經成為企業不可控的壓力。

From 2010 to 2011, in Guangzhou alone, the annual overall rental growth of shops was 5%~10%, and that in popular areas was more than 15%.128 The cost of rent has become an uncontrollable pressure for enterprises.

2014 年上海市中心的大賣場日租金約在每平方米 5~6 元，繁華地段的品牌專賣店日租金從每平方米 3, 40 元到 7, 80 元不等。[129]

In 2014, the daily rent of the hypermarket in central Shanghai was about ¥5 — 6 per square meter, and the daily rent of brand stores in the bustling area ranged from ¥30-40 to ¥70-80.[129]

商業地產服務和投資公司世邦魏理仕（CBRE）2014 年 4 月發佈的報告顯示，2014 年第一季度北京零售物業首層租金同樣本比較環比上漲 2.4%，達到每天每平方米 43.2 元。特別是餐飲企業租金壓力已經突破安全線（15%），佔到了營業額的 20%。據天虹商場的公告，支付其他與經營活動有關的現金

---

128 電商蠶食傳統領地 十家零售商或行將末路 . 投資者報 E-commerce Erodes Traditional Commerce, Ten Retailers May Face Crisis. Investor' s Journal

129 高房租擠壓商業利潤 零售業恐現關店潮 . 經濟參考報 High Rents Squeeze Business Profits, the Retail Industry is Under the threat to Close the Store. Economic Information Daily

130 商業地產租金五年翻兩倍 致零售業毛利率跳水 . 一財經 Commercial Real Estate Rent Doubled in Five Years Causing the Gross Margin of the Retail Industry to Dive.

由上年同期的 9.99 億元上升至 13.39 億元，同比增長 34%。主要原因是水電費、房屋租賃費、新店增加及人工成本的增加。[130]

According to a report released in April, 2014 by CBRE, a commercial property services and investment company, the rent of the first floor of retail properties in Beijing in the first quarter of 2014 rose by 2.4% from the previous month to daily ¥43.2 per square meter. In particular, the rental pressure of catering companies has broken through the safety line of 15%, accounting for 20% of turnover. According to the announcement of Tianhong Shopping Mall, the cash paid for other business activities increased from ¥999 million in the same period of the previous year to ¥1.339 billion, a year-on-year increase of 34%. The main reason for this is the increase in utilities, housing rental fees, labor costs and number of new stores.

表 Table 5-2 2013 年第四季度連鎖企業分業態的典型企業房租同比增幅 Year-on-year typical enterprises rent increase of chain enterprises in the fourth quarter of 2013

業態 Format	房租同比增幅（%）Year-on-year increase of rent(%)
百貨店 Department store	20.9
便利店 Convenience store	21.4
普通超市 Normal supermarket	24.4
大型超市 Hypermarket	27.5
專賣店 Specialty store	22.8
專業店 Specialized store	19.8
總計 Total	23.6

資料來源：中國連鎖協會 Source: China Chain-Store & Franchise Association

表 Table 5-3　2013 年第四季度連鎖企業分業態的典型企業職工薪酬同比增幅
Year-on-year typical employee's salary increase of chain enterprises
in the fourth quarter of 2013

業態 Format	從業人員同比增幅（%）Year-on-year increase of number of employees(%)	職工薪酬同比增幅（%）Year-on-year increase of employee's salary
百貨店 Department store	0.7	8.6
便利店 Convenience store	2.5	13.8
普通超市 Normal supermarket	3.4	13.1
大型超市 Hypermarket	2.3	8.1
專賣店 Specialty store	2.2	12.1
專業店 Specialized store	-2.8	4.0
總計 Total	1.7	9.0

資料來源：中國連鎖協會 Source: China Chain-Store & Franchise Association

在利潤和各項主要指標下滑壓力之下，不少零售商以關店止損，比如農工商集團、羅森便利店、7-11 便利店等。2012 年中國連鎖百強榜顯示，百聯集團門店同比減少 8.2%，五星電器門店同比減少 9.7%，海航商業門店同比減少 4.3%，利群集團門店同比減少 12%，百盛商業門店同比減少 7.7%，三江購物門店同比減少 3.3%，北京京客隆門店同比減少 4%，農工商集團門店同比減幅達 19%，而湖南友誼阿波羅門店減幅更高達 27.3%。據中華全國商業資訊中心發佈的統計資料顯示，排名全球前三的外資零售企業沃爾瑪、家樂福、樂購2012 年在華開店幅度同比平均降低27%，且都出現了關店的現象。

Under the pressure of falling profits and major indicators, many retailers, like the NGS Group, Lawson Convenience Store, and 7-11 Convenience Store have shut up stores to stop loss. The 2012 China's top 100 list of chain stores showed that the number of Bailian Group's stores decreased by 8.2% year-on-year, Five Star

Appliance by 9.7%, HNA Retailing by 4.3%, Liqun Group by 12%, and Parkson Retailing by 7.7%, Sanjiang Shopping by 3.3%, Jingkelong in Beijing by 4%, NGS Group by 19% and Friendship&Apollo in Hunan by 27.3%. According to the statistics released by the China National Business Information Center in 2012, even the top three foreign retailers Wal-Mart, Carrefour, and Tesco underwent a year-on-year growth decline of 27% in China, and all of them shut up some of their stores.

表 Table 5-4　2012 年部分連鎖百強企業門店減少情況 Store losses in some of the top 100 chain enterprises in 2012

企業名稱 Name of the enterprise	2012 年較 2011 年門店減少速度（%）Store loss rate in 2012 compared to 2011
百聯集團 Bailian Group	8.2
五星電器 Five Star Appliance	9.7
海航商業 HNA Retailing	4.3
利群集團 Liqun Group	12.0
百盛商業 Parkson Retailing	7.7
三江購物 Sanjiang Shopping	3.3
北京京客隆 Jingkelong in Beijing	4.0
農工商集團 NGS Group	19.0
湖南友誼阿波羅 Friendship&Apollo in Hunan	27.3

2013 年 10 月 24 日，沃爾瑪宣布在華發展的 3 年計畫，其中重要內容之一就是透過市場評估關閉中國門店 36 家，佔在華門店數量的近 9%。可的便利店也對全國門店進行了調整，僅在揚州一地就已關閉 20 多家門店。家樂福、瀋陽伊勢丹、無錫大洋百貨、上海光一百貨、天津亞瑪達電器等實體零售企業的關店也越來越頻繁。

On October 24, 2013, Wal-Mart announced its three-year plan for development in China. One of the important contents was to close 36 stores in

China through market evaluation, accounting for nearly 9%. Kedi Convenience Store has also carried out the adjustment, more than 20 stores were closed in Yangzhou alone. The store closing of Carrefour, Isetan in Shenyang, Dayang Department Store in Wuxi, Guangyi Department Store in Shanghai, Yamada Denki in Tianjin and other physical retail enterprises became more frequent.

2015、2016 年，實體零售商業績下滑的趨勢仍然沒有得到遏制。以百貨及購物中心為例，據統計，2015 年 63 家百貨門店關閉，涉及 17 個省市自治區的 14 個品牌，其中福建省就有 9 家百貨門店關店。以關店品牌來看，萬達百貨關店 35 家、天虹百貨關店 5 家、瑪莎百貨關店 5 家、金鷹百貨關店 5 家。據統計，2016 年上半年全國 50 家重點大型零售企業零售額同比累計下降 3.1%，並且所有品類零售額均表現為同比下降。

In 2015 and 2016, the decline in the performance of physical retailers has not been curbed. Taking department stores and shopping centers as examples, according to statistics in 2015, 63 department stores were closed, involving 14 brands in 17 provinces, municipalities and autonomous regions, including 9 department stores in Fujian Province alone. As the brands of the closed stores, there were 35 Wanda Department Stores, 5 Rainbow Department Stores, 5 Marks & Spencer and 5 Golden Eagle Department Stores. According to statistics, in the first half of 2016, the retail sales of 50 key large-scale retail enterprises nationwide decreased by 3.1% year-on-year, and the retail sales of all categories showed a year-on-year decline.

2018 年，在中國大陸經營了 23 年的國際零售巨頭家樂福和騰訊、永輝簽訂了中國潛在投資意向條款，黯然退出中國市場。此前的五年間（2012

年-2017年），家樂福在中國大陸地區的銷售份額由 55.83 億歐元下降到 46.19 億歐元，下降 9.64 億歐元，折合人民幣 75.93 億元，降幅為 17.26%。

In 2018, international retail giant Carrefour, which have been operating in mainland China for 23 years, signed the terms of potential investment intentions in China with Tencent and Yonghui Supermarket and quitted the Chinese market. In the previous five years (2012-2017), Carrefour's sales share in mainland China fell from € 5.583 billion to € 4.619 billion, going down by € 964 million, equivalent to ¥7.593 billion, a decrease of 17.26%.

表 Table 5-5 2013 年連鎖企業分業態的典型企業毛利率及淨利率變化 2013 Changes in gross profit margin and net interest rate of typical enterprises in chain format

業態 Format	毛利率（%）		毛利率差	淨利率（%）		淨利率差
	2012 年第四季度	2013 年第四季度		2012 年第四季度	2013 年第四季度	
百貨店 Department store	20.4	20.2	-0.2	3.7	3.6	-0.1
便利店 Convenience store	19.3	20.2	0.8	2.2	2.4	0.2
普通超市 Normal supermarket	18.1	17.7	-0.4	4.3	4.1	-0.2
大型超市 Hypermarket	16.1	15.9	-0.2	1.6	1.4	-0.3
專賣店 Specialty store	24.8	24.3	-0.5	5.1	4.6	-0.5
專業店 Specialized store	22.8	22.4	-0.4	5.0	5.1	0.1
總計 Total	19.9	19.8	-0.2	3.9	3.7	-0.2

資料來源：中國連鎖協會 Source: China Chain-Store & Franchise Association

4· 傳統零售商面臨的是全球性危機
Traditional retailers face a global crisis

不僅在中國，而且在全球很多國家和地區，傳統零售業面臨網路零售的挑戰和擠壓是當前實體零售業面臨的普遍問題。

Not only in China, but also in many countries and regions around the world, the challenges and squeezing from online retailing are common problems for traditional physical retail industry.

以美國為例，2013 年美國的網路購物數量增長了 14%，而全國整體的零售業銷售只增長了 3%。9 月，共計有超過 1.2 億購物者湧向網路商城，創下了 14.6 億美元的消費紀錄。據 ComScore 公司報告，2013 年美國假日購物季的網上購物支出增長了 10%，僅在「網購星期一」一天，全球各地在亞馬遜網站上訂購的商品就超過了 3680 萬件，很多零售業者也報告了其網站的銷售業績強勁增長。與此相對照的，是實體零售業的不甚景氣。據 ShopperTrak 統計，2013 年假日購物季，美國零售業者迎來的客流量是 3 年前的約一半。
131

In the United States, for example, the amount of online shopping increased by 14% in 2013, while the overall retail sales in the country increased by only 3%. In September, a total of more than 120 million shoppers flocked to the online

---

131 Jeremy Bogaisky. 零售業危機：實體店必須做出的改變. 福布斯中文網. 2014-02-27
Jeremy Bogaisky. Retail Crisis: Changes that Physical Stores Must Make. Forbes China. Feb 27, 2014

stores, setting a record of $1.46 billion in consumption. According to ComScore, online shopping spending in the US holiday shopping season increased by 10% in 2013. On the 「Online Shopping Monday」 alone, more than 36.8 million items were ordered on Amazon.com worldwide. Many retailers also reported the sales performance of their website has grown strongly. In contrast, the physical retail industry was in depression. According to ShopperTrak, in the holiday shopping season of 2013, the traffic was about half of that three years ago.

2016 年 11 月，M&S 瑪莎百貨宣告退出中國市場，關閉全部 10 間門店，同時退出法國市場，關閉 7 間所有門店。國際市場關店數量將達到 113 間，同時裁員 2100 人。瑪莎百貨擬關閉的國際市場還包括比利時、愛沙尼亞、匈牙利、立陶宛、荷蘭、波蘭、羅馬尼亞和斯洛伐克。

In November 2016, Marks & Spencer announced its withdrawal from the Chinese market and the French market, closing all 10 and 7stores respectively. The number of closed stores in the international market reached 113 and 2,100 employees were laid off. The international markets that Marks & Spencer intended to close include Belgium, Estonia, Hungary, Lithuania, the Netherlands, Poland, Romania and Slovakia.

另以購物中心的發展為例。20 世紀以來，購物中心的繁榮使得 Shopping Mall 成為美國經典文化的一部分。1956—2005 年，超過 1500 家大型購物商場在美國各地拔地而起；然而，自 2006 年開始，美國境內再無新開張的購物中心。據美國地產分析公司 CoStar Group 2013 年的調查報告，5 年內，美國境內 15% 的購物中心將停業關閉，10 年內購物中心將會消失 50%，「50 年

後購物中心對人們消費的統治將最終結束」。[132]

Another example is the development of shopping malls. Since the 20th century, prosperity of shopping malls has made it a part of American classic culture. From 1956 to 2005, more than 1,500 large shopping malls were launched across the United States. However, since 2006, no new shopping malls have appeared in the country. According to a 2013 survey by US real estate analyst firm CoStar Group, within 15 years, 15% of shopping malls in the US will be closed down and 50% will disappear within 10 years. 「The rule of shopping malls over people's consumption will eventually come to an end」.[132]

為此,面臨危機的實體零售業必須做出改變。

To this end, the physical retail industry must make changes.

---

[132] 美國購物中心變「鬼城」網路購物重創美國零售業態. 贏商網. 2014-06-06 American Shopping Center is Changed into A"Ghost City", Online Shopping Hit the US Retail Business. Winshop.com. June 6, 2014

零售商的應對 2. Retailers' reponse

　　顯然，市場的變化、消費格局的變化、消費者的選擇決定了傳統零售業不可能因循原有的發展路徑，必須迎接挑戰。危機是壓力也是動力，是傳統零售業改革、創新、發展的契機，是零售業全行業實現跨越的先機。

　　Obviously, the changes in the market and the consumption pattern, and the choice of consumers have determined that the traditional retail industry cannot follow the original development path and must meet the challenges. The crisis is both pressure and driving force. It is also an opportunity for the reform, innovation and development of the traditional retail industry, and to achieve leap development.

　　近年眾多國際國內零售企業發力電商。如擁有大量的實體店的資生堂公司從 2012 年起開始線上線下佈局，線上上，他們積極謀求異業合作，互相導流。資生堂在網上還有美容諮詢、在家美容檢測服務、直營網店和實體店導航等，利用互聯網＋實體店的優勢與顧客間實現互動。

　　In recent years, many international and domestic retailers have launched e-commerce. For example, Shiseido, which has a large number of physical stores, has been working on this online and offline since 2012. It actively seeks cross-industry cooperation and mutual guidance online. There are also beauty consultation, home beauty testing services, direct online stores and physical store navigation on the website, using the advantages of the 「Online+physical store」 to interact with customers.

　　永旺則採取與軟銀、雅虎合作的模式。顧客可以在雅虎的網站上下載優惠券，在永旺實體店門口掃描出來後就可以在永旺實體店使用。這種「永旺＋軟銀＋雅虎」模式的亮點是：零售商可以透過這種方式收集到顧客資料，並透過雅虎引流到店，從而使雅虎和永旺能共用顧客資源。

　　AEON cooperates with SoftBank and Yahoo. Customers can download coupons on Yahoo and scan them at the AEON store to use them. The highlight of this 「AEON+Softbank+Yahoo」 model is that retailers can collect customer data and divert customers to the store through Yahoo, so that Yahoo and AEON can share customer resources.

　　2012 年中國連鎖百強企業中，已有 62 家開展了網路零售業務，比 2012 年初統計的 59 家增加 3 家。2013 年，在連鎖百強企業中已有 67 家開展網路零售業務。許多零售企業表示，願與網路零售企業競合發展，北京朝陽大悅城公開支持「抄貨號」，願做「雙 11」試衣間。

　　Among the China top 100 chain enterprises in 2012, 62 have launched online retail business, an increase of 3 from 59 in the beginning of 2012. In 2013, 67 have launched online retail business. Many retailers have expressed their willingness to compete and cooperate with online retailers. Joy City in Chaoyang District, Beijing has publicly supported the number-copying behavior and is willing to be the fitting room for Double Eleven.

表 Table 5-6　2013 年連鎖百強企業分業態的典型企業網上零售業務淨利率變化情況 Changes in the net interest rate of typical online retail business of China top 100 chain enterprises in 2013

業態 Format	第三季度網路零售同比增幅（%）Year-on-year growth of online retail sales in the third quarter(%)	第四季度網路零售同比增幅（%）Year-on-year growth of online retail sales in the fourth quarter(%)
百貨店 Department store	24.8	19.8
便利店 Convenience store	17.7	16.0
超市 Supermarket	78.3	82.3
大型超市 Hypermarket	11.1	11.6
專賣店 Specialty store	32.3	22.1
專業店 Specialized store	58.4	89.0
總計 Total	31.3	45.3

資料來源：中國連鎖協會 Source: China Chain-Store & Franchise Association

表 Table 5-7　2013 年中國連鎖百強企業開展網路零售業務情況 Online retail business of China top 100 chain enterprises in 2013

序號 No.	企業簡稱 Enterprise	網點名稱 Branch	網址 Site URL
1	蘇寧雲商 Suning Commerce Group	蘇寧易購 Suning E-commerce	http://www.suning.com
		蘇寧紅孩子母嬰商城 Suning Redbaby	http://redbaby.suning.com
		蘇寧繽購美妝商場 Suning Binggo	http://binggo.suning.com
2	國美電器 Gome Electrical Appliances	國美線上 Gome Online	http://www.gome.com.cn
		國美線上天貓官方旗艦店 Gome Online Tmall Flagship Store	http://gome.tmall.com
		大中電器 Dazhong Electronics	http://www.dazhongdianqi.com.cn
3	沃爾瑪 Wal-Mart	1 號店 No.1 Store	http://www.yhd.com
		山姆會員店 Samclub	http://www.samsclub.cn/home

		銀泰網 Yintai	http://www.yintai.com
4	銀泰商業 Yintai	銀泰百貨京東精品旗艦店 Yintai-JD Flagship Store	http://yintai.jd.com
		銀泰百貨天貓精品旗艦店 Yintai-Tmall Flagship Store	http://yintai.tmall.com
5	百勝餐飲集團 Yum! Brands Inc.	肯德基宅急送 KFC Delivery	http://www.4008823823.com.cn
		必勝宅急送 Pizza Hut Delivery	http://www.4008123123.com
6	麥當勞 Mcdonald's	麥樂送 網上訂餐 McDeliveries	https://www.4008-517-517.cn
7	王府井集團 Wangfujing Group	王府井網上商城 Wangfujing Online	http://www.wangfujing.com
8	山東省商業集團 Shandong Commercial Group	銀座網 Yinzuo100	http://www.yinzuo100.com
		銀座食品旗艦店 Yinzuo Food	http://yinzuoshipin.jd.com
9	大商集團 Dashang Group	大商網 66buy	http://www.66buy.cn
		麥凱樂網上商城 Mykal Online	http://shop.qdmykal.com
10	新世界百貨 New World Department Store	新世界百貨網上商城 New World Department Store Online	http://www.xinbaigo.com
11	大潤發 RT-Mart	飛牛網 Feiniu	http://www.feiniu.com
12	廣州屈臣氏 Watsons in Guangzhou	屈臣氏官網商城 Watsons Online	http://www.watsons.com.cn
		屈臣氏天貓旗艦店 Watsons Tmall	http://watsons.tmall.com
		屈臣氏京東官方旗艦店 Watsons JD	http://watsons.jd.com
13	天虹商場 Rainbow Department Store	網上天虹 Rainbow	http://www.tianhong.cn
14	步步高集團 BBK Group	步步高商城 BBK	http://www.bubugao.com
15	迪信通 D.Phone	迪信通移動生活商城 Dixingtong	http://shop.dixintong.com
16	錦江麥德龍 Metro Supermarket in Jinjiang	麥德龍 MetroMall	http://www.metromall.com.cn

17	武漢武商集團 Wushang Group in Wuhan	武商網 Wushang	http://www.wushang.com
18	中百集團 Zhongbai Holdings Group	中百商網 Zon100	http://www.zon100.com
19	利群集團 Liqun Group	利群商城 Liqun Shop	http://www.liqunshop.com
		利群醫藥資訊 Lqyaopin	http://www.lqyaopin.com
20	中石化銷售公司 Sinopec	易捷網 Ejoy365	http://www.ejoy365.com
21	南京中央商場 Nanjing Central Shopping Centre	南京中央商場天貓旗艦店 Njzysc-Tmall	http://njzysc.tmall.com
		雲中央 CloudCentre	http://www.600280.com
22	聯華超市股份有限公司 Lianhua Supermarket Holdings Co., Ltd	聯華易購 LhMart	http://www.lhmart.com
			http://www.lhok.com
23	上海友誼集團 Shanghai Friendship Group	百聯股份網上商城 Blzoom	http://www.blzoom.com
24	石家莊北國人百 Shijiazhuang Beiguo-Renbai Group	北國如意購 Ruyigou	http://www.ruyigou.cn
25	百盛商業集團 Parkson Retail Group	百盛網 Parkson	http://www.parkson.com.cn
26	歐尚（中國）Auchan China	歐尚網購 Auchan	http://www.auchan.com.cn/Default.aspx
27	江蘇五星電器有限公司 Jiangsu Five Star Appliance Co., Ltd	五星電器網上商城 5star	http://www.5star.cn
28	合肥百貨大樓 Hefei Department Store Group	百大易商城 bdysc	http://www.bdysc.com
29	宏圖三胞 HISAP100	宏圖三胞·慧買網 Huimai100	http://www.huimai100.com

30	農工商超市（集團）NGS Supermarket Group	便利通網 CHBLT	http://www.chblt.com
31	廣百股份 Grandbuy Co., Ltd	廣百薈 iGrandbuy	http://www.igrandbuy.com
		廣百天貓官方旗艦店 Grandbuy	http://grandbuy.tmall.com
32	家樂園商貿有限公司 Jialeyuan Group	家樂園速購 Subuy	http://www.subuy.com
33	永旺 Aeon	永旺網上商城 Aeon E-shop	http://eshop.qdaeon.com
34	廣州友誼集團 Guangzhou Friendship Group	友誼網樂購 Cgzfs	http://mall.cgzfs.com
35	北京首商集團 Beijing Capital Retailing Group	西單商場 i 購物 igo5	http://www.igo5.com
36	煙臺市振華百貨集團 Yantai Zhenhua Group	振華商城 Zhenshang Eshop	http://eshop.zhenshang.com
37	文峰大世界 Wenfeng Great World	文峰大世界網上商城 Wenfeng Great World Online	http://www.wfdsj.com.cn/shop
		愛上文峰 iWengfeng	http://www.iwenfeng.com
38	長春歐亞集團 Changchun Eurasia Group	歐亞 e 購 oysd	http://www.oysd.cn
39	三江購物俱樂部 Sanjiang Shopping	三江購物網 sangjiang	http://www.sanjiang.com
40	新一佳 Xinyijia	新一佳網上超市 xyj-shop	http://www.xyj-shop.com
		新一佳生活超市淘寶店 xyj365	http://xyj365.taobao.com
41	伊藤洋華堂 Ito Yokado	伊藤洋華堂網路超市 yiteng365	http://www.yiteng365.com http://shop.iy-cd.com
42	遼寧興隆大家庭 Liaoning Xinglong Happy Family	網上興隆 xlgoo	http://www.xlgoo.net
43	邯鄲市陽光百貨 Handan Sunshine Department Store	陽光天天購 ygttg	http://www.ygttg.com

44	青島維客 Qiangdao Weekly Group	點點網 weeklydd	http://www.weeklydd.com
45	一丁集團 Yiding Group	一丁集團網上商城 itu4	http://www.itu4.com
46	人人樂 RenRenLe	人人樂購網上商城 rrlgou	http://www.rrlgou.com
47	湖南友誼阿波羅 Hunan Friend&Apollo	友阿奧特萊斯 youaoutles	http://www.9448.net
48	重慶商社（集團）Chongqing General Trading Group	世紀購 sigo365	http://www.sjgo365.com
49	廣東嘉榮超市 Guangdong Jiarong Supermarket	喜伴啦啦網上超市 sparlala	http://www.sparlala.com
50	徽商集團旗下 Huishang Group		
	1. 徽商農家福 1.Nongjiafu	農家福農資網上商城 hsnjf	http://old.hsnjf.com/mallcenter.asp
	2. 商之都 Commercial Capital	徽之尚商城 hzsmall	http://www.hzsmall.com
51	青島利客來 Qingdao Likelai	立刻送商城 shoplikelai	http://www.shoplikelai.com
52	浙江人本超市 Zhejiang C&U Supermarket	人本網上超市 rbcs	http://www.rbcs.cn
53	長沙通程式控制股 Changsha Tongcheng Group	愛尚通程 dolton	http://www.dolton.cn
54	濟南華聯商廈集團 Jinan Hualian Commercial Group	華聯易購 hlyigou	http://www.hlyigou.com
55	TESCO 樂購 Tesco	e 樂購 elegou	http://elegou.cn.tesco.com

56	山東新星集團 Shandong Xinxing Group	新星商城 xinxing001	http://www.xinxing001.com
57	江蘇新合作常客隆 Jiangsu New Cooperation CKL	常客隆網上商城 csckl	http://www.csckl.com/qjd.php
		家易樂 csckl	http://csckl.com/ytj/pc
58	全福元商業集團 Shangdong Quanfuyuan Commercial Group	全福元商城 qfy365	http://www.qfy365.com
59	山東濰坊百貨 Shandong Weifang Department Store	小蜜蜂購物網 92xmf	http://www.92xmf.com
		中百天貓官方旗艦店 Zhongbai Tmall Flagship Store	http://zhongbai.tmall.com
60	山西省太原唐久超市 Shangxi Taiyuan Tangjiu Supermarket	唐久京東大賣場 tjcvs.jd	http://tjcvs.jd.com
61	北京樂語世紀通訊設備 Beijing FunTalk Communications	樂語天貓官方旗艦店 FunTalk-Tmall	http://funtalk.tmall.com
62	大參林醫藥集團 Dashenlin Medical Group	大參林大藥房天貓旗艦店 Dashenlindyf-Tmall	http://dashenlindyf.tmall.com
63	雄風集團有限公司 Xiongfeng Group	雄風電器天貓官方旗艦店 Samsungxf-Tmall	http://samsungxf.tmall.com
64	阜陽華聯集團 Fuyang Hualian Group	華聯易購淘寶店 hlyigou	http://fyhlyg.taobao.com
65	北京菜市口百貨 Beijing Caishikou Department Store	菜百首飾京東網店 caibai.jd	http://caibai.jd.com
66	全家便利店 FamilyMart	FamilyMart 全家天貓旗艦店 FamilyMart Tmall Flagship Store	http//familymart.tmall.com
		全家 FamilyMart 淘寶店 FamilyMart Taobao	http://shop101170741.taobao.com

| 67 | 家家悅 Jiajiayue Supermarket | U 箱超市——家家悅 uxiangjiajiayue | http://uxiang.com/index. htm?cityId=1314 |

資料來源：中國特許經營協會 2014 年 4 月 21 日發佈 Source: China Chain Store&Franchise Association. Apr 21,2014

在「雲消費」時代，面對撲面而來的零售革命大潮，零售企業是如何應對的？

In the era of cloud consumption, how do retail companies deal with the tide of the retail revolution?

（一）蘋果公司做了什麼？ What did Apple do?

近年來，蘋果公司在整體戰略上實現了從純粹的消費電子產品生產商向以終端為基礎的綜合性內容服務提供者的轉變。主要在產品和技術、行銷、商業模式、零售店體驗方面進行了一系列創新。

In recent years, Apple has transformed its strategy from a purely consumer electronics manufacturer to a terminal-based, integrated content service provider. A number of innovations have been made primarily in product and technology, marketing, business models, and retail store experience.

1 · 產品和技術創新
Innovation in product and technology innovation

蘋果公司從 iPod、iMac、iPhone 到 iPad 不斷地推陳出新，引領潮流，

從最初單一的電腦公司，逐步轉型成為高端電子消費品生產和服務企業。最重要的是蘋果公司抓住行動電話市場的發展趨勢，推出了 iPhone 系列產品。2007 年 1 月，蘋果公司首次公布進入 iPhone 領域，正式涉足手機市場。蘋果公司在 MP3 市場上依靠 iPod ＋ iTunes 大獲成功後，緊接著在手機市場依靠 iPhone+APP Store 的組合，透過在產品、性能、作業系統、管道和服務方面的差異化定位，一舉擊敗其他競爭對手。2011 年 2 月，蘋果公司打破諾基亞在移動通信領域連續 15 年銷售量第一的壟斷地位，成為全球第一大手機生產廠商，2017 年蘋果手機出貨量約 2.34 億部。

From the iPod, iMac, iPhone to iPad, Apple continues to innovate and lead the trend, from a pure computer company in the beginning to a high-end consumer electronics production and service company. The most important thing is that Apple has seized the development trend of the mobile phone market and launched the iPhone series. In January 2007, Apple first announced its entry into the iPhone field and officially entered the mobile phone market. After its success in the MP3 market with iPod and iTunes, Apple beat other competitors in a differentiated position in terms of products, performance, operating systems, channels and services with the combination of iPhone and APP Store in the mobile phone market. In February 2011, Apple broke Nokia's monopoly position in the mobile communications field for 15 consecutive years, becoming the world's largest mobile phone manufacturer. In 2017, Apple's mobile phone shipments were approximately 234 million units.

2・ 營銷創新 Innovation in marketing

蘋果公司的「飢餓行銷」策略讓很多消費者被它牽著鼻子走，同時也

為蘋果產品聚集了一大批忠實粉絲。從 2010 年 iPhone 4 開始到 iPad 2 再到 iPhoneX，蘋果產品全球上市呈現出獨特的傳播曲線：發佈會一上市日期公布一等待一上市新聞報導一通宵排隊一正式開賣一全線缺貨一黃牛漲價。與此同時，蘋果公司一直採用「捆綁式行銷」的方式帶動銷售量，從 iTunes 對 iPod、iPhone、iPad 和 iMac 的一系列捆綁，讓使用者對其產品產生很強的依賴性。

Apple's strategy of hunger marketing has led many consumers by the nose, and has also gathered a large number of loyal fans for Apple products. From the iPhone 4 in 2010 to the iPad 2 and then to the iPhoneX, the global launch of Apple products presents a unique communication curve: the press conference-the release date- waiting-news reports-overnight queues-officially selling-out of stock on all fromts- high price by the scalpers. At the same time, Apple has been using the bundled marketing to drive sales, such as iTunes for iPod, iPhone, iPad and iMac, to make users have a strong dependence on their products.

3 · 商業模式創新 Innovation in business model

（1）iPod 的商業模式創新 ——iTunes 網上音樂商店
Innovation of iPod business model-iTunes the online music store

蘋果公司最初透過「iPod+iTunes」的組合開創了一個新的商業模式，將硬體、軟體和服務融為一體。iPod 的成功是線上音樂產業上下游合作的結果，即價值鏈整合，它實現了 PC、消費電子和音樂三者的集成，將 iPod 播放機、iTunes 音樂下載、Macintosh 視頻播放軟體有機結合起來，為客戶打造了播放、

下載和視頻等客戶價值鏈系統.

Apple initially created a new business model through the combination of iPod and iTunes that combines hardware, software and services. The success of the iPod is the result of the upstream and downstream cooperation of the online music industry, namely the integration of the value chain, which realizes the integration of PC, consumer electronics and music, and combines the iPod player, iTunes music download and Macintosh media player software, creating a value chain for the customers.

（2）iPhone 的商業模式創新——電信運營商合作捆綁銷售與「iPhone ＋ APP Store」 Innovation of iPhone's business model-telecom operators cooperate with bundled sales and the combination of iPhone and APP Store

iPhone 採取與電信運營商合作捆綁銷售 iPhone 產品的商業模式，利潤分成。同時採用 iPhone ＋ APP Store 的商業模式，蘋果公司的 APP Store 對所有開發者開放，任何公司或個人的有想法的 APP 都可以在 Apple Store 上銷售，銷售收入與蘋果七三分成，除此之外沒有任何的費用。這極大地調動了協力廠商開發者的積極性，同時也豐富了 iPhone 的用戶體驗，適應了 iPhone 手機使用者對個性化軟體的需求。

iPhone adopts a business model that bundles and sells iPhone products in cooperation with telecom operators, and shares the profit. At the same time, it adopts the business model of the combination of iPhone and APP Store and opens its APP Store to all developers. Any interesting APP developed by any company or individual can be sold on the Apple Store. The sales revenue is on a 70/30 split with no exra fees. This greatly mobilizes the enthusiasm of third-party developers and enriches the experience for iPhone user, which meets the needs of iPhone users for personalized software.

4 · 零售創新 Innovation in retail

（1）人才聘用 Staffing

在選擇優秀的市場行銷人才方面，約伯斯聘請了 Gao 服裝的董事長兼 CEO 米奇•德雷克斯勒和經驗豐富的零售大師羅恩•詹森。羅恩•詹森曾經幫助 Target 品牌實現從入駐沃爾瑪賣場的落選者向高端供應者的轉變。

In selecting excellent marketing talents, Jobs hired Mickey Drexler, Chairman and CEO of Gao Clothing and Ron Johnson, an experienced retail guru. Ron Johnson has helped Target become a high-end supplier from a losing candidate to Wal-Mart.

（2）零售店的區別定位 Differential positioning of retail stores

蘋果公司的零售店想要出售一種「生活方式」，顧客可以在這裡感受蘋果的數位生活方式——離開的時候很有可能會帶走一臺機器。

Apple's retail stores attempt to sell a "lifestyle", where customers can experience Apple's digital lifestyle and leave with a product.

（3）商店選址 Location of the store

蘋果公司早期決策非常關鍵的一步是將商店設在人流量大的地區，事實證明，這個決策是一大突破。但是最初這個決策卻遭到了普遍的批評，因為人流量大的地方租金非常貴。

A very critical step in Apple' s early decision-making was to locate the stores in crowded areas, which proved to be a major breakthrough. But the decision was widely criticized at first because rent was too high in places with high traffic.

（4）商店設計 Store design

在 iPod 產品推出市場之前，蘋果公司的所有產品是兩種筆記型電腦以及兩種桌上型電腦。如何讓這 4 種產品來充實 6000 平方英尺的蘋果零售商店真是一個巨大的挑戰；蘋果公司決定用消費者體驗來充實它。商店只出售合適的產品；商店的設計特別注重細節，商店的顏色是淺色調，良好的燈光效果使產品看起來非常耀眼奪目；商店在佈局上分成了 4 個區域，每個區域都以專門的「方案解決區域」為中心。

Before iPod was introduced to the market, all of Apple's products at that time were two kinds of laptops and two kinds of desktops. How to make these four kinds of products to fill up the 6,000-square-foot Apple retail store was a huge challenge and Apple decided to fill it up with consumer experience. The store only sold the right products and the store design paid special attention to details such as light color and the good lighting, which made the products look dazzling. The store was divided into 4 areas, each with a special 「Solution Area」 as the center.

（5）天才吧 Genius Bar

蘋果零售商店最重要的創新是天才吧，它為消費者提供了實踐培訓和技術支援。在天才吧，消費者可以與維修人員面對面地進行問題檢修，或者順便將有故障的設備拿到當地的商店，而不需要特意送修。據蘋果公司估算，2006 年平均每週有 100 多萬消費者光顧天才吧。

The most important innovation in the Apple retail store is the Genius Bar, which provides hands-on training and technical support to consumers. In the Genius Bar, consumers can do a face-to-face troubleshooting with maintenance personnel, or take the faulty devices to a local store without intentionally sending them to repair. It is estimated by Apple that more than 1 million consumers visited the Genius Bar every week in 2006.

（6）零售服務 Retail service

蘋果零售商店嚴格控制員工與消費者互動的方式，對現場技術支援人員的用語進行規範培訓，考慮門店的每個細節，細緻到樣機上的預載圖片和音樂。[133]

Apple strictly controls the way employees in retail stores interact with consumers and conducts standard language training for on-site technical support staff. Besides, Apple considers every detail of the store like pre-loads pictures and music on the demonstration devices.

（7）iBeacon 技術應用 Application of iBeacon technolog

2013 年 12 月起，蘋果公司開始在全美的 254 家蘋果零售店中部署 iBeacon（智慧信標）技術。 Beacon 技術是一個實現室內資料傳輸的解決方案，該技術能以廉價的硬體透過低功耗藍牙的方式向網路內的移動設備捕捉和推送資訊，從而充當一個小型資訊基站，幫助人們完成室內導航、移動支付、店內導購和人流量分析等活動。蘋果公司將 iBeacon 技術應用到零售業

---

133 蘋果──創新與未來的引領者. 聖才學習網 Apple-the leader of innovation and the future. www.100xuexi.com

之中，可以向消費者提供更明確、更具針對性的資訊，進而實現更為精確的銷售和移動購物。

Since December 2013, Apple has been deploying iBeacon (Smart Beacon) technology in 254 Apple retail stores across the United States. Beacon technology is a solution for indoor data transmission. It can capture and push information to mobile devices in the network through low-power Bluetooth in low-cost hardware, thus serving as a small information base station to help people complete activities such as indoor navigation, mobile payments, in-store shopping guides, and traffic analysis. Apple's application of iBeacon technology to the retail industry can provide consumers with clearer and more targeted information to realize more accurate sales and mobile shopping.

圖 Figure 5-4 蘋果 iBeacon 手機介面 Mobile interface of iBeacon of Apple

（二）沃爾瑪做了什麼？ What did Wal-Mart do?

　　早在 2000 年，沃爾瑪便成立了網上商城。2010 年，沃爾瑪成立新業務部門 Global.com，對沃爾瑪電子商務業務進行全面監管；隨後成立沃爾瑪實驗室，緊跟互聯網新趨勢、新技術，制訂公司的社交和移動商務戰略，將實體店和電子商務更好地整合到一起。2011 年，沃爾瑪又投入 5500 萬美元收購了社交媒體技術提供商 Kosmix，並重組包括美國、英國、日本和加拿大等國在內的電子商務部門，以便更好地整合門店和網路零售業務。截至 2011 年，沃爾瑪已經在電子商務領域投入超過 1.5 億美元，並在世界很多國家擁有相對獨立的電子商務平臺，如美國、英國、加拿大、墨西哥和巴西。其中美國沃爾瑪 Walmart.com 的點擊率僅僅次於亞馬遜，而英國沃爾瑪 Asda.com 也較為成熟。

　　As early as 2000, Wal-Mart had set up its first online store. In 2010, Wal-Mart set up a new business unit, Global.com, to oversee Wal-Mart's e-commerce business. Subsequently it established the Wal-Mart Lab to keep up with new Internet trends and new technologies, and develop social and mobile commerce strategies to better integrate physical stores with e-commerce. In 2011, Wal-Mart invested another $55 million in the acquisition of Kosmix, a social media technology provider, and reorganized its e-commerce divisions in the United States, the United Kingdom, Japan and Canada, to better integrate its retail stores and online retail business. As of 2011, Wal-Mart has invested more than $150 million in e-commerce and has relatively independent e-commerce platforms in many countries around the world, such as the United States, the United Kingdom,

Canada, Mexico and Brazil. Among these platforms, click rate of Walmart.com in the United States ranks second only to Amazon. Asda.com in Britain is also relatively mature.

在中國，2011 年 5 月份沃爾瑪宣布入股在倉儲物流和配送系統上表現出色的國內電子商務公司——1 號店，隨後將山姆會員店網上商城從深圳拓展到北京，之後，沃爾瑪更是宣布要在上海設立沃爾瑪全球電子商務中國總部，全面負責其全球電子商務在中國市場的運營。沃爾瑪入股 1 號店後，其業務增長迅速。2011 年 1 號店銷售額達 27.2 億元，2012 年銷售額較上年增長 2.5 倍，無線業務增長 26 倍，銷售額達到 68 億元；同時 2012 年 1 號店透過提升供應鏈運營效率，在採購、倉儲、配送等環節相較前一年降低了 37% 的成本。做為首家擁有海外直採資格的電商，進口食品是 1 號店的王牌品類，2012 年其進口食品年度增長 407%。其中進口牛奶品類在 1 號店的品牌就覆蓋 18 個國家 60 個品牌，成功帶動整個 1 號店進口食品成為線上零售平臺第一。[134]

In China, in May 2011, Wal-Mart announced that it would invest in No. 1 Store, a domestic e-commerce company which performed well in the warehousing logistics and distribution system. Then it expanded the Sam's Club online stores from Shenzhen to Beijing. After that, Wal-Mart announced that it would established Wal-Mart Global E-Commerce China Headquarters in Shanghai, fully responsible for the operation of its global e-commerce in the Chinese market. After Wal-Mart became a shareholder of No. 1 Store, its business grew rapidly. In 2011, the sales

---

[134] 1號店2012銷售額增長2.5倍. 人民網. 2013-03-14No. 1 Store Sales Increased by 2.5 Times in 2012. People.cn. March 14, 2013

volume of No. 1 Store reached ¥2.72 billion. In 2012, the sales volume increased by 2.5 times compared with the previous year. The wireless business of the year grew by 26 times and the sales volume reached ¥6.8 billion. At the same time, by improving the efficiency of supply chain operation, No.1 Store reduced the cost of purchasing, warehousing, distribution and other links by 37% compared with the previous year. As the first e-commerce company with overseas direct investment qualifications, No. 1 Store's ace is the imported food. In 2012, sales volume of imported food grew by 407%. Among the category, the brand of imported milk products in the No. 1 Store covered 60 brands in 18 countries, which successfully made No.1 Store rank the top in imported food among online retail platforms.

近年沃爾瑪還與 Facebook 合作開發「My Local Walmart」，沃爾瑪可以和其粉絲進行互動，更好地瞭解客戶需求，加強內容行銷，並推送附近門店的優惠資訊據美國電商門戶。據 Internet Retailer 統計，沃爾瑪移動 App 使用者以更高的頻率光臨沃爾瑪，在超市停留時間較普通顧客高 40%，年平均消費額高於線下消費者 38%（Nielsen 資料）。

In recent years, Wal-Mart has also partnered with Facebook to develop 「My Local Walmart」 which enables Wal-Mart to interact with its fans to better understand customer needs, enhance content marketing, and push offers from nearby stores. According to statistics by Internet Retailer, Wal-Mart mobile app users visit Wal-Mart at a higher frequency, they stay in the supermarket 40% longer than the general customers and spend 38% more than the offline consumers (Data by Nielsen).

（三）H&M 做了什麼？ What did H&M do?

創建於 1947 年的「老字型大小」品牌 H&M 是全球知名的快時尚品牌，這個品牌永保年輕活力的祕訣不僅來自於其自身對時尚的不懈追求，而且也源於其善於把握主流消費，採用與時俱進、不斷創新的行銷手法。近年，很好地利用社交媒體促進行銷已成為 H&M 新的「時尚祕笈」。

Founded in 1947, H&M is both a time-honored and world-renowned fast fashion brand. The secret of this brand's youthful vitality is not only its own unremitting pursuit of fashion, but also its adeptness of grasping mainstream consumption and using advancing and innovative marketing tools. In recent years, the use of social media to promote marketing has become H&M's new 「fashion secrets」.

1‧ 善用優勢社交平臺，凝聚目標消費者 Make good use of social platforms and foucus on target consumers

一方面，H&M 透過運營 Facebook、Twitter、Pinterest 等社交平臺，發佈新品的最新時尚資訊，將線上線下融為一體；另一方面，H&M 積極拓展新的社交領域 Google+，將內容定位為富有靈感、獨特，以分享新奇、獨特的圖片和視頻為主保持使用者黏性和忠誠度，並透過對第一手資料的分析，制訂精細化的行銷策略，目前，其在 Google+ 的品牌關注度榜單中位列第三位，高於 Facebook 和 Twitter。

On the one hand, through the social platforms such as Facebook, Twitter and Pinterest, H&M publishes the latest fashion information of new products and integrates online and offline. On the other hand, H&M actively enters Google+, a new social domain and positions the content as something inspirational and unique. Also, it shares novelty and unique images and videos to maintain user stickiness and loyalty. Through the analysis of first-hand data, it develops a refined marketing strategy. Currently, H&M' s brand attention on Google+ ranks third, higher than Facebook and Twitter.

圖 Figure5-5　Google+ 上的 H&M 頁面 H&M page on Google+

2‧ 善用意見領袖，充分發揮互動的力量 Make good use of opinion leaders and give full play to interaction

H&M 邀請貝克漢為代言人，不僅請他出演廣告片，而且將其品牌效應用在實際行銷中。2012 年，為提高紐約、洛杉磯和三藩市門店的客流量，將潛在的線上消費者吸引到線下門店當中，H&M 將貝克漢的人像設置在城市的不同地方，貝克漢的粉絲透過 H&M 提供的線索，利用網路找到分布在各

地的人像，並與其合影拍照，分享到各個社交網路，提高了品牌知名度，增加了實體店的客流量。

H&M invited David Beckham as the spokesperson, not only asking him to appear in commercials, but also using the brand effect in actual marketing. In 2012, in order to increase the traffic of stores in New York, Los Angeles and San Francisco and atract potential online consumers to offline stores, H&M placed Beckham's portraits in different parts of the city. Through the clues provided by H&M, Beckham' s fans used the network to find portraits distributed around the country, took photos with them, posting the photos on various social networks, thus increasing the brand awareness and the traffic of physical stores.

2013 年，H&M 還透過開展名為「50 States of Fashion」的行銷活動，推廣其在美國新上線的電商網站。活動主要是鼓勵美國消費者將自己的照片上傳到 Instagram 主頁，然後對上傳的參與人進行投票，選出每個州的形象代言人。被選出的形象代言人不僅可以獲得 1000 美元的代金券，而且可以去紐約參加時裝週。這一活動得到了廣大消費者的積極回應。

In 2013, H&M also promoted its new e-commerce website in the US through a marketing campaign called 「50 States of Fashion」. The main activity was to encourage American consumers to upload their photos to the homepage of Instagram, then vote on the participants and select the image spokesperson for each state. The selected image spokesperson could not only get a $1,000 voucher, but also attend Fashion Week in New York. The activity had received positive response from consumers.

圖 Figrue5-6　粉絲互動增強消費黏性 Consumer stickiness is enhanced by interaction with fans

3．把握客戶需求，以優質內容創造良好體驗 Grasp customer needs and create a good experience with quality content

H&M 定位為平民化的時尚品牌。主要消費群體是一群對時尚極度敏感的年輕人。因此，H&M 把握客戶需求，將品牌內容延伸到時尚的各個領域，根據不同地域消費者的喜好，創造各種不同的有趣話題，比如自行車時尚、布魯克林的時尚未來派等。讓消費者有更多的關注點，有更多共同的時尚興趣點。[135]

---

135 米米 . H&M 的時尚祕笈：利用社交媒體俘獲消費者的心 [J]. 行銷智庫，2014（04）：4. Mimi. H&M's Fashion Tips: Using Social Media to Capture Consumers' Hearts [J]. Marketing Think Tank, 2014(04): 4.

H&M is positioned as a popular fashion brand. The main consumer group is young people who are extremely sensitive to fashion. Therefore, H&M grasps the needs of customers and extends the brand content to all areas of fashion. It creates various interesting topics according to the preferences of consumers in different regions, such as bicycle fashion and fashion futurism in Brooklyn, thus letting consumers have more concerns and more common fashion points of interest.

（四）蘇寧做了什麼？ What did Suning do?

蘇寧公司以電器業務起家，近年來，在「去電器化」的道路上邁出了自己的步伐。2010 年蘇寧公司正式上線蘇寧易購，蘇寧電子商務開始飛速發展。2011 年，公司制訂未來 10 年「科技轉型、智慧服務」戰略。2012 年，蘇寧公司提出超電器化戰略，將線上線下融為一體，進行全品類擴張。線上下，2012 年蘇寧新開 Expo 超級旗艦店 13 家、樂購仕生活廣場 8 家。線上上，2012 年蘇寧易購在原有圖書品類的基礎上，上線了虛擬產品、酒水、食品、保險、團購、電子書等非電器品類。2013 年 2 月 19 日，蘇寧公司名稱變更為蘇寧雲商集團股份有限公司，並闡述了「雲商」新模式的內涵：店商＋電商＋零售服務商。其核心是以雲技術為基礎，整合開放蘇寧前臺後臺、融合開放蘇寧線上線下，服務全產業、服務全客群。這標誌著蘇寧雲商零售模式全面落成。這不僅是蘇寧未來實現跨越發展的新方向，也在一定程度上代表

---

136 張輝東 . 摘掉電器帽子 蘇寧要做雲商 [N]. 長沙晚報 · 2013-02-22. Zhang Huidong. No Longer Electrical Appliance Provider: Suning Comes into Cloud Commerce [N]. Changsha Evening News, Feb 22, 2013.

了中國零售行業轉型發展的新趨勢。[136]

　　Suning started from electrical appliance business. In recent years, it has taken its own steps on the road of 「de-electricization」. In 2010, Suning Company officially launched its Suning E-commerce and began to develop rapidly. In 2011, the company formulated the strategy of 「technology transformation and smart service」 in the next 10 years. In 2012, Suning Company proposed the strategy of super-electricalization to integrate online and offline and carry out an all-category expansion. For offline bussiness, Suning opened 13 new Expo flagship stores and 8 Tesco Life Plazas in 2012; for online business, Suning E-commerce launched non-electrical items such as virtual products, drinks, food, insurance, group purchases, and e-books on the basis of the book categories in the same year. On February 19, 2013, Suning Company changed its name into Suning Commerce Group Co., Ltd., and explained the connotation of the new model of cloud business: store+e-commerce+retail service. Its core was based on cloud technology, integrating and opening the Suning front-end and back-end, online and offline, serving the whole industry and the entire customer base. This marked the completion of the retail model of Suning E-commerce. This was not only the new direction for Suning to achieve leapfrog development in the future, but also represented a new trend in the transformation and development of China's retail industry to a certain extent.[136]

　　由此，蘇寧公司結合市場需求，正式推出線上線下結合的雲商模式，透過組織變革、實施線上線下同價，開始了 O2O 的全面融合。原有的家電、3C、消費類電子、生活電器等品類已開始全面實施線上線下的融合。與之相適應，2013 年 9 月，蘇寧公司推出雙線開放平臺「蘇寧雲臺」，10 月戰略投

資 PPTV，並在 11 月舉辦首屆 O2O 購物節，在矽谷成立研究院，進軍海外。
[137] 2014 年 2 月，蘇寧公司又獲得國際快遞業務經營許可，成為國內電商企業中第一家取得國際快遞業務經營許可的企業。

As a result, in combination with market demand, Suning Company officially launched the cloud business model of combining both online and offline business, and began the comprehensive integration of O2O through organizational change and keeping the online and offline prices the same. The original home appliances, 3C, consumer electronics, and consumer electronics began to fully implement online and offline integration. In line with this, in September 2013, Suning launched a two-line open platform "Suning Yuntai", strategically invested in PPTV in October, held the first O2O shopping festival in November, and established a research institute in Silicon Valley, successfully entering overseas areas.137 In February 2014, Suning Company obtained the international express business license, becoming the first domestic e-commerce company to obtain such license.

蘇寧雲商主要透過 3 件大事實現轉軌：[138]

Suning Commerce Group mainly achieved transition through three major events:[138]

---

137 【O2O 案例】蘇寧：O2O 戰略方向、機會、挑戰．中國電子商務研究中心 [The O2O Cases] Suning: Strategic Direction, Opportunity, Challenge of O2O. China Electronic Commerce Research Center

138 蘇寧轉型進入深水區 O2O 帶來機會和挑戰．商業價值 Suning Transforms into Deep Water Area O2O Brings Opportunities and Challenges. Business Value

1・破除組織壁壘 Break down organizational barriers

蘇寧公司原先採用三級矩陣式管理體系，其最上一級為 14 大管理中心。經過調整，現在組織結構為：線上線下 2 大開放平臺、3 大經營事業群、28 個事業部、60 個大區組成了蘇寧雲商的新架構。

At first, Suning used a three-level matrix management system, of which the top level was 14 major management centers. After adjustment, the current organizational structure is composed of 2 online and offline open platforms, 3 major business groups, 28 business divisions and 60 large regions.

在組織架構方面：蘇寧雲商新增連鎖平臺經營部、電子商務經營總部、商品經營總部。

In terms of organizational structure: Suning Commerce Group added the chain platform operation department, the e-commerce headquarter and the commodity management headquarter.

電子商務經營總部下設 8 大事業部：網購、移動購物、本地生活、商旅、金融產品、數位應用、雲產品和物流。業務類型包括實體商品經營、生活服務、雲服務和金融服務。物流事業部納入電子商務經營總部，支援小件商品全國快遞服務。

The e-commerce headquarter consists of 8 business divisions: online shopping, mobile shopping, local life, business travel, financial products, digital applications, cloud products and logistics. Business types include physical commodity operations, life services, cloud services and financial services. The Logistics Division is incorporated into the e-commerce headquarter to support national express delivery services for small items.

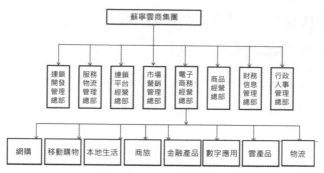

圖 Figure5-7 蘇寧雲商集團組織結構圖 Organization chart of Suning Commerce Group

2・ 破除價格壁壘 2.4.2 Break the price barrier

蘇寧公司原來一直以實體門店為主,線上線下不同價。現在打通了線上線下,破除價格壁壘,實行線上線下價格一致。為此,主要做了以下嘗試:

Suning Company has always centered on physical stores, with different prices online and offline. Now it has eliminated the price difference broke the price barrier, and implemented the same prices both online and offline. To this end, the main attempts made by Suning are as follows:

2012 年「8·18」電商價格戰期間,蘇寧公司在北京以 3C 品類為突破口,試點線上線下同價。此次試水,使蘇寧在銷售等方面的總體增長達到 4 - 5 倍。

During the 「8·18」 e-commerce price war in 2012, Suning Company took the 3C category as a breakthrough and piloted the same prices online and offline in Beijing. The pilot work enhanced the overall growth in sales and other aspects 4 to

5 times.

2013 年 6 月 8 日開始，蘇寧公司全國所有門店、樂購仕門店商品與蘇寧易購實現同品同價，這是全國首例大型零售商全面推行線上線下同價。2013 年 11 月 11 日 O2O 購物節持續 4 天，不僅實現了商品、價格、服務、支付等重要環節全部線上線下同步，而且將電商大戰推向了一個高潮。

Starting from June 8, 2013, Suning's all stores, Laox stores and Suning Yigou stores throughout the country achieved the same prices, which was the first case of a large-scale retailer in the country fully implementing the same prices online and offline. On November 11, 2013, the O2O shopping festival lasted for 4 days, which not only realized the synchronization of important links such as commodities, prices, services and payment online and offline, but also pushed the e-commerce war to a climax.

3. 破除體驗壁壘 Break the experience barrier

蘇寧公司過去的實體門店一直以專業賣場形式、以銷售產品為主。轉型雲商後蘇寧公司對外公布開放平臺，將原先純粹銷售功能的店面，升級為集展示、體驗、物流、售後服務、休閒社交、市場推廣為一體的新型互聯網化門店。比如，全店開通免費 WiFi、實行全產品的電子價籤、佈設多媒體的電子貨架，利用互聯網技術收集分析各種消費行為，推動實體零售創新。同時，蘇寧公司大力改造超級店，詮釋全新的 O2O 商業模式，實現線上蘇寧易購和線下實體店的融合，實現產品豐富度與購物體驗的全面提升，在銷售、服務、體驗上形成嶄新且更具活力的商業模式。

In the past, Suning's physical stores were mainly specialty stores. After the transformation into cloud business, Suning announced to open the platform to upgrade the original stores of pure sales into a new Internet-based store integrating display, experience, logistics, after-sales service, leisure and socialization and marketing. For example, all stores provide free WiFi, electronic price tags of all products and electronic shelves with multimedia, using the Internet to collect and analyze various consumer behaviors and promote innovation in physical retail. At the same time, Suning Company vigorously renovates their Expo Superstores in order to interpret the new O2O business model, realizing the integration of online and offline stores and the comprehensive improvement of product richness and shopping experience, and forming a new and vital business model in sales, service and experience.

2013 年中秋節期間，蘇寧上海五角場 EXPO 超級店做為第一家改造超級店全新開業。該店位於蘇寧生活廣場 1 － 5 層，營業面積 1.5 萬平方米，商品 SKU 近 10 萬，以彙集全品類、融合全管道、服務全客群為目標，著力打造「一站式」體驗的購物中心。

During the Mid-Autumn Festival in 2013, as the first renovated super store, Suning EXPO Superstore in Wujiaochang, Shanghai opened. The store is located on the 1st to 5th floors of Suning Life Plaza, with a business area of 15,000 square meters and a product SKU of nearly 100,000. It aims to create a 「one-stop」 shopping centre by bringing together all categories, integrating all channels and serving the entire customer base. center.

圖 Figure5-8　蘇寧超級店現場早教迪士尼英語 On-site early education of Disney English in Suning Expo Superstore

（五）盒馬鮮生做了什麼？
What did Freshhema do?

　　盒馬鮮生是阿里旗下新零售的標竿項目，也是阿里巴巴對線下超市完全重構的新零售業態。這種新業態既不是超市，不是便利店，不是餐飲店，也不是菜市場，但又是超市＋便利店＋餐飲店＋菜市場＋物流 +APP，根據阿里巴巴集團 CEO 張勇對新零售的詮釋，盒馬鮮生是在大數據驅動下完成人、貨、場的重構，產生化學反應，形成新的消費價值和體驗的零售業態和品牌。據華泰證券 2016 年 12 月研報，盒馬上海金橋店 2016 年營業額約 2.5 億元，坪效約 5.6 萬元，遠高於同業平均水準（1.5 萬元）。2017 年末盒馬鮮生已有22 家店鋪。據盒馬公司 CEO 侯毅表示，盒馬營業時間超過半年的門店基本實現盈利。

Freshhema is the pilot project of Ali's new retail and is also a new retail format of Alibaba's complete renovation of offline supermarkets, which is not a supermarket, a convenience store, or a vegetable market, but a combination of the above, logistics and APP. According to the new interpretation of retail by Zhang Yong, CEO of Alibaba Group, Freshhema is a retail format and brand that completes the reconstruction of people, goods and fields under the drive of big data, which forms new consumption values and experiences. According to the research report of Huatai Securities in December 2016, Freshhema Shanghai Jinqiao Store had a turnover of about ¥250 million in 2016 and an area efficiency of about ¥56,000, which was much higher than the industry average of ¥ 15,000. At the end of 2017, there were 22 Freshhema stores. According to Hou Yi, CEO of the company, stores that have been operating for more than half a year were generally profitable.

本書主要介紹盒馬鮮生 3 方面較獨特的零售創新。

This book mainly introduces the unique retail innovation of Freshhema in three aspects.

1 · 大數據支撐的線上線下一體化
Online and offline integration supported by big data

盒馬網路科技有限公司與其說是一家零售企業,還不如說是一家業務龐雜的科技創新企業。公司業務範圍涵蓋了從電腦網路到進出口、票務代理、餐飲服務等方方面面,盒馬鮮生總部員工中,有一半是軟體技術開發人員。

擁有強大的資料支援能力，使其能精準定位消費者、精準選址、精準選擇消費者需求最集中的商品。

Freshema Network Technology Co., Ltd. is not so much a retail enterprise, but rather a technology innovation company with a complex business. The company's business scope covers everything from computer networks to import and export, ticketing agents, catering services, etc. Half of the employees are software developers. With strong data support capabilities, the company can accurately locate consumers and sites, and select the products with the most concentrated consumer demand.

在消費物件的選擇上，透過大數據分析，盒馬公司認知到，有一定經濟基礎的 80/90 後，特別是 80/90 後女性用戶是盒馬主要的消費群體。男性用戶在線上線下融合的採購場景中熱情較高。線上下圍繞家庭場景消費時，多以男性用戶主導買單行為。從消費能力看，中高端消費者佔比較高。為此，盒馬消費對象主要定位於 80、90 後的年輕消費群，集中在女性，偏重於中高端消費。

In the choice of target consumers, through analysis on big data, Freshhema realizes that the 80s' and 90s' users with a certain economic base, especially female users, are the main consumer group. Male users show more passion in the online-and-offline procurement scenes. When consuming in the family scene offline, male users usually dominate the decision to buy. From the perspective of consumption power, middle and high-end consumers account for a relatively high proportion. To this end, Freshhema' s target consumer group is mainly positioned in the 80s' and 90s' , especially female, focusing on high-end consumption.

在選址上，以阿里大數據為指導，盒馬鮮生能較精準地瞄準目標使用者的生活圈、消費圈，所選的商場多為中高檔精品生活廣場，周邊有寫字樓、中高端社區等配套功能。附近樓盤價格偏高，居民消費水準偏中上。盒馬鮮生北京十里堡店、雙井店的選址正是如此。

In the site selection, with the guidance of Ali Big Data, Freshhema aims at the target users' life circle and consumption circle more accurately. The selected shopping malls are mostly medium and high-grade life squares, surrounded by office buildings, high-end communities and other supporting functions. The prices of nearby estates are high and the consumption level of residents is on the upper hand. The location of Freshhema Beijing Shilibao Store and Shuangjing Store is selected by this principle.

在商品選擇上，盒馬鮮生透過 APP，Wi-Fi 探頭、電子射頻等多項技術，建立了銷售端的客戶資訊庫，再結合淘寶大數據來找到消費者需求最集中的商品來滿足消費者需求。盒馬鮮生建立了銷售端的客戶資訊庫，能夠獲取顧客的年齡、性別、購物偏好、購買頻次、評價等多項資料。他們將這些資料提供給門店運營團隊及買手團隊，對門店需要補什麼貨、什麼時候補貨、如何精準送達等提供針對性的指導。

In terms of product selection, Freshhema has established a customer information database at the sales end through APP, WI-FI probe, electronic RF and other technologies, and combined with Taobao Big Data to find the goods with most concentrated consumer demand to meet consumer demand. Freshhema has established a customer-side information base at the sales end, which can obtain data such as the customer's age, gender, shopping preferences, purchase frequency, and reviews. They provide such data to the store operation team and the buyer team and targeted guidance on when and what goods to replenish, and how to deliver them accurately.

正是以大數據為支撐實現線上線下一體化，盒馬鮮生呈現出一系列亮眼的經營資料：用戶月購買次數達到 4.5 次，坪效為傳統超市的 3-5 倍；門店線上訂單佔比超過 50%；APP 的線上轉化率 35% 左右，是傳統電商的 10 到 15 倍。[139]

It is the support of big data to achieve the online and offline integration that Freshhema presents a series of eye-catching business data: on average, every single user purchases 4.5 times a month, the area effectiveness is 3-5 times that of the traditional supermarket. The ratio of online orders is over 50%. APP online conversion rate is about 35%, which is 10 to 15 times that of traditional e-commerce.[139]

2. 聚焦「吃」的場景 Focus on the scenes of "eating"

場景聚焦是新零售變革的重要方向。圍繞目標群體，盒馬鮮生打破了傳統零售業態以商品為中心的經營理念，轉換為以消費者實際生活需求為中心的場景體驗理念。

Focusing on the scenes is an important direction for the new retail revolution.

139 盒馬鮮生究竟鮮在哪裡 . 新浪財經 .2017.07.27What is good in Freshhema. Sina Finance. July 27, 2017

Around the target group, Freshhema has broken the traditional commodity-centric operation theory and changed it into a scene experience concept centered on the actual needs of consumers.

如針對商品品類的劃分，不同於普通超市根據商品屬性的常規分類劃分商品品類，盒馬鮮生選擇根據場景進行分類，目的是產生聯動銷售作用，同時有針對性的迎合消費需求。如在早餐場景中，盒馬鮮生或是圍繞中式的餛飩、麵條、餃子、粥等單品構建相關商品體系；或是針對西式的牛奶、麵包、三明治調整商品結構。在晚餐場景中，考慮到消費者在週一至週五需要快捷，盒馬鮮生會推出半成品蔬菜、日日鮮、成品滷菜等單品；但在週末時段，盒馬鮮生則會主推高檔海鮮、牛肉、進口水果等改善生活的商品體系。

Different from ordinary supermarkets, which divide the commodity categories according to the general classification of commodity attributes, freshhema divided the classification of commodities according to the scenes, the purpose of which is to generate linkage sales, and at the same time, to cater to consumer demand. For example, in the breakfast scenes, Freshhema builds a related commodity system with the Chinese style of glutinous rice, noodles, dumplings, porridge and other foods, or adjusts the structure of goods on account of Western-style milk, bread and sandwiches. In the dinner scene, taking into account the need for consumers to eat fast from Monday to Friday, Freshhema sells semi-finished vegetables, dairy products, finished stewed dishes, etc. However, during the weekend, it promotes high-end seafood, beef and imported fruits and other commodity systems that improve living.

盒馬鮮生店內採用餐飲連動的模式，消費者下單購買海鮮後，可以到加工櫃檯稱重、選擇加工方式現場烹製，還能在現場採買各類現場加工餐飲，從品質不俗的豪華的牛排大餐到家常的黏玉米，採買後在餐臺與家人、友人聚餐。在盒馬鮮生「逛吃」已經成為眾多消費者的樂趣。據媒體反映，盒馬鮮生各店開業後均受到吃貨們的追捧，許多消費者從選購到取餐，用時超過兩個小時，如此也難以擋住吃貨們的熱情。

Freshhema store adopts the model of catering linkage. After consumers purchase the seafood, they can go to the processing counter to weigh and choose the processing method to cook the seafood on site. They can also buy all kinds of on-site processed food on site, from the high-quality luxury steak to the homemade sticky corn. After payment, they can gather with family and friends at the dining table. 「Walking and eating」 has become great fun for lots of consumers. According to the media, each Freshhema store is sought-after by the foodies. Many consumers spend more than two hours on purchasing and waiting, but their passions can still not be blocked.

3. 如何實現低售價高毛利
How to achieve low selling price and high gross profit

低零售價、高毛利率是零售企業的經營目標，但對於經營者又談何容易，透過綜合運用現金直採、線上線下聯採、運營管控等措施，盒馬鮮生在生鮮經營中已摸索出自己的路，在保障高毛利率的同時有效地將零售價控制在合理的水準。

Low selling price and high gross profit are the business objectives of retail enterprises, but at the same time it is no easy to achieve. Through comprehensive measures such as direct procurement in cash, online and offline procurement, operation control, Freshhema has carved its own path in fresh food business. It effectively guarantees the high gross profit margin while keeping the retail price at a reasonable level.

在採購上，盒馬鮮生採用直採體系，如針對波士頓龍蝦等大海鮮品類，實現全球直採；針對常規商品、自自有品牌等品類，執行全國統採。同時還根據門店所在區域，引進當地供應商，增強生鮮品類，當地語系化商品的差異性與性價比。

In terms of procurement, Freshhema adopts the direct procurement system. For example, the company directly procures major seafood such as Boston lobsters from foreign countries and uniformly procures general and self-owned commodities nationwide. At the same time, according to the area where the store is located, local suppliers are introduced to enhance the diversity of fresh foods, and variance and cost-effectiveness of localized products.

在生鮮品的購銷上，盒馬鮮生以技術、標準和資料支援，盡可能降低損耗，控制成本。以「日日鮮」品類為例，盒馬鮮生在選擇供貨基地時通常會輸出相關標準，比如要求供應商在種植地附近建立生產車間，車間配備冷鏈溫控、接入 WiFi 探頭、預包裝生產線等多種設施。而冷鏈運輸車則必須有溫度感應器、GPS 等裝置，以此實現將蔬菜收割到門店上架的間隔控制在 18 個小時之內。同時，因為「日日鮮」只賣一天，所以盒馬省去了大量的保鮮、

包裝環節，在壓縮成本方面效果顯著。

In the purchase and sale of fresh products, Freshhema is supported by technology, standards and data to minimize losses and control costs. Take the 「Daily Fresh」 category as an example. Freshhema always sets the relevant standards when selecting the supply base. For example, the supplier is required to establish a production workshop near the plantation site and the workshop must be equipped with cold chain temperature control, Wi-Fi probes, pre-packaging production lines and other facilities. The cold chain transporter must be equipped with a temperature sensor, GPS and other devices to ensure the interval from harvesting to sale is within 18 hours. At the same time, 「Daily Fresh」 category is on sale for only one day, so that Freshhema saves a lot of links of fresh-keeping and packaging, which significantly reduces costs.

在防損層面，盒馬鮮生則會利用精準推送等方式處理積壓單品。例如一旦某天訂貨有問題，突然有兩百克裝的幾份豬肉賣不掉。普通超市通常採用打折處理，但是到店客群畢竟有限，就會出現損耗。盒馬鮮生則會用大數據篩選有豬肉消費意向的顧客，推送一條優惠資訊給他，基本上幾分鐘就能夠處理完。[140]

In terms of loss prevention, Freshhema uses the precision push to get rid of inventory. For example, something is wrong with the orders one day and some

---

[140] 盒馬「惡補」商品力：大數據選品、產地直採、自有品牌，它與「傳統零售企業」差距越來越小 . 今日零售 .2018-01-07 Freshhema Fortifies the Commodity Power Badly: Big Data Selection, Direct procurement, Own Brands, Its Gap with Traditional Retailers is Getting Smaller and Smaller. Today's Retail. Jan 7, 2018

200-gram pork doesn' t sell. In such case, ordinary supermarkets usually provide discounts. However, because of limited number of customers, there still will be losses. For Freshhema, it will use big data to screen customers with the intentions to consume pork and push a special offer to them, which can be processed in a matter of minutes.[140]

（六）上品折扣做了什麼？
What did Shopin do？

上品折扣（Shopin）是中國都市型百貨折扣連鎖品牌，在北京地區有王府井、首體、五棵松、中關村、朝陽門等 8 家實體店和 1 家電子商務網站上品折扣網，包括 600 餘個國內外知名品牌、近 10 萬款商品。由於商品大多是國內外名牌專櫃正品，其目標市場定位於以年輕人為主的、追求時尚和品牌、講究提高生活品質和崇尚理性購物的大眾消費群體。

Shopin is an urban chain department store in China. In Beijing, there are 8 physical stores in Wangfujing, Shouqu, Wukesong, Zhongguancun and Chaoyangmen, etc. and an online discount network, on which sells more than 600 well-known brands at home and abroad, and nearly 100,000 goods. Since most of the goods are authentic ones, the market is targeted at the mass-consumer group that focuses on young people, pursues fashion and brand, pays attention to improving the quality of life and advocating rational shopping.

1·建立零庫存、虛實一體化的電子商務模式 Establish an e-commerce model of zero inventory and integration of the virtual and the real

2009 年 5 月，上品折扣官方購物網站——上品折扣網成功上線，這是中國第一家百貨連鎖折扣店的官方購物網站。

In May 2009, Shopin's official shopping website, Shopin Online was successfully launched. This is the official shopping website of the first chain department store in China.

2010 年 4 月，上品折扣網系統升級，多店購物系統上線。上品折扣網依託現有上品折扣實體店連鎖體系，線上、一體化、全方位整合上品折扣多店、全品類、全品牌、全庫存商品資源，最大化吸納實體賣場資源，讓賣場在進行實體行銷的同時發揮出「網路庫房」的優勢。從而實現了線上線下統一資料庫、統一庫存管理，上品折扣網與上品折扣連鎖系列門店同步銷售商品，網站上所有的商品均來自於上品折扣實體店，每一個網路訂單中的貨品均由各個系列門店供貨，創造了全新的實體店 & 網路同步行銷的經營模式，也是中國首家由實體店供貨的品牌折扣網路商場，真正實現了零庫存、虛實一體化的電子商務模式。

In April 2010, Shopin Online underwent asystem upgrade and the multi-store shopping system was launched. Shangpin Online relies on the chain system of physical stores and integrates its multi-store, all-category, full-brand and all-inventory commodity resources of in an online, integrated and all-round manner, maximizing the absorption of resources in physical stores and allowing the store to play the advantage of "network warehouse" while conducting physical marketing, which realizes the unified database and unified inventory management both online and offline. Shopin Online sells good concurrently with physical stores, which means all the goods on the former are from the latter, creating a new business model of physical store plus network synchronization marketing. Besides, Shopin Online is also the first online discount mall in China, goods of which are supplied by physical stores, truly realizing the e-commerce model of zero inventory and integration of the virtual and the real.

2· 用自主品牌 PDA 進行商品數據採集和銷售
Collect product data and sell products using self-owned PDA

2011 年，上品折扣開發了自主品牌 PDA 用於銷售環節，借助移動互聯網終端來加強使用者體驗。上品折扣 PDA 主要用於解決商品資料的採集和現場的物品銷售。上品折扣的 6000 名銷售員，每個人的資料都對接到同一個系統中。

In 2011, Shopin developed a self-owned PDA for sales in order to enhance the user experience with terminals of mobile Internet. The PDA is mainly used to collect commodity data and sell goods on site. Of all the 6,000 salespersons, each one's data is connected to the system.

3· 與支付寶、微信的戰略合作
Strategic cooperation with Alipay and WeChat

2012 年 8 月 23 日，上品折扣與支付寶聯合推出一項新穎的商場 O2O 購物服務。2013 年 12 月，上品折扣又攜手微信支付開展促銷活動。在中關村店舉行的 50 元購買大牌商品的促銷活動中，雙方基於移動互聯網和 O2O 應用在支付結算、行銷推廣等方面展開了合作，是國內首家使用微信支付的百貨商場。活動持續 15 天，日均交易超過 1000 筆，日交易額 24 萬元，上品折扣微信支付總交易額近 100 萬元。

On August 23, 2012, Shopin and Alipay jointly launched a new shopping mall O2O shopping service. In December 2013, Shopin, together with WeChat Payment, carried out promotional activities. In the promotion of buying big-name products with RMB¥50 held in Zhongguancun store, the two sides started cooperation on payment and settlement, marketing promotion, etc. based on mobile Internet and O2O applications, and the store was also the first department store in China to use WeChat payment. The activity lasted for 15 days, the number of daily average transactions exceeded 1000, daily transaction volume reached ¥240,000, and the total transaction volume of the Shopin WeChat payment was nearly ¥1 million.

4· 開設全國首家微信體驗店
Opened the first WeChat experience store in China

2014 年 4 月 25 日，上品折扣全國第一家微信體驗店在杭州下沙區開業，真正打通線上線下，實現一體化經營。

On April 25th, 2014, the first WeChat experience store of Shopin in China was opened in Xiasha District, Hangzhou, and it truly achieved integrated operation of online and offline.

上品折扣微信體驗店裡沒有一個收銀臺，實體門店、網上商店和微信掃碼購物等終端管道整合在同一套庫存系統，永不打烊的「虛擬購物牆」24 小時展示著上品折扣的精選商品和相應的二維碼，消費者可以透過互動螢幕，直接用微信掃碼購買所有商品，不用去收銀臺排隊，而且顧客的購物場景可擴展到任何地方。

There is no cashier at the WeChat experience store. The terminal channels such as physical stores, online stores and WeChat code-scanning shopping are integrated in the same inventory system, and the virtual shopping wall that never rests is always displaying goods and the corresponding QR codes 24 hours a day so that consumers can use the interactive screen to directly purchase the goods by scanning the codes without going to the cashier, which means that shopping scene can be extended to any place.

在運營創新和會員體系方面，上品折扣微信體驗店推出了「微秒殺」活動。透過價格優惠和商品的稀缺性調動消費者的興趣，並且形成微信掃碼的支付習慣。

In terms of operational innovation and membership system, Shopin WeChat experience store launches the「micro-seckill」campaigns. Through the discounts and the scarcity of goods, such campaign arouses consumers' interest and forms the payment habit of scanning WeChat code.

上品折扣微信體驗店鼓勵消費者每天透過微信掃碼進行簽到，並為此制訂了專門的會員成長體系。同時，透過「上品折扣杭州」的微信公眾帳號，

消費者可以獲得從查詢、下單、支付到提貨、發貨、物流、退換貨在內的功能和消息回饋，瞭解商品的每一步動向。購物環節中的分享也可以透過微信完成。消費者可以把掃碼生成的訂單分享給親友，邀請對方幫自己參考或請對方幫忙付款都可以。如果對方感興趣，也可以自己透過微信直接購買。[141]

Shopin WeChat experience store encourages consumers to sign in by scanning WeChat code every day, and has developed a special membership growth system for this purpose. At the same time, through the WeChat official account of 「Shopin Hangzhou」, consumers can enjoy the functions and receive feedback of inquiry, order, payment, delivery, logistics, return and exchange, and track every step of the purchase. Shopping sharing can also be done via WeChat. Consumers can share the orders generated by the Wechat code with relatives and friends, invite them to give advices or ask them to pay. If the invited party shows interest, he or she can also palce the same order directly through WeChat.[141]

圖 Figure 5-9　消費者在商品折扣微信體驗店中體驗微信支付購物 Consumers experience WeChat payment shopping in the WeChat experience store

---

141 上品折扣微信體驗店杭州開業 未來商店雛形初現. 驅動中國 Shopin WeChat Experience Store Started in Hangzhou: Future Store is Beginning to Appear. Qudong.com

（七）萬達廣場做了什麼？
What did Wanda Plaza do?

　　萬達廣場是由著名的地產企業——萬達集團在國內各地投資興建的綜合性商業項目，一般包括購物中心、娛樂中心和城市公寓。近年來，萬達廣場發展速度較快，到 2013 年年底全國已經開業的萬達廣場達 85 個；2014 年計畫開業 24 個，到 2014 年年底，開業的萬達廣場達到 109 個，持有物業面積規模超過 2203 萬平方米，成為全球最大的不動產企業。

　　Wanda Plaza is a comprehensive commercial project invested by Wanda Group, a well-known real estate company, in various parts of the country, generally including shopping centers, entertainment centers and urban apartments. In recent years, Wanda Plaza has developed rapidly. By the end of 2013, there have been 85 Wanda Plazas in the country. In 2014, 24 mrore were planned to open. By the end of 2014, the number reached 109, and the size of the property held exceeded 22.03 million square meters, making it the world's largest real estate company.

　　2001 年以來的 10 餘年間，萬達廣場的建築形式和業態結構經歷幾次升級，已形成業態多元，包括購物、休閒、餐飲、文化、娛樂等多種功能，集大型商業中心、商業步行街、五星級酒店、商務酒店、寫字樓、高級公寓等多種商業設施於一體的大型城市綜合體。

　　In the more than 10 years since 2001, Wanda Plaza's architectural form and business format have undergone several upgrades, and it has formed a variety of formats, including shopping, leisure, catering, culture, entertainment and other functions, integrating large-scale commercial centers, commercial pedestrian streets, five star-rated hotels, business hotels, office buildings, and high-end apartments into a large-scale urban complex.

做為百餘家萬達廣場物業的管理者，萬達集團旗下萬達商業地產公司透過初步實施購物中心大數據策略，對全國龐大體量的購物中心實施有效管理，由此向「智慧廣場」建設邁出實質性的步伐。

As the property manager of more than 100 Wanda Plazas, Wanda Commercial Real Estate Co., Ltd., a subsidiary of Wanda Group, implemented the strategy of shopping center big data to implemented effective management of the huge shopping malls in China, thus establishing the 「Smart Plaza」.

1. 開展 7 個方面的大數據管理
Carry out big data management in seven aspects

（1）對租賃的全過程進行資料化管理
Data management of the entire process of leasing

對商戶從進場開始到退場整個過程中所有團隊的變化、進出貨的變化及其各個時間段、各個季節的銷售情況，以及租賃的全過程進行資料化管理。

Implement data management of all the team changes, the changes of the incoming and outgoing shipments, the sales of each time slot and season, and the whole process of the lease from the start to finish of merchants.

（2）對所有品牌建檔管理 File management of all brands

根據消費者年齡層、消費額及客流曲線進行品牌定位，將品牌精確分類，

為未來大數據的分析提供依據和分類，將購物中心佈局調整至最合理狀態。

Conduct brand positioning according to the age groups, consumption amount and passenger flow curve and accurately classify the brands, which provides the basis and classification for the analysis of big data in the future. In addition, adjust the layout of the shopping center to the most reasonable state.

（3）對城市的所有資訊進行統計

Statistics on all information about the city

盡可能拿到諸如政府真正的統計資料、區域內的人口、區域內的 GDP 等，以及相關政策在這個區域內的動向等資料資訊。

Get as much data, such as the government's real statistics, the population, the GDP and relevant future policies within the region, as possible.

（4）POS 交易紀錄 Transaction records on POS

所有在萬達廣場經營的商戶都安裝 POS 機。幾乎所有商戶在不分時段、不同位置、不同業態的銷售資料，最終可以合併到大數據的資料庫中進行處理。

All merchants operating in Wanda Plaza have equipped with POS machines. Sales data of almost all merchants in different time slots, locations and formats can be merged into the database of big data for processing.

（5）客流監控採集 Passenger flow monitoring acquisition

從 3 個層段分析消費者，建立廣場策略。第一層段是統計進出廣場客流量；第二個層段是分區域、分業態進行客流資料統計（萬達廣場做了人臉識別攝像頭的統計，識別率非常高）；第三個層段是在每一家經營的店鋪做客流資料的統計。

Wanda Plaza analyzes consumers from three levels and establish a plaza strategy. The first level is the inbound and outbound passenger flow statistics; the second is the passenger flow data statistics in region and format (Wanda Plaza has the face recognition camera statistics with high recognition rate); the third is the passenger flow data statistics made in each store.

（6）消費者 WiFi 跟蹤 WiFi tracking of consumers

在整個廣場搭建大 WiFi 和大會員體系，透過 WiFi 體系捕捉廣場裡面所有的智慧手機用戶，瞭解其行走路線、他所關注的商品和他的消費習慣，對接會員體系，由此掌握消費者需求及相關產品喜好。

Wanda Plaza builds large WiFi and membership systems in the whole plaza, capturing all the smartphone users in the plaza through the WiFi system to understand the walking route, the products they care about and their consumption habits. Wanda Plaza docks with the membership system to grasp the consumer demand and related product preferences.

（7）建立大會員體系 Establish a large membership system.

綜合所有有效資料合併到大數據的資料庫裡進行處理，這是萬達廣場建立大數據管理的基礎，也是萬達廣場全數據模式的基礎。

The integration of all valid data into the database of big data for processing is the basis for the establishment of big data management in Wanda Plaza, and also the basis of Wanda Plaza's full data model.

2· 分片區、分業態實施大數據策略

Implement big data strategies by region and format

（1）分片區管理 Management by region

近百家萬達廣場分屬不同的省份、不同的城市，經濟結構、消費結構不盡相同。根據大數據思維，萬達商業地產公司將其分為七大片區：東北、華北、西北、華中、華東、華南和西南，並針對不同區域市場予以差異化管理。

There are nearly one hundred Wanda Plazas in different provinces and cities, and their economic structure and consumption structure are not the same. According to big data thinking, Wanda Commercial Properties Co., Ltd. divides itself into seven major regions: Northeast China, North China, Northwest China, Central China, East China, South China and Southwest China, and differentiates them for in management.

（2）大業態管理 Big format management

萬達商業地產公司將萬達廣場業態分為四大類：服裝、餐飲、精品（與生活服務相關項目）和體驗（娛樂體驗類項目），並根據不同業態進行針對性的資料分析研究和管理。

Wanda Commercial Properties Co., Ltd. divides formats of Wanda Plaza into four categories: clothing, catering, quality goods (life-related services) and experience (entertainment experience projects), and conducts targeted data analysis and management according to different formats.

萬達廣場體系內現有 497 個服裝品牌，涵蓋 15 個品類，店鋪有 3005 個，總經營面積約 82 萬平方米。在對服裝零售系統進行資料分析的基礎上，萬達

商業地產公司得出近年服裝零售產業的發展趨勢：快時尚、淑女和戶外休閒將成為未來一段時間增長的主力軍；量販休閒少女裝和設計師品牌由於進入的門檻較低，出現了比較明顯的下滑，但不代表這個業態是不可以選擇的，對具體品牌要做慎重選擇。

There are 497 clothing brands in the Wanda Plaza system, covering 15 categories, with 3005 stores and a total operating area of about 820,000 square meters. Based on the data analysis of the clothing retail system, Wanda Commercial Properties Co., Ltd. has drawn the development trend of the clothing retail industry in recent years: categories like fast fashion, ladies and outdoor leisure will become the main force for future growth; girlish casual and designer brands, due to the low entry barriers, have experienced a significant decline. However, it does not mean that the format is undesirable as we must make careful choices for specific brands.

萬達廣場體系內現共有 1034 個餐飲品牌，涵蓋 65 個品類或者是菜系，共有 2459 個店鋪，總經營面積達 92 萬平方米。在對餐飲服務系統進行資料分析的基礎上，萬達商業地產公司認為其餐飲系統整體坪效較好。按區域對增長突出的店鋪進行分析，得出的結論是西餐、茶餐、日式料理出現了正增長，這體現了現代消費者的消費趨勢；而前幾年在全國都非常火爆的韓式料理和越南菜出現了雙向負增長。休閒餐飲日漸受到消費者青睞，具有鮮明民族特色的菜系生命週期偏短。

There are 1,034 catering brands in the Wanda Plaza system, covering 65 categories or cuisines, with a total of 2,459 stores and a total operating area of 920,000 square meters. Based on the data analysis of the catering service system,

Wanda Commercial Properties Co., Ltd. believes that its catering system has a good overall area-effectiveness. According to the analysis of the shops with outstanding growth by region, Western food, tea meals, and Japanese cuisine have seen positive growth, which reflects the modern consumption trend. In the past few years, the Korean and Vietnamese cuisines were very popular in the country. However, there has been a two-way negative growth in both of them. Casual catering is increasingly favored by consumers, and the life cycle of cuisines with distinctive ethnic characteristics is short.

萬達廣場體系內現共有生活精品類品牌 468 個，分為 14 個品類，共有 1794 個，總經營面積達 26 萬平方米。對其分析的結果是：絕對坪效比較突出，但是個別品類已經開始出現了負增長。在對各個區域的資料分析之後，發現生活化的東西越來越符合消費者的消費趨勢和消費習慣；個人護理、創意禮品和時尚表開始出現了坪效和銷售額的雙增長；前幾年進入購物中心的數碼店和眼鏡店則出現了負增長。

There are 468 lifestyle boutique brands in the Wanda Plaza system, which are divided into 14 categories, with a total number of 1794 and a total operating area of 260,000 square meters. The analysis result is that the absolute area-effectiveness is prominent while a few individual categories have begun to show negative growth. After analyzing the data of various regions, it is found that the lifestyle goods become more and more consistent with consumption trends and consumption habits. Personal care, creative gifts and fashion watches have begun to show growth in both the area-effectiveness and sales. Digital stores and optical shops entering the mall experiences negative growth.

萬達廣場體系內現共有體驗類品牌 239 個，涵蓋 20 個品類，經營面積 110 萬平方米。近期資料分析顯示：兒童相關業態（特別是兒童培訓、兒童攝影、教育、遊樂等）均有比較好的表現，坪效和銷售額兩個方面都有比較好的增長，顯示出小手拉大手的經濟依然有非常大的發展空間，也應該是購物中心下一步經營的趨勢。[142]

There are 239 experience brands in the Wanda Plaza system, covering 20 categories with an operating area of 1.1 million square meters. Recent data analysis shows that children-related formats (especially training, photography, education, amusement, etc.) have relatively good performance with good growth in both area-effectivenes and sales, showing the parent-child economy still has a very large space for development, and should be the next step in the operation of shopping centers.[142]

（八）大悅城做了什麼？
What did Joy City do?

大悅城集大型購物中心、甲級寫字樓、服務公寓、高檔住宅等為一體，是中糧集團城市綜合體的核心品牌。集聚購物、娛樂、觀光、休閒、餐飲等多種功能，以年輕、時尚、潮流、品位為定位的大悅城，已成為中國高品質

---

142 解析萬達廣場如何實施購物中心大數據策略 . 中國連鎖經營協會官方微博 .2014-05-28

城市生活的新標誌。目前，除了在北京朝陽、西單，瀋陽中街、上海蘇河灣、天津內環開設的 5 家大悅城，2014 年、2015 年兩年煙臺大悅城、成都大悅城也即將揭幕。

Joy City is a core brand of COFCO Urban Complex, which integrates large shopping malls, Grade-A office buildings, service apartments and high-end residences. Gathering shopping, entertainment, sightseeing, leisure, catering and other functions, Joy City, which is young, fashionable, trendy and taste-oriented, has become a new symbol of high-quality urban life in China. At present, in addition to the five Joy Citys opened in Chaoyang and Xidan in Beijing, Zhongjie in Shenyang, Suhewan in Shanghai and Tianjin Inner Ring, Yantai Joy City and Chengdu Joy City were unveiled in 2014 and 2015 repectively.

2013 年「雙十一」期間，全國大悅城當日的銷售業績同比增長了 22%、客流量增長了 26%、單日交易筆數增長了 29%。在電商創造的節日當天，大悅城的經營業績成為實體商業的亮點。一個 2007 年才開出第一家購物中心的企業，為什麼會在短短的幾年間在中國購物中心領域形成響亮的品牌？應該說，大悅城的成功離不開其針對消費者需求所實施的全面的「大悅城體驗」。每年大悅城都會請尼爾森、零點等公司開展細緻的消費者滿意度調查，將消費者體驗細緻到近百個指標上，並力求透過對這些指標的改進來完善購物體驗。

During the 「Double Eleven」 period in 2013, the sales performance of Joy City nationwide increased by 22%, passenger traffic increased by 26%, and the number of single-day transactions increased by 29% year-on-year. At the festival created by e-commerce, Joy City's business performance became the highlight of all physical business. How come a company that opened its first shopping mall in 2007 forms a best-known brand in the field of Chinese shopping mall in just a few years? It should be said that Joy City's success is inseparable from its comprehensive 「Joy City Experience」 for consumer demand. Every year, Joy City invites companies such as Nielsen Company and Horizon Research Group to conduct detailed consumer satisfaction surveys, detailing the consumer experience to nearly one hundred indicators to improve the shopping experience by improving these indicators.

大悅城體驗主要體現在以下 4 個維度：

Joy City experience is mainly reflected in the following four dimensions:

1. 建築及環境體驗
Architectural and environmental experiences

大悅城設計別具匠心。其對空間舒適性、節奏感和柔軟度的把控有獨到之處，如高聳的建築挑空及更加開闊軒敞的公共空間、跨越多層的飛天梯、更多豐富趣味的景觀等，在最大化購物便利性和舒適度的同時，打造室內動線節點上的地標型景觀，為購物中心賦予了「悅」的溫度。

Design of Joy City design shows originality. Its thinking on space comfort, rhythm sensation and softness is unique, such as towering atriums, wider public spaces, multi-storey flying ladders and interesting landscapes. Such design maximizes shopping convenience and comfort while creating a landmark-like landscape on the indoor pedestrian flow nodes, giving Joy City a「Joy-ous」temperature.

在環境上，大悅城一直致力於營造舒適創意的購物空間和環境氛圍。如位於天津大悅城北區 5 層主題街區的「騎鵝公社」，由天津大悅城和創意平臺「瘋果」聯手打造，總建築面積 2000 平方米的街區，目前已成為天津最具創意的市級地標性創意市集。

In terms of the environment, Joy City has always been committed to creating a comfortable and creative shopping space. For example, the Cheer Market, located in the 5th floor of the North District of Joy City in Tianjin, is jointly created by Tianjin Joy City and the creative platform FengGuo, with a total area of 2,000 square meters, and has become the most creative city-level market in Tianjin.

透過對 2012 年大悅城購物環境的滿意度進行二級指標細分後，大悅城發現，「背景音樂適合程度」指標的滿意度比值偏低，為此，大悅城著手與專業的背景音樂管理公司合作，特別訂製了與大悅城品牌調性相一致的背景音樂曲庫，並透過同一個網路後臺向各大悅城賣場同步發佈統一的背景音樂，同時還特別為聖誕、新年、春節、情人節等打造節慶主題背景音樂，以及店慶、促銷等不同的推廣活動主題所用不同曲風的背景音樂，營造了輕鬆、歡樂的音樂氛圍。

After subdividing the secondary indicators of satisfaction degree of Joy City shopping environment in 2012, Joy City found that the satisfaction rate of the background music was low. To this end, Joy City cooperates with a professional background music management company to customize a background music library that is consistent with the tonality of Joy City, and releases unified background music to each Joy City store through the same network background. For festivals like Christmas, New Year, the Spring Festival and Valentine's Day, and different promotion activities such as celebrations and promotions, special theme musics are made to create a relaxed and joyful music atmosphere.

大悅城還嘗試將生態農場搬進購物中心。天津大悅城實景搭置近 200 平方米的「農場」，種植了茄子、草莓、人參果、朝天椒、番茄等 20 多個品類、70 多盆果蔬，展示了植物的不同生長階段。在「農場」中間還有 1 座微型牧場，3 隻小香豬、多隻松獅兔和倉鼠，與消費者定期見面。消費者用手機掃描「農場」內隨處可見的二維碼，就可以快速瞭解關於植物的詳細資訊。許多孩子和家長被小動物吸引，不少人等著抱小豬合影。由此，大悅城幫助人們體驗到久違的回歸田園的自然之感。

Joy City also tries to move the ecological farm into the shopping center. Tianjin Joy City has set up nearly 200 square meters of「farm」, where more than 70 pots of fruits and vegetables in more than 20 categories such as eggplant, strawberry, ginseng fruit, pod pepper and tomato, are planted, showing different growth stages of plants. In the middle of the「farm」, there is also a mini ranchland with three small musk-pigs, many chow-rabbits and hamsters, all of which meet regularly with consumers. Consumers can quickly find out more about plants by scanning QR codes everywhere in the farm with their mobile phones. Many children and parents are attracted by small animals, waiting to have a photo with a small pig. In this way, Joy City helps people experience the long-lost return to the pastoral nature.

2． 品牌體驗 Brand experience

大悅城的品牌體驗包括保持品牌商戶組合在市場的引領性、以豐富的業態組合滿足全方位「一站式」的消費需求、引入新興業態挖掘消費者的潛在需求、打造室內主題街區提升逛街的趣味體驗和淘寶樂趣。

Brand experience of Joy City includes maintaining the leading position of the brand merchants combination in the market, satisfying the all-round 「one-stop」 consumer demand with rich format combinations, introducing emerging formats to tap the potential needs of consumers, and creating indoor theme districts to enhance fun experience of shopping and finding great stuff.

天津大悅城 400 餘個品牌中有 20% 為天津市場獨有，如 INDITEX 旗下的 Massimo Dutti、日本人氣品牌 MOUSSY、創意香水品牌氣味圖書館、貓的天空之城概念書店、外婆家、港麗餐廳等，都是天津的城市青年客群最追捧的品牌商家。在「騎鵝公社」主題街區， 34 家店鋪包括 22 個原創設計師品牌。「70 後」熟悉的陶瓷杯，「80 後」熱愛的明信片，供「90 後」展現個性的不插電舞臺，「鎮上限速」、「單眼皮路」等個性標牌，「騎鵝公社」裡的每個角落都散發著誘人的個性和文藝氣質。

In Tianjin Joy City, 20% of more than 400 brands are unique only in the Tianjin market, such as Massimo Dutti under INDITEX, MOUSSY from Japan, creative perfume brand Scent Library, concept bookstore Mao Mi Cafe, Grandma's Home and Charme Restaurant, etc., all of which are the most sought-after brand merchants in Tianjin's urban youth group. In Cheer Market, there are 34 stores, including 22 original designer brands. Ceramic cups familiar to the post-70s, postcards loved by the post-80s, the unplugged stages for the post-90s to show their personality, and personalized signs like "town limit speed" and "single eyelid road" can all be seen in every corner of Cheer Market, which exudes a seductive personality and literary temperament.

來自日本的專為女性消費群體提供料理課堂服務的新興業態——ABC Cooking Studio 首次進入中國市場便與大悅城合作，在上海大悅城成功開設了在中國的首家店並獲得非常高的市場認可度。開業後連續 5 個月月銷售額持續增長，增長率高達 86.8%。之後又進駐了北京朝陽大悅城，銷售坪效情況一直持續在娛樂服務業態中的第一名，並帶動了相鄰商戶的銷售增長平均約 17%。大悅城透過引入新興業態和創新體驗型的品牌商戶，為消費者帶來的是更加新奇有趣的購物體驗，為商戶帶來的是更好的經營業績。幫助商戶去一同開拓新的市場，也為大悅城自身帶來更高的租金收益和商業價值。

ABC Cooking Studio, a new format from Japan, which provides cooking class services for female consumer groups, cooperated with Joy City at the first time to enter the Chinese market and, successfully opened the first store in Shanghai Joy City and achieved very high market recognition. In the next five months after opening, the monthly sales continued to grow, with a growth rate of 86.8%. Later, ABC Cooking Studio entered the Chaoyang Joy City in Beijing. The area-effectiveness of sales has continued to be the first in the entertainment service industry, and spurred the sales growth of neighboring merchants about 17% on averaged. Through the introduction of the emerging format and innovation in experience brand merchants, Joy City brings a more new and interesting shopping experience to consumers, and also better business performance to merchants.

Helping merchants to explore new markets together also brings higher rental income and business value to Joy City itself.

3 · 服務體驗 Service experience

　　大悅城相信，細緻化的、統一化的、完善的顧客服務是商業經營層面最重要的取勝因素。為此，大悅城千方百計地透過細緻化的、一致性的顧客服務、量身訂製的專屬會員服務打造服務體驗，並將新的科技技術──如室內定位系統、反向尋車系統、集中點餐排號系統等運用在大悅城為顧客所提供的服務中。

Joy City believes that meticulous, unified and perfect customer service is the most important factor in the business management. To this end, Joy City has done everything possible to create a service experience through detailed, consistent customer service and tailor-made exclusive membership services, and integrated new technology technologies such as indoor positioning system, reverse car-seeking system and food ordering system in the services provided by Joy City for customers.

　　在會員忠誠度維護方面，大悅城更加注重對會員的文化歸屬感培養，透過大悅城官方網站、大悅城智慧手機 APP、大悅城微信平臺、實體店會員中心 4 個管道的打通，建立全方位的 360° 會員服務平臺，讓會員感受到 24 小時的服務覆蓋，並提供更加訂製化的專屬服務，以及透過優惠折扣、定期舉行會員專屬活動等，提升會員對大悅城的文化歸屬感。

In terms of member loyalty maintenance, Joy City pays more attention to the cultivation of members' sense of culture belonging. Through the four channels of Joy City official website, Joy City smartphone APP, Joy City WeChat platform and member center of physical store, Joy City establishes a 360-degree member service platform, letting members feel the 24-hour service coverage. It also provides more customized and exclusive services, and enhances members' sense of culture belonging of Joy City by discounts and regular member-only activities.

4· 營銷體驗 Marketing experience

大悅城的行銷體驗包括對全國超過 75 萬會員的精準行銷、大悅城的新媒體行銷和不斷推陳出新的主題推廣活動。如幾米、憤怒的小鳥、HELLO KITTY、哆啦 A 夢、麥兜、漫威英雄聯盟等主題景觀展，萬聖節、美食節、聖誕檔期等年度固定的主題推廣互動，以及「大悅城 JOY 24 小時傳遞正能量」等品牌公關活動等，大悅城每年舉辦大型主題活動 50 餘場、各類小型活動百餘場。如 2014 年 4 月 17—6 月 22 日期間，北京朝陽大悅城舉辦的「100 哆啦 A 夢祕密道具博覽」。100 隻 1:1 比例的哆啦 A 夢攜帶願望實現機、竹蜻蜓飛行器、隨手取物提包等百種祕密道具現身朝陽大悅城，吸引了大批看著哆啦 A 夢長大的年輕人及其他們的小家庭。

Marketing experience of Joy City includes precision marketing for over 750,000 members nationwide, new media marketing and ongoing theme promotion. For example, there are thematic exhibitions of Jimmy the cartoonist, Angry Birds, HELLO KITTY, Doraemon, McDull and Marvel Heroes, promotions with fixed themes like Halloween, Gourmet Festival and Christmas, and brand public relations activities like 「JOY 24-hour positive energy passing」. Joy City holds more than 50 large-scale theme activities and more than 100 small-scale events every year. For example, from April 17 to June 22 in 2014, the 「100 Doraemon Secret Props Expo」 was held in Beijing Chaoyang Joy City. A hundred 1:1 Doraemons, carrying a variety of secret props such as the wish-fulfillment machine, the bamboo copter, and the wonderbag, appeared in Chaoyang Joy City, attracting a large number of young people who grew up watching Doraemon with their families.

大悅城富於創意的品牌活動，適應年輕人的生活方式、娛樂方式，讓年輕人感受到快樂。透過舉辦活動，讓更多年輕人愛上大悅城，享受大悅城的體驗文化。[143]

Joy City's creative brand events adapt to the lifestyle and entertainment of young people and make young people feel happy. By organizing events, more young people fall in love with Joy City and enjoy the experience culture of it.[143]

（九）吳裕泰做了什麼？
What did Wu Yutai did?

創業於清光緒年間的吳裕泰茶莊，是中國著名的茶品經營老字型大小，經過百年發展，現在已形成擁有近 300 家門店的中型專業連鎖企業。做為一家品牌商業老字型大小，吳裕泰並未故步自封，而是在零售創新方面走出了

---

143 大悅城品牌的核心競爭力：大數據與體驗經濟. 北方網. 2013-11-27

自己的路子。

Wu Yutai Tea House, founded in the Guangxu Period of the Qing Dynasty, is a famous time-honored tea brand in China. After development for a century, it has now been formed into a medium-sized professional chain enterprise with nearly 300 stores. As a time-honored brand, Wu Yutai did not stand still, but walked out of his own path in retail innovation.

1. 創新花茶時尚，研發茶衍生品　Innovate the flower tea fashion, research and develop tea derivatives

吳裕泰公司（以下簡稱吳裕泰）主營茶葉、茶製品，傳統以銷售自拼茉莉花茶為主要特色，吳裕泰茉莉花茶「香氣鮮靈持久，滋味醇厚回甘，湯色清澈明亮，耐泡」，被消費者親切地稱為「裕泰香」。

Wu Yutai Tea House Company (hereinafter referred to as Wu Yutai) is mainly engaged in tea and tea products. It sells self-planted jasmine tea as its main feature. The jasmine tea 「has a long-lasting aroma, a mellow and sweet taste, a clear and bright soup, and is resistant to foam」. Consumers affectionately call the tea "Yutai Tea".

做為茶品經營的老字型大小，吳裕泰在面對遍及世界五大洲 160 多個國家近 30 億飲茶人的同時，看到年輕消費群體更崇尚時尚茶飲的消費趨勢和特徵。所以，吳裕泰在繼續保持茉莉花茶經營優勢的同時，積極研發新茶，力求使茶產品呈現多元化，確立了從內容到形式創新茶文化的吳裕泰發展方向。為此，吳裕泰在茉莉花茶的基礎上不斷創新花茶時尚，研發茶衍生品。2012

年，吳裕泰採用分離窨製技藝研製的花香綠茶翠谷幽蘭上市，僅 1 年時間銷售同比上漲 130%，而同期茉莉花茶銷售僅提升 10%。圍繞花茶方向，吳裕泰還推出了系列花草茶、紅茶、烏龍茶產品，如主打養生概念的桂圓紅棗茶，提神醒腦的沁心薄荷茶，馨香甘美的玫瑰紅茶、桂花紅茶、梔子紅茶，典雅清新的桂花烏龍茶等。美茶配美器，每款茶的包裝均有獨到之處，三角獨立茶包便捷時尚，受到更多年輕消費者的歡迎。

As a time-honored tea brand, Wu Yutai not only sells its tea to nearly 3 billion tea drinkers in more than 160 countries on five continents, but also notices the consumption trends and characteristics of fashionable teas for young people. Therefore, while continuing to maintain the advantages of jasmine tea, Wu Yutai actively researches and develops new teas, striving to diversify tea products and establishing the development direction of innovative tea culture from content to form. To this end, Wu Yutai continues to innovate the flower tea fashion based on jasmine tea and develop tea derivatives. In 2012, the floral green tea Cuiguyoulan (Orchid in a green valley), which was developed using the separation and scenting technology, went on sale, sales volume of which increased by 130% in the second year, while that of jasmine tea only increased by 10%. In the category of flower tea, Wu Yutai also launched a series of herbal tea, black tea, oolong tea products, such as the longan-jujube tea featured health, mint tea featured the ability of refreshing, as well as sweet and fragrant rose black tea, osmanthus black tea, gardenia black tea and elegant osmanthus oolong tea. Good teas are also served with good tea sets. Besides, each kind of tea packaging has its own unique feature. Individual tea bag in triangle is convenient to use and fashionable, and is welcomed by young consumers.

2009 年，在吳裕泰推出茉莉花茶冰淇淋時，曾一度受到很多質疑，但經過幾年的市場檢驗，證明消費者喜歡茶葉店做出來的冰淇淋，老字型大小茶莊做冰淇淋絕非不務正業。2013 年，吳裕泰前門、雍和宮等 4 家店，僅茶冰淇淋一項就賣了 103 萬個，價值 160 萬元。每年耶誕節期間，在王府井店外消費者穿著羽絨服排著長隊購買冰淇淋已成為一景。

In 2009, when Wu Yutai launched the jasmine tea ice cream, it was once questioned a lot. However, after several years of market testing, it proved that consumers like the ice cream made by the tea house. The time-honored tea house was not fooling along. In 2013, Wu Yutai' s four stores including Qianmen store and Lama Temples store sold 1.03 million pieces of tea ice cream worth ¥1.6 million. Every year, consumers wait outside the Wangfujing store wearing down jackets to buy ice cream at Christmas, which has become a regular phenomenon.

2·開展多管道經營，積極探索電子商務 Conduct multi-channel management and actively explore e-commerce

2013 年 5 月 18 日，吳裕泰在天貓商城上的官方旗艦店正式開業，標誌著吳裕泰全面開始了網路行銷體系。之後，吳裕泰陸續登錄 1 號店、京東、當當、亞馬遜等其他電商平臺。吳裕泰電商產品透過專業研發共上市 10 個系列 85 款產品，上線 6 個月銷售額就達到了 400 萬元，而且直接進到連鎖店購買產品的轉化率達到 4%，由此使吳裕泰品牌知名度在全國範圍得以迅速提升，使其在茶葉行業中的排名得到迅速提升。

On May 18, 2013, Wu Yutai opened its official flagship store on Tmall, marking Wu Yutai's comprehensive network marketing system. Later, Wu Yutai successively settled in other e-commerce platforms such as No. 1 Store, Jingdong, Dangdang, and Amazon. Wu Yutai's 85 e-commerce products of 10 series went on sale through professional R&D, the sales volume reached ¥4 million yuan in 6 months, and the conversion rate of purchasing directly in the chain stores reached 4%, thus making Wu Yutai brand popular throughout the country and its ranking in the tea industry rapidly improved.

3· 融合傳統與時尚，打造體驗化零售終端 Integrate tradition and fashion to create an experiencing retail terminal

吳裕泰是 2010 年上海世博會特許商品生產商和零售商。 2013 年吳裕泰上海旗艦店「裕泰東方」茶薈館正式亮相浦東「世博園」商業區。此後「裕泰東方」又進駐北京前門，其店鋪設計風格軒敞溫馨，色彩沉穩，以清簡文化為特色，使消費者享有溫暖、自然、舒適的感受。經典的原葉茶、各類的時尚飲品、健康美味的蛋糕、DIY 體驗，均打造出獨特的時尚消費模式。

Wu Yutai is a licensed product manufacturer and retailer for the Shanghai World Expo 2010. In 2013, Yutai Oriental the tea club, Wu Yutai's flagship store in Shanghai, officially appeared in the Expo Park business district in Pudong. Since then, Yutai Oriental has been stationed in the front door of Beijing. The store design is open and warm, and the color is calm. It features a simple culture and makes consumers feel warm, natural and comfortable. The classic original tea, all kinds of fashionable drinks, healthy and delicious cakes, and DIY experience, all create a unique and fashionable consumption model.

圖 Figure 5-10 「裕泰東方」——提供茶文化 DIY 現場體驗 Yutai Oriental–providing DIY live experience of tea culture

4· 發展自媒體，開展精準化營銷

Develop the self-media to conduct precision marketing

近年來，吳裕泰嘗試利用微博、微信、做《茗鑑》電子雜誌等自媒體手段及二維碼等技術開展精準化行銷。2011 年，開通官方微博；2012 年，開通官方微信，投拍《928 就愛吧—微電影三部曲》，受到業界的肯定；2013 年，搭建「碼上傳播行銷平臺」，新浪微博「吳裕泰中國」拓展粉絲地域，將新粉絲轉化為老茶客，每個月都有線上粉絲互動活動，不斷提升粉絲活躍度，增強粉絲黏性。吳裕泰還邀請粉絲到吳裕泰展位和門店參與茶文化體驗活動，實現線上線下的緊密連接。[144]

In recent years, Wu Yutai has tried to use Weibo and WeChat, launch the electronic magazine Tea Mirror and two-dimensional code to carry out precision

---

144 孫丹威. 靠觸網把握先機 憑創新直面挑戰. 第八屆京商論壇 Sun Danwei. Grasping the Opportunity by Turning to the Internet, Facing the Challenge with Innovation. The 8th Beijing Business Forum

marketing. In 2011, it opened its official Weibo. In 2012, it opened its official WeChat account and developed 928 Love: The Micro-Film Trilogy, which was recognized by the industry. In 2013, it established the code-based communication marketing platform,where Sina Weibo account Wu Yutai China expanded its fan scope, turning new fans into old tea customers. Also, it has online fan interaction activities every month, constantly improving the fan activity and enhancing the fan stickiness. Wu Yutai also invited fans to its stands and stores to participate in the tea culture experience activities to achieve a close connection online and offline.[144]

圖 Figure5-11 吳裕泰網上行銷 Online marketing of Wu yutai

三

問題與對策——零售商的出路
3. Problems and countermeasures-the way out
for retailers

（一）主要問題
The major problems

面對撲面而來的零售革命浪潮，儘管已有相當多的國內傳統零售企業已經堅定不移地開始「雲消費」時代的零售創新之旅，但還有很多企業處於觀望、等待或無所適從之中。現階段傳統零售企業在零售革命性創新中所面臨的主要問題是：

In the face of the wave of retail revolution, although quite a few domestic traditional retailers have unswervingly started the retail innovation in the era of cloud consumption, many companies are still waiting to see, or don' t know what to do. The main problems faced by traditional retailers in retail revolutionary innovation at this stage are:

1· 創新觀念問題 Innovation concept

傳統零售商都是在多年市場環境下「一刀一槍」幹出來的。多年以來，

「終端為王」的經營理念深入人心，控制終端、圈地發展是其主流擴張模式，他們對傳統行銷、管道、服務模式均形成了相對固化的傳統思維，也相信這種模式是經市場檢驗極為成功的。要突破這種思維模式和思維教條，有一定難度。

Traditional retailers have done their job in the market environment for many years. Over the years, the business philosophy of 「terminal rules」 has been deeply rooted in people's minds. The mainstream expansion model includes terminal controlling and enclosing land development, such relatively solid traditional thinking on traditional marketing, channels and service models has long been believed to be successful tested by the market. It is quite difficult to break through such thinking pattern.

同時，部分零售企業不成功的互聯網零售創新，也讓很多其他企業在零售的革命性發展方面更為謹慎。由於對風險與挑戰的準備不足，或過於放大風險，使得他們在創新面前裹足不前。

Moreover, the unsuccessful Internet retail innovations made by some retailers have also made many other companies more cautious in the revolutionary development of retail. They hesitate to innovate because they are under-prepared for or overestimate risks and challenges.

2· 創新時機問題 Innovation opportunities

當前，網上零售模式演進的速度已超過大多數人的預期，各種新觀念、新創意、新模式不斷出現，從 B2C 與 C2C 模式到網上商城，再到網路團購，

再到新型的 O2O 與 SOLOMO 模式，各種零售創新理念不斷湧現，網上零售競爭日趨激烈，品牌電商日趨成熟，其觀念、模式，甚至在品牌運營、物流系統建設等方面的專業化程度均遠遠超過多數傳統零售商。而大多數傳統零售企業已經錯過了進入網路的最佳時期，在商業模式轉型過程中，存在大量不確定性因素與風險，制約了部分傳統零售企業的轉型和發展。

At present, the evolution speed of online retailing is beyond most people' s expectation. Various new concepts, ideas, and models are emerging, from B2C and C2C models to online malls, then to online group purchases and new O2O and SoLoMo models are constantly emerging, making online retail competition increasingly fierce and brand e-commerce more mature. As a result, its concept, model, and even professionalization in brand operation and logistics system construction far exceeds that of most traditional retailers, which have missed the best time to enter the network. In the process of business model transformation, there are a lot of uncertain factors and risks, which restrict the transformation and development of some traditional retailers.

3. 全局創新問題 Overall innovation

中國傳統零售企業依賴傳統的經營管道，在嘗試革命性轉型過程中，必然產生大量的衝突與矛盾。例如，戰略重構問題、業務流程再造問題、組織創新問題、零售門店控制成本問題、新技術選擇和應用問題等。這不僅要打破原來熟悉的運作框架，還要對新的框架進行戰略選擇，要投入大量的人、財、物，面臨系列矛盾和衝突。為此，許多傳統零售企業在這一進程中往往

故步自封或患得患失，喪失了發展的時機。

Chinese traditional retailers rely on traditional business channels, and in the process of t revolutional transformation, there will inevitably be a lot of conflicts and contradictions, such as issues of strategic refactoring, business process reengineering, organizational innovation, cost control of retail stores and new technology choice and application. This is not only to break the familiar operational framework, but also to make strategic choices for the new framework and invest a large number of people, money, and materials, and to face series of contradictions and conflicts. To this end, many traditional retailers tend to stand still or hesitate in this process, losing the opportunity for development.

4‧ 創新技術問題 Innovation technology

先進的零售科技正在成為傳統零售企業創新發展的載體：一方面，需求供應鏈等一系列成熟的零售技術可以幫助精簡後臺功能和物流系統，降低成本，大幅提高工作效率。另一方面，虛擬導購、互動櫥窗、LBS 地理位置、二維碼等新興科技承載物在互聯網、移動互聯網、實體商店上的綜合運用，增強了用戶體驗。此外，微信、微博等新興社交媒體技術正在成為傳統零售商與消費者聯繫的紐帶。總而言之，傳統企業急需創新現有技術，迎接互聯網的挑戰。

Advanced retail technology is becoming the carrier of innovation and development of traditional retailers. On the one hand, a series of mature retail technologies such as demand supply chain can help streamline background functions and logistics systems, reducing costs and greatly improving work efficiency. On the other hand, the comprehensive use of virtual technology shopping, interactive show window, LBS geographic location, QR code and other emerging technology carriers in the Internet, mobile Internet, and physical stores enhances the user experience. In addition, emerging social media technologies such as WeChat and Weibo are becoming the link between traditional retailers and consumers. All in all, traditional companies urgently need to improve existing technologies to meet the challenges of the Internet.

5· 創新人才問題 Innovation talents

　　傳統零售企業加快零售革命創新面臨的主要「瓶頸」之一是缺少具有新的專業能力、創新發展能力的人才。以電子商務為代表的零售革命，線上線下的融合需要大量具備較高的網路構建、網站開發、商務運營、物流組織配送、客戶服務等專項技術的人才，需要新的更為扁平化的組織架構。傳統零售企業由於制度原因，在薪酬、激勵、晉升等方面缺乏系統的制度設計，在人才引進、人才待遇、留住人才等方面存在較大的障礙。在 2010 廈門網路零售發展戰略高峰論壇上，沃爾瑪、家樂福等國際零售巨頭紛紛表示，進軍互聯網零售存在智力與人才「瓶頸」。[145]

---

145 姚國章. 從沃爾瑪看傳統零售商的電子商務發輾轉型 [J]. 南京郵電大學學報（社會科學　版　）. 2011·13（01）. Yao Guozhang. The Transformation of E-Commerce of Traditional Retailers, Taking Wal-Mart as the Example[J]. Journal of Nanjing University of Posts and Telecommunications (Social Sciences Edition). 2011,13(01).

One of the main bottlenecks faced by traditional retailers in accelerating revolutionary retail innovation is the lack of talents with capabilities of new profession and innovative development. For the retail revolution represented by e-commerce, the integration of online and offline requires a large number of talents with high-level specific techniques like network construction, website development, business operations, logistics organization and distribution and customer service, as well as a new and flat organizational structure. Due to institutional reasons, traditional retailers lack systematic design in terms of salary, incentives, promotion, etc., and there are major obstacles in talent introduction, talent treatment, and retaining talents. At the 2010 Xiamen Online Retail Development Strategy Summit Forum, international retail giants such as Wal-Mart and Carrefour have said that there is a bottleneck in intelligence and talent to enter the Internet retail.[145]

（二）零售商的對策
Countermeasures made by retailers

「雲消費」時代，發展危機與挑戰並存，零售業面臨新的發展契機。我們認為，「雲消費」時代也是零售業盈利模式重構的時代，零售商要主動適應變化，善於把握現代主流消費模式，適應消費者新型生活方式需求，特別要在以下四個方面做出努力，做出特色，做出亮點。

In the era of cloud consumption, development crisis and challenges coexist, and the retail industry is facing new opportunities for development. We believe that the era of cloud consumption is also an era of restructuring of the retail industry's profit model. Retailers should take the initiative to adapt to the change, be good at grasping the modern mainstream consumption model and adapt to consumers' needs of new lifestyles. Special features and highlights must be made in the following four aspects.

1. 賣場的體驗化 Experience of the store

傳統零售業追逐有效營業面積內每平方米的利潤，「雲消費」時代商家更應追求賣場的體驗化，應從坪效導向轉向顧客體驗導向。

The traditional retail industry pursues the profit per square meter in the effective business area. In the era of cloud consumption, merchants should pursue the experience of the store, and shift from the area-effectiveness-orientation to the customer experience orientation.

顧客體驗是在顧客消費過程中建立的一種純主觀的感受。良好的顧客體驗是吸引新顧客、留住老顧客、增加顧客黏性的基礎。

The customer experience is a purely subjective feeling established during the consumption process. A good customer experience is the basis for attracting new customers, retaining old customers and increasing customer stickiness.

北京金源新燕莎 Mall 有一家叫「CHAPTER 7 章」的品牌概念店。這家店宣導的是一種生活方式和生活態度，推崇「服裝＋家居＋有機食品＋有意思的東西＝另一種生活」的經營理念。人們走進這家店，並不僅僅是為了購買其誕生於英國的童裝品牌 Mitti、關注中國都市獨立女性生活現狀和生存方

式的 OTT、男裝品牌 VOL3，以及來自 YYO，引領健康生活方式的有機食品等，而更多的是被其自然、動物、各類服飾、玩具巧妙搭配，渾然天成的環境及其所宣導的文化所吸引。在這裡，購物成為第二位的東西，新鮮、有趣、好玩才是吸引人們流連忘返的主因，這就是體驗化賣場的魅力。

Beijing New Yansha Shopping Mall has a brand concept store called 「CHAPTER 7」. The store advocates a lifestyle and life attitude of 「clothing + home + organic food + interesting things = another life」. People walk into the store not only to buy the British children's wear brand Mitti, OTT, which concerns about the status and survival of independent women in China's cities, the men's wear brand VOL3, and the organic foods brand YYO that leads a healthy lifestyle, but also are attracted by clever collocation of nature, animal, clothing and toys, together with the culture that the mall advocates. Here, shopping becomes the secondary. Feeling of freshness and fun is the main reason to attract people to linger. This is the charm of experience store.

圖 Figure 5-12 CHAPTER 7 章概念店 Concept store 「CHAPTER 7」

2 · 功能的社交化 Socialization of function

在傳統意義上，商業是購物的場所。20世紀90年代中後期，北京的百貨店內才出現頂層餐飲。人們逐漸適應了百貨＋餐飲的模式，適應了百貨＋餐飲＋超市的模式。在國內，人們認識購物中心的歷史更短，北京首家真正意義的購物中心始於1998年1月18日開業的新東安市場。隨著購物中心的日益增多，人們對其「一站式」服務，購物、餐飲、娛樂三分天下的經營模式有了切身的認識。購物中心的發展方向已不再侷限於購物的中心，而更多地成為社交體驗的場所，達到生活方式化和休閒娛樂目的地化。

In the traditional sense, business is the shopping places. It was the mid-to-late 1990s that top-level catering appeared in department stores in Beijing. People gradually adapted to the models of department store + catering and department store + catering + supermarket. In China, the history of shopping centers is shorter. The first true-sense shopping mall in Beijing was Bejing APM opened on January 18, 1998. With the increasing number of shopping centers, people have a personal awareness of the "one-stop" operation model of shopping, catering and entertainment. The development direction of the shopping mall is no longer limited to the shopping mall itself, but to become a place for social experience, lifestyle and entertainment.

生活方式化就是將購物中心變為一種生活方式中心，變為體驗式的消費文化中心。它更關注的是消費者在消費過程中的身心感受，強調以舒適友好的環境設計，營造多元化的體驗式消費空間。

By lifestyle, it means to turn the shopping mall into a lifestyle center and an experience consumer culture center. It is more concerned about the consumer's physical and mental experience in consumption, emphasizing a comfortable and friendly environment design to create a diversified experience consumption space.

在香港有一家以五行為環境主題的圓方購物中心絕對值得一提。這家購物中心引入了中國風水學中的五行——金、木、水、火、土，以五行元素做為各區域的標誌性主題，將五行元素透過現代的詮釋，將色彩、感覺、內涵、公共藝術和購物中心體驗化環境有機結合。每一五行元素都有自己的個性，如「金」象徵尊貴的品位，與世界一線品牌店和高檔餐廳相得益彰。這家購物中心透過這種渾然天成的創意設計，對每一個細節的精益求精，敞亮舒適的消費環境，將人們的生活方式融入其中，讓消費成為一種時尚精緻的體驗過程，由此在香港林立的購物中心中獨樹一幟。

In Hong Kong, there is the Elements Shopping Mall themed the five elements, which is definitely worth mentioning. This shopping mall introduces the five elements of Feng Shui, the Chinese geomancy-Metal, Wood, Water, Fire, and Earth. The five elements are used as the iconic theme for each region. Through modern interpretation, they are combined with the colors, feelings, connotations, public art and the experience environment of the shopping mall. Each of the five elements has its own personality, like Metal symbolizes the noble taste, which complements the world's first-line brand stores and high-end restaurants. Through natural and creative design, the shopping mall is perfect for every detail. The open and comfortable consumption environment integrates people's lifestyles in it and makes consumption a fashionable and exquisite experience. All makes the mall a unique one among so many malls in Hong Kong.

圖 Figure 5-13　香港圓方購物中心 Elements Shopping Mall in Hong Kong

　　休閒娛樂目的地化就是將購物中心變為一種能輕鬆遊玩，可以經常光顧甚至能待上幾天的旅遊場所。美國最大的購物中心──美國摩爾購物中心（MALL OF AMERICA）就是一個典型的休閒娛樂目的地。

　　By entertainment, it means to turn the shopping center into a recreation place that people can easily have fun in or even stay in for a few days. MALL OF AMERICA, the largest shopping mall in the United States, is a typical recreational destination.

　　美國明尼蘇達州布魯明頓市的美國摩爾購物中心是全美最大的購物娛樂體驗中心，建築面積約 39 萬平方米，共分上下 4 層，據說能裝下 2 座胡夫金字塔。商城內共有約 520 家商店，其中包括 4 家大型百貨公司、40 多家餐館，商城中央有佔地面積達 2.8 萬平方米的「史努比營」主題遊樂園，還有 1 座由 14 個影廳組成的電影院和 8 家夜總會。顧客來到這裡不僅是購物，更是旅遊、探險和放鬆。

Located in Bloomington, Minnesota, MALL OF AMERICA is the largest shopping and entertainment experience center in the United States. It has a building area of approximately 390,000 square meters and is divided into four floors. It is said that it can hold two Pyramids of Khufu. There are about 520 stores in the mall, including 4 large department stores and more than 40 restaurants. In the center of the mall lies the 「Snoopy Camp」 theme park with an area of 28,000 square meters, a cinema with 14 halls and 8 nightclubs. Customers come here not only for shopping, but also for travel, adventure and relaxation.

在體驗化場景的背後,是精細的管理和體貼的服務。商城內外共安裝了 125 部閉路電視監視器 24 小時監控;停車場上的 130 個呼叫器和商城裡的 44 部求助電話可以直接聯絡到安全中心;12750 個車位的停車場全部免費,還為準媽媽和殘障人士提供專用停車場;帶小孩來逛商城的人也不用擔心會不方便,在一層入口處的服務臺,花 3 美元就可以租 1 輛童車用 1 天。行動不便的人在這裡交 5 美元的押金就可以免費租 1 輛輪椅。

Behind the experience scenes, there is the sophisticated management and considerate services. A total of 125 CCTV monitors are installed inside and outside the mall for 24-hour monitoring. There are 130 beepers on the parking lot and 44 emergency telephones in the mall, all of which are directly connected to the security center. All the 12,750 parking lots are for free, among which specialized lots for pregnant women and disabled people are also available. People with children to visit the mall can rent a stroller for one day for $3 at the service desk at the entrance of the first floor. People with limited mobility can rent a wheelchair for free by paying a $5 deposit here.

這一切帶來的是絕佳的口碑和效益。美國摩爾購物中心不僅改變了美國西北部地區人們的購物習慣，也成了全球聞名的「旅遊景點」。據報導，目前，購物中心出租率達到 99%，年客流量達 4250 萬人次。不少顧客把這裡當成了休閒度假的好去處，全家人遠道而來，在附近的旅館裡住上兩三天，盡情體驗逛商店的樂趣。遊客在此每花費 1 美元，就相應地要在摩爾購物中心之外為汽油、住宿、餐飲等花 2－3 美元。這樣不僅每年給明尼蘇達州帶來的收入高達 16 億美元，而且也給布魯明頓這座小城帶來了巨大變化。布魯明頓的就業率一直很高，僅旅館業就因此增加了 50% 的床位，總數達到了 7600 個，相當於明尼蘇達首府雙城（明尼阿波利斯市和聖保羅市）的總和。**146**

All these measures and services brought great word of mouth as well as benefits. MALL OF AMERICA not only changed the shopping habits of people in the northwestern United States, but also became a world-famous tourist attraction. According to reports, at present, the occupancy rate of shopping centers has reached 99%, and the annual passenger flow has reached 42.5 million. Many customers regard the mall as a good place to spend a relaxing holiday. The whole family come from afar and stay at a nearby hotel for two or three days to enjoy the fun of shopping. For every $1 spent by tourists, it is necessary to spend an extra $2 or $3 for gasoline, accommodation and catering outside the mall. This not only increases the financial revenue of Minnesota to $1.6 billion a year, but also brings a huge change to the small town of Bloomington. The employment rate in Bloomington has long been high, and the hotel industry alone has increased the number of beds by 50%, reaching a total of 7,600, equivalent to the sum of the twin cities (Minneapolis and São Paulo)of Minnesota.**146**

圖 Figure 5-14　美國摩爾購物中心——與海洋神祕接觸的購物中心 MALL OF AMERICA-a shopping mall that mysteriously contacts with the ocean

圖 Figure 5-15　美國摩爾購物中心——有「史努比營」主題遊樂園的購物中心 MALL OF AMERICA-a shopping mall with the "Snoopy Camp" theme park

3‧ 運營的智慧化 Intelligent operation

在商業零售領域，智慧化技術不但為企業提供了新的管理、運營、行銷工具，更重要的是，智慧化技術為真正意義上的精細化管理、個人化推薦以及提高賣場的體驗性提供了可能。智慧化技術從根本上改變了企業的管理運

---

146 美國摩爾——開創 "娛樂零售" 新概念. 全球品牌網 MALL OF AMERICA-Creating A New Concept of Entertainment Retail. Global Brand

營模式。

In the field of commercial retail, intelligent technology not only provides enterprises with new tools of management, operation and marketing, but more importantly, it provides the possibility of refined management, personalized recommendations and improving store experience in the true sense. Intelligent technology fundamentally changes the management and operation model of the enterprise.

商業企業運營智慧化主要體現在以下幾個方面：

The intelligent operation of enterprises is mainly reflected in the following aspects:

（1）底層建設——超前的資訊系統架構無縫契合企業商業模式轉型 The underlying architecture construction-the advanced information system architecture seamlessly matches the transformation of the business model

資訊系統在「雲消費」時代，是企業決策、管理、運營、服務中樞，其重要性不言而喻。而資訊系統底層架構是企業資訊系統的根基，其設計的超前性與可擴展性決定了未來企業智慧化發展的能力，是企業資訊系統建設的重中之重。由於資訊系統通常都是由專業的程式師構建，企業難以真正識別資訊系統的底層架構，因此對企業來說形成了巨大的「陷阱」。

In the era of cloud consumption, information system is the center of enterprise decision-making, management, operation and service, importance of which is self-evident. The underlying architecture of information system is the foundation of enterprise information system. The advancement and scalability of its design determine the ability of enterprise intelligent development in the future, enjoying the top priority of enterprise information system construction. Since information systems are usually built by professional programmers, it is difficult for enterprises to truly identify the underlying architecture of information systems, thus creating a huge "trap" for enterprises.

蘇寧易購從 2013 年 12 月初開始，出現大面積的下單後遲遲未發貨，網友紛紛在微博上對此表示憤怒：「蘇寧滾出電商圈，下單付款 25 天不發貨，退款了，目前還有 6000 多塊沒退。」「蘇寧（12 月）12 日下單，今天還沒收到貨，看來要到 2014 年才能收到了。售後電話始終忙線，蘇寧電商之路至少要再爬幾年坡。」蘇寧對此給出的解釋是系統升級令人傷不起。

Since the beginning of December 2013, Suning Yigou has been delaying delivery of a large amount of orders. Netizens expressed their anger on Weibo and posted comments like 「Suning should get out of the e-commerce circle, it has delayed delivery for 25 days. And there is still more than ¥6,000 left to refund.」「I placed an order on Suning on December 12th and has not received the goods until today. It seems that I will not receive my goods until 2014. The after-sales hotline is always busy, Suning Yigou still has a long way to go.」 Suning explained that the delay was because of the system upgrade, which was apparently not satisfactory.

其起因在於 2013 年年底，蘇寧將使用了 7 年之久的 SAP 系統全面切換為 LES 系統，同步升級的還有蘇寧易購的前臺下單系統、訂單管理系統以及客戶管理系統……在升級過程中，資料導入之複雜、系統切換之艱鉅，使蘇寧的 IT 系統受到巨大考驗，同時也使蘇寧運營體系經受了巨大的信任危機。

其背後的根源在於系統底層架構的滯後性。

The origin of the case is that at the end of 2013, Suning changed the SAP system used for 7 years to the LES system, together with the front-end ordering system, order management system and customer management system. During the upgrade process, Suning's IT system has been greatly tested by the complexity of data importing and system switching. Suning's operating system has also experienced a huge crisis of trust. The root cause lies in the lag of the underlying architecture of the system.

（2）智慧升級──從資料庫到資料倉庫，從經營統計到資料採擷
Intelligent upgrade-from database to data warehouse, from operating statistics to data mining

傳統企業對經營的全盤掌握與分析有賴於對銷售資料的報表統計。但隨著資料幾何級的倍增、大數據時代的來臨，商業智慧平臺需求將從「以報告為中心」（Reporting-Centric）轉變為「以分析為中心」（Analysis-Centric）。對商業企業來說，即在於如何快速而正確地分析和理解資料，如何將資料分析轉變為對消費者的個人化服務。

Traditional enterprises' overall mastery and analysis of the operation depends on the reports and statistics of sales data. But with the explosion in data and the advent of the big data era, the demand for business intelligence platforms shifts from Reporting-Centric to Analysis-Centric. For enterprises, this means how to analyze and understand data quickly and correctly, and transform data analysis into personalized services for consumers.

IDC 在 2014 年 5 月發佈的一項針對亞太地區的相關調查報告中指出，近七成企業高管認為，複雜的資料管理是他們目前最艱鉅的工作之一，一旦這些資料丟失，企業將面臨巨大的損失。然而，在整個亞太地區，在所有的交易資料中，只有 73% 的資料得到了有效的分析。許多企業有太多報表、太多資料系統，這讓企業的決策者難以從中迅速挖掘出有效資訊。而針對消費者個體的個人化行銷，則更有賴於智慧的「商業探索」，從而説明客戶迅速找到資料間的關聯，更快地進行分析以及更精準地進行行銷。簡單地說，從具有記錄、統計功能的資料庫向具有智慧挖掘與自我解析的資料庫升級，成為企業運營智慧化的重要一步。

In a survey report for the Asia-Pacific region released by IDC in May 2014, nearly 70% of corporate executives believe that complex data management is one of their most difficult tasks. Once these data are lost, enterprises will suffer huge loss. However, in the entire Asia Pacific region, only 73% of all transaction data has been effectively analyzed. Many companies have too many reports and data systems, which makes it difficult for decision makers to quickly dig out effective information. Personalized marketing for individual consumers is also dependent on intelligent 「business exploration」 to help customers quickly find data associations, analyze faster and market more accurately. In short, upgrading from a database with records and statistics to a database with intelligent mining and self-analysis has become an important step in the intelligent operation of enterprises.

（3）運營突破——從單品、客群分析到精準客戶訂製 **Breakthroughs in operation–from single product, customer group analysis to precise customer**

customization

　　有了超前的系統架構基石以及強大的資料採擷能力後，企業在終端面對消費者時就具有了運營智慧化突破的可能。過去企業將市場進行細分，透過對幾類目標顧客的分析從而為某一類細分人群提供商品或服務。今天，商業智慧讓這一傳統的企業經營方式做到極致。透過大數據的支援，企業能夠實現對每一個顧客進行個性化的商品、服務推薦，甚至根據顧客的需求為顧客訂製獨一無二的產品或服務。

With the advanced system architecture foundation and powerful data mining capabilities, it is possible for breakthroughs in operation when enterprises face consumers. In the past, enterprises subdivided the market and provided goods or services to certain subdivisions of people through analysis of several types of target customers. Today, business intelligence makes the best of this traditional way of doing business. Supported by big data, enterprises can realize personalized products and service recommendations for each customer, and even customize unique products or services according to customers' needs.

　　零售行業中的巨頭塔吉特就透過大數據分析走在了零售革命的前列。每位顧客初次到塔吉特刷卡消費時，都會獲得一組顧客識別編號，內含顧客姓名、信用卡卡號及電子郵件等個人資料。日後凡是顧客在塔吉特消費，電腦系統就會自動記錄消費內容、時間等資訊。再加上從其他管道取得的統計資料，塔吉特便能形成一個龐大的資料庫，運用於分析顧客喜好與需求。依靠分析顧客資料，塔吉特的年營業收入從 2002 年的 440 億美元增長到 2010 年的 670 億美元。這家成立於 1961 年的零售商能有今天的成功，資料分析功不

可沒。

The retail giant Target Corp. is at the forefront of the retail revolution through big data analysis. Each customer receives the customer identification number containing personal data like customer name, credit card number and email, for the first time when he or she pays by card. Later on, whenever the customer consumes in Target, the computer system will automatically record information such as content and time. Together with the statistics obtained from other channels, Target can form a huge database for analyzing customer preferences and needs. By analyzing customer data, Target's annual revenue increased from $44 billion in 2002 to $67 billion in 2010. The retailer, founded in 1961, succeeded due to data analysis.

4· 組織的扁平化 Flat organization

組織的扁平化就是以客戶介面導向重構組織結構與運營流程。與傳統組織體系相比，這裡突出兩大變化：一是要增強企業的客戶服務能力。要從過去以產品分類進行組織劃分向以客戶分類進行組織劃分轉變，真正做到以客戶為中心，圍繞客戶需求整合產品與服務資源，頂層設計從企業架構上增強客戶服務能力。二是改變企業管理結構與運營流程。一定要改變企業管理結構，簡化管理層級，實現企業扁平化管理；對企業運營流程做出相應改變，實現對企業顧客的全流程統籌化管理。

The flattening of the organization is to restructure the organizational structure and operational processes with customer interface orientation. Compared with the traditional organizational system, two major changes are highlighted. First, it is necessary to enhance the customer service capabilities of the enterprise and shift the division by product classification in the past to division by customer classification. It is also necessary to be truly customer-centric, integrate product and service resources according to customer needs, and enhance customer service capabilities from enterprise architecture. Second, it should change the enterprise management structure and operation process. It is necessary to change the enterprise management structure and simplify the management level to realize the flat management of the enterprise. Also, it should make corresponding changes to the business operation process to realize the overall management of the customers.

騰訊公司 2012 年宣布進行公司組織架構調整，把原有業務系統制變為事業群制。經過調整，騰訊公司將現有業務重新劃分為企業發展事業群（CDG）、互動娛樂事業群（IEG）、移動互聯網事業群（MIG）、網路媒體事業群（OMG）、社交網路事業群（SNG）；整合原有的研發和運營平臺，成立新的技術工程事業群（TEG）；並成立騰訊電商控股公司 (ECC) 專注於電商業務。2014 年 5 月 6 日，騰訊公司又宣布成立微信事業群（WeiXin Group，WXG）。騰訊公司在組織架構上的這一番大「動作」，是為了保持創新的活力和靈動性，而進行的由「大」變「小」，把自己變成整個互聯網大生態圈中的一個具有多樣性的生物群落。我們可以認為，騰訊公司此舉是在組織體系建設上創建了類似生物型的組織，讓企業組織自我進化，使龐大的企業輕裝上陣。以對騰訊公司貢獻卓越的微信事業群團隊而言，這個團隊本身就是騰訊公司內部自我創新、自我競爭的產物，這個團隊有 150 多人，

技術人員佔了絕大部分，他們主要從事前臺開發、後臺開發、測試等工作；
還有 20 名左右產品和設計人員，從事研發、市場、設計一體化等工作。在騰
訊公司內部，這個團隊沒有績效考核指標；但從全國數以億計的微信用戶來
看，他們為騰訊公司創造了未來。

In 2012, Tencent announced the restructuring of its organization and changed the original business system into a business group system. After adjustment, Tencent has reclassified its existing business into the Corporate Development Group (CDG), the Interactive Entertainment Group (IEG), the Mobile Internet Business Group (MIG), the Online Media Business Group (OMG), and the Social Network Business Group (SNG). It integrated the original R&D and operation platform into a new technology engineering business group (TEG) and established Tencent E-commerce Company (ECC) to focus on e-commerce business. On May 6, 2014, Tencent announced the establishment of the Weixin Group (WXG). Tencent's 「big action」 in organizational structure is to maintain the vitality and agility of innovation, and to change from 「big」 to 「small」 and turn itself into one of the communities in the entire Internet ecosystem. We can think that Tencent's move is to create a bio-like organization in the organization system, let the organization evolve itself and make the huge enterprises lightly loaded. The WeChat business group team itself is the product of Tencent's internal self-innovation and self-competition. This team has more than 150 people, and technicians account for the vast majority. They are mainly engaged in front-end development, back-end development, testing, etc. Moreover, there are about 20 product researchers and designers who engage in integration work of research and development, marketing and design. Within Tencent, there are no performance appraisal indicators for the team. But from the perspective of hundreds of millions of WeChat users across the country, they have created a future for Tencent.

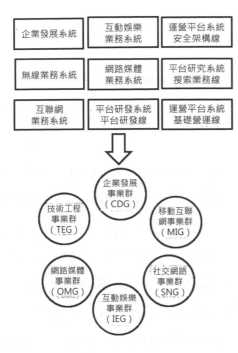

Enterprise Development System\
Interactive Entertainment System\
Security Structure of Operation Platform System
Wireless Buisness System\ Network Media Service System\ Searching Business of Platform Research System Internet Business System\ Platform Development of Platform Development System\ Basic Operation of Operation Platform System

企業發展事業群（CDG）、互動娛樂事業群（IEG）、移動互聯網事業群（MIG）、網路媒體事業群（OMG）、社交網路事業群（SNG）、技術工程事業群（TEG）:the Corporate Development Group (CDG), the Interactive Entertainment Group (IEG), the Mobile Internet Business Group (MIG), the Online Media Business Group (OMG), the Social Network Business Group (SNG), Technology Engineering Business Group (TEG)

圖 Figure 5-16　騰訊公司組織架構調整前後對比 Before and after Tencent's organizational structure adjustment

5· 營銷的口碑化 Word of mouth marketing

在「雲消費」時代，行銷的概念也在改寫。受網路影響，2013 年中國傳統廣播電視廣告收入的增幅持續大幅下降，增幅僅為 2.52%，較 2012 年 13% 的增幅降低了近 11%。[147] 據上海市工商局發佈的 2013 年度《上海廣告市場狀況報告》，2013 年度上海市電視、廣播、報紙和期刊四大傳統媒體廣告營

收總體下降，同比減少 7.2%。[148] 與此同時，上海互聯網廣告媒介經營單位 2013 年廣告營收達 39.8 億元，同比增長 117.6%。在電視、廣播、報紙、期刊和互聯網五大大眾傳播媒介的廣告營收份額中，互聯網媒體廣告的份額由 2012 年的 16.4% 上升至 2013 年的 31.5%，展現出強勁的增長勢頭。據研究，近年中國互聯網廣告業務處於持續增長態勢，2013 年百度廣告收入已超過中央電視臺，成為中國最大的廣告公司。而同年谷歌公司 600 億美元的收益幾乎都由網路廣告貢獻，其廣告營收已經超過了全美所有報紙或雜誌的廣告營收總和。[149]

In the era of cloud consumption, the concept of marketing is also being rewritten. Affected by the network, the growth rate of China's traditional radio and television advertising revenue continued to drop sharply in 2013, an increase of only 2.52%, which was nearly 11% lower than the 13% increase in 2012. [147] According to the 2013 Shanghai Advertising Market Status Report released by the Shanghai Municipal Administration of Industry and Commerce, the 2013 Shanghai traditional advertising revenues of TV, radio, newspapers and periodicals generally declined, down 7.2% year-on-year.[148] At the same time, the advertising revenue of Shanghai Internet advertising media business units reached ¥3.98 billion in 2013,

---

147 傳統媒體廣告收入下降與新媒體融合任重道遠 . 中國廣播網 . 2014-03-05 Traditional Media Advertising Revenue Declined and Its Integration with the New Media Has a Long Way to Go. cnr.cn. March 5, 2014

148 上海市工商局：2013 年上海四大傳統媒體廣告營收下降 7.2%. 2014-03-29 Shanghai Municipal Administration of Industry and Commerce: In 2013, Shanghai's Four Traditional Media Advertising Revenues Fell by 7.2%. March 29, 2014

149 谷歌今年營收將超全美報刊廣告收入總和 . 新浪科技 . 2013-11-14 Google's Revenue This Year Will Exceed the Total Revenue of Newspapers and Magazines in the United States. Sina Technology. Dec 14, 2013

a year-on-year increase of 117.6%. In the advertising revenue share of the five major mass media of television, radio, newspapers, periodicals and the Internet, the share of Internet media advertising increased from 16.4% in 2012 to 31.5% in 2013, showing a strong growth momentum. According to research, China's Internet advertising business has continued to grow in recent years. In 2013, Baidu's advertising revenue exceeded that of CCTV and became China's largest advertising company. In the same year, Google' s $60 billion in revenue was almost exclusively contributed by online advertising, and its advertising revenue exceeded the total advertising revenue of all newspapers or magazines in the United States.[149]

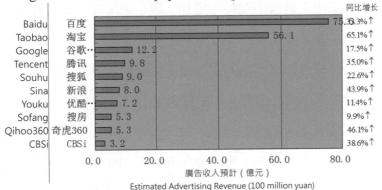

圖 Figure 5-17 2013Q2 中國網路廣告市場媒體營收規模 2013Q2 Top10 China online advertising market media revenue scale
資料來源：億邦動力發佈的《中國互聯網廣告核心資料》 Source: China Internet Advertising Core Data released by Yibang Power

據英國《每日郵報》的一篇報導，普通人每天平均每 6.5 分鐘就會看一眼手機。以每日清醒時間 16 小時計算，人們一天要看 150 次手機。[150] 隨著互聯網技術特別是移動互聯技術的發展，在「雲消費」環境下，口碑化行銷已成為行銷的重要選擇，商家的發展要更多地借助自媒體傳播。

According to a report in Daily Mail, average people check the phone every 6.5 minutes a day, which means people check phones for 150 times a day within waking time of 16 hours.[150] With the development of Internet technology, especially mobile internet technology, word-of-mouth marketing has become an important choice for marketing in the cloud consumption environment. The spead of businesses should rely more on self-media.

小米手機創造了口碑化行銷的範例。規模並不大的小米公司在小米論壇、微博、微信、百度知道、QQ 空間等社會化媒體擁有近百人的團隊負責新媒體運營。其中有 30 多名微博客服人員，每天處理私信 2000 多條，提及、評論等 4, 5 萬條。透過在微博上互動和服務，讓小米手機深入人心。通常某顧客買了一支小米手機經常死機，在微博上吐槽一下，15 分鐘就會得到微博客服的專業回覆。這種自媒體行銷方式改變了傳統的傳播習慣和規律，使企業更緊密地聯繫消費者、代入消費者，使買賣雙方共同受益。2013 年，小米手機的微博帳號已經有 200 多萬的粉絲，每天微信上的使用者互動資訊達 3 萬多條；小米論壇註冊用戶已經近 1000 萬，每天有 100 萬用戶在裡面討論；小米手機 QQ 空間認證帳號的粉絲數超過了 1000 萬，小米手機在 QQ 空間做活動時，往往很容易就有幾萬轉發。正是擁有並很好地經營著如此龐大的自媒體傳播平臺，使小米手機取得了非凡的成功，成為業界的領路者。

Xiaomi has created an example of word-of-mouth marketing. Xiaomi, a small company, has a team of nearly 100 people in social media such as Xiaomi Forum,

---

150 每人每天看多少次手機？生命時報 . 2013-02-22　How Many Times does One Person Check the Phone Everyday? Life Times. Feb 22, 2013

Weibo, WeChat, Baidu, and QQ Space, responsible for new media operations. Among them, there are more than 30 micro-blog service personnel, handling more than 2,000 private messages every day, mentioning and commenting for forty or fifty thousand. Through the interaction and service on Weibo, Xiaomi is deeply rooted in the hearts of the people. Very often, a customer whose Xiaomi phone crashes complains on Weibo, in 15 minutes he or she will get a professional reply from service staff. Such self-media marketing approach has changed the traditional communication habits and patterns, enabling companies to more closely contact consumers, so that buyers and sellers can both benefit. In 2013, Xiaomi' s Weibo account had more than 2 million fans, and there were more than 30,000 user interactions on WeChat every day. Xiaomi Forum has nearly 10 million registered users, and number of online users reaches 1 million every day. The number of fans of QQ space verified account exceeds 10 million. When Xiaomi holds events in QQ space, it is easy to have tens of thousands of forwardings. It is the possession and good operation of such a large self-media communication platform that makes the Xiaomi achieve extraordinary success and become the leader of the industry.

「雲消費」時代零售業的發展方向

Chapter Six
The Development
Direction of the
Retail Industry in
the Era of Cloud
Consumption

第六章

「雲消費」時代零售業的發展方向

Chapter Six The Development Direction of the Retail Industry in the Era of Cloud Consumption

「雲消費」引領現代零售革命，顛覆了固有的商業經營思想和經營模式。不破不立，零售商必須進一步突破傳統的思維教條，順應零售革命的大勢所趨，重構全新的盈利模式，實現以消費需求為核心的線上線下一體化的運營創新。

Cloud consumption leads the modern retail revolution, subverting the inherent business management thinking and business models. Without destruction there is no construction. Retailers must further break through the traditional dogmatic thinking, conform to the general trend of the retail revolution and reconstruct a new profit model to realize online-offline operation innovation with consumer demand as the core.

我們從六方面看「雲消費」時代零售業的發展方向：

The book divides development direction of the retail industry in the era of cloud consumption into six aspects.

一

創新個性化商業
1. Innovation in personalized business

　　傳統商業經營以連鎖化、標準化、規範化為努力方向；在「雲消費」時代，獨特性才是商業企業生存的根本。消費者要消費獨特的、個性的商品，選擇到獨特的、個性的商店消費。

　　The traditional business operation makes efforts in the chainization, standardization and normalization. In the era of cloud consumption, the uniqueness is the basis for the survival of enterprises. Consumers consume unique and personalized products and choose to shop in unique and personalized stores.

（一）傳統商業經營──連鎖化、標準化、規範化
Traditional business operation-chainization, standardization and normalization

　　傳統商業經營講求連鎖化、標準化、規範化。

　　Traditional business operation stresses on chainization, standardization and normalization.

　　連鎖化是指經營同類商品或服務的若干個企業，以一定的形式組成一個

聯合體，在整體規劃下進行專業化分工，並在分工基礎上實施集中化管理，把獨立的經營活動組合成整體的規模經營，從而實現規模效益。這是一種高度組織化、系統化的商業組織形式和經營制度，具備六個統一的特點：統一採購、統一配送、統一標識、統一行銷策略、統一價格、統一核算。

Chainization refers to a number of enterprises that operate similar goods or services form a consortium in a certain form, carry out specialized division of labor under the overall planning, and implement centralized management on the basis of division of labor. The consortium combines independent business activities into an overall scale operation tp achieve economies of scale, which is regarded as a highly organized and systematic business organization and management system with six unified characteristics: unified procurement, distribution, identification, marketing strategy, price, and accounting.

標準化是指依據科學技術和實踐經驗的綜合成果，對技術、管理等活動中具有多樣性、相關性重複事物，以特定的程式和形式頒發的統一規定。商業經營活動中的標準化更側重於標準化的管理，即以標準化原理為指導，將標準化貫穿於管理全過程；以增進系統整體效能為宗旨，提高工作品質與工作效率。

Standardization refers to the uniform provisions in specific procedures and forms for repeated things with diversity and relevance in technology and management activities. Standardization in operation is more focused on standardized management, that is, guided by the principle of standardization, to carry out standardization throughout the management process, and improve work quality and work efficiency for the purpose of improving the overall efficiency of the system.

規範化通常指使事物發展變化符合規定的程式和標準。商業經營中的規範化通常指根據企業章程及業務發展需要，合理地制訂組織規程、基本制度以及各類管理事務的作業流程，以形成統一、規範和相對穩定的運營管理體系。

Normalization usually refers to procedures and standards of development and changes of things. Standardization in business operation usually refers to the rational formulation of organizational procedures, basic systems, and various management transactions in accordance with the company's charter and development needs, in order to form a unified, standardized and relatively stable operation management system.

連鎖化、標準化、規範化具有典型的工業時代特徵。

Chainization, standardization and normalization have typical characteristics of the industrial age.

連鎖經營最早出現在美國，後來逐漸在世界各地得到迅速推廣。

The chain operation first appeared in the United States and was promoted rapidly around the world.

第一家頗具規模的連鎖商店是 1859 年喬治•F. 吉爾曼和喬治•亨廷頓•哈特福特在紐約創辦的大美國茶葉公司。在 6 年時間內，該公司發展了 26 家正規店，全部經銷茶葉。1869 年，更名為大西洋和太平洋茶葉公司」。到 1880 年，其發展規模已經達到 100 多家分店。

The first large-scale chain store was The Great American Tea Company founded in 1859 by George F. Gilman and George Huntington Hartford in New York. In six years, the company developed 26 regular stores, all of which engaged in tea. In 1869, it changed its name to The Great Atlantic and Pacific Tea Company. By 1880, it developed more than 100 branches.

在同一時期另一家透過連鎖經營取得成功的公司是勝家縫紉機公司。該公司於 1865 年開始採用「特許經營」分銷網路的方式進行產品銷售，收到很好的效果，迅速打開了產品銷路。20 世紀 60 年代，勝家縫紉機公司在全球已有 3 萬多家專賣店和經銷點，形成了強大的銷售網路，成為縫紉機行業當之無愧的領導者。

During the same period, another company that succeeded through the chain operation was the Singer Sewing Machine Company. The company began to sell its products in a franchising distribution network in 1865 and quickly found market. In the 1960s, the company had more than 30,000 specialty stores and distribution points around the world, forming a strong sales network and becoming a well-deserved leader in the sewing machine industry.

1953 年開始連鎖經營的麥當勞在全世界 128 個國家和地區已擁有 3.1 萬多家餐廳，每隔 15 小時就有一家麥當勞餐館開業。1965 年 4 月 15 日，麥當勞公司股票上市，每股為 22.5 美元，20 年後該公司股價約為原來的 175 倍。

McDonald's, which began chain operation in 1953, has more than 31,000 restaurants in 128 countries and regions around the world. On average, a new McDonald's restaurant opens every 15 hours. On April 15, 1965, McDonald's shares were listed at $22.5 per share. After 20 years, the share price was about 175 times higher.

1962 年，在美國不起眼的阿肯色州創立的沃爾瑪，經過 50 餘年的發展，已經成為世界上最大的連鎖零售企業。在全球 27 個國家開設了超過 1 萬家商場，下設 69 個品牌，全球員工總數 220 多萬人，每週光臨沃爾瑪的顧客 2 億人次。

In 1962, Wal-Mart, founded in Arkansas in the United States, has become the world's largest chain retailer after more than 50 years of development. It has opened more than 10,000 stores in 27 countries around the world, with 69 brands and more than 2.2 million employees worldwide. About 200 million customers visit Wal-Mart every week.

日本連鎖商業的發展引人注目。以 7-11 便利店為例，1974 年 5 月，日本的第一家 7-11 本土便利商店在東京都江東區開張。到 2003 年時，7-11 本土便利商店的總店數達到 1.6 萬多家，遍及全日本城鄉各地。

The development of Japanese chain commerce is eye-catching. Take the 7-Eleven as an example. In May 1974, Japan' s first 7-Eleven opened in Koto-ku, Tokyo. By 2003, the number of 7-Eleven stores had reached more than 16,000, covering all parts of Japan.

20 世紀 90 年代中期以後連鎖經營做為一種新型流通方式在中國發展迅猛，尤其是以食品、零售、餐飲業、服務業等行業最具代表性。2013 年，中國連鎖百強企業銷售規模已達到 2.04 萬億元，約佔全國社會消費品零售總額的 8.7%；百強企業門店總數達到 94591 家，其中中石化易捷連鎖便利店達到

2.33 萬家，百勝集團的連鎖門店也超過了 6000 家。

After the mid-1990s, chain operation, as a new type of circulation, developed rapidly in China, especially in the food, retail, catering, service and other industries. In 2013, the sales scale of China's top 100 chain enterprises reached ¥2.04 trillion, accounting for 8.7% of the total retail sales of consumer goods nationwide. The total number of top 100 enterprises reached 94,591, of which Sinopec Ejoy reached 23,300, and Yum! Brands Inc. more than 6,000.

（二）「雲消費」時代的商業經營——個性化、特色化
Business operation in the era of cloud consumption-personalization and specialization

「雲消費」時代是張揚消費者個性、銷售者個性的時代，個性化、特色化是張揚創造力、創新力的基礎。

The era of cloud consumption is an era of publicizing the personality of both consumers and sellers. Personalization and specialization are the basis for publicizing creativity and innovation.

在「雲消費」時代，消費者追求高品位、高品質的生活已成為消費潮流和趨勢。這種由消費者的消費意識而確立的主體地位，為個性化產品的產生、發展提供了機遇。企業賣的不只是產品，而是根據消費者的特殊需求，不斷推出按需訂製的個性化商品，甚至是超越消費者期待的商品，滿足消費者對

個性化商品的多元化需求甚至是極致需求。

In the era of cloud consumption, consumers' pursuit of high-grade, high-quality life has become a trend of consumption. This subject status, established by the consumption consciousness, provides an opportunity for the generation and development of personalized products. Enterprises sell not only products, but also constantly introduce customized products on demand according to the special needs of consumers or even beyond the expectations of consumers, to meet the diversified needs of consumers for personalized products and even the acme demand.

多年來，傳統的百貨業態在不斷創新。我們看到，這些年城市百貨業創新的步伐一直沒有停滯，從單一的綜合百貨發展到高檔百貨、精品百貨、生活百貨、主題百貨等，業態在不斷細分，百貨店的連鎖化也在不斷推進。截至 2010 年年底，北京的王府井百貨集團已在全國 15 個城市開業運營 22 家大型百貨商場。此外，百貨與購物中心的結合、百貨與生活廣場的結合、百貨與品牌地產的結合，統一經營、購銷分離的經營模式、買手製的引進等，都讓我們看到了百貨業創新的步伐。而這些，在當下的「雲消費」時代還遠遠不夠。適應個性化的消費，提供個性化、特色化的服務，是百貨業新一輪創新的路徑。

Over the years, the traditional department store format has been constantly innovating. We have seen that the pace of innovation in the department store industry in these years has never stopped, from a single comprehensive department store to high-end, boutique, lifestyle, theme department stores, etc., the business is constantly being subdivided, and the chainization of department stores is also constantly advancing. As of the end of 2010, Beijing Wangfujing Department Store

Group has opened 22 large department stores in 15 cities across the country. In addition, the combination of department stores and shopping malls, the combination of department stores and lifestyle squares, the combination of department stores and brand real estate, operation models of unified operation and the separation of purchase and sales business, as well as the introduction of buyer system have all showed the pace of innovation in the department store industry. And these are not enough in the current era of cloud consumption. Adapting to personalized consumption and providing personalized and specialized services is a new round of innovation in the industry.

　　我們不妨借鑑日本丸井百貨公司（以下簡稱丸井百貨）的發展案例。丸井公司創業於 1931 年，是一家有 80 餘年歷史的老牌百貨店，但同時也是日本年輕人最喜歡的百貨店。這兒聚集了各式各樣時下年輕人喜愛的商品與品牌，如日本國產手袋 SAZABY、G -SHOCK 的最新款式，反町隆史在日劇《海灘男孩》中所戴的黑框藍鏡片太陽眼鏡、Hamlet Lang 的泡綿背包等，還有不少強調創意與顛覆的設計師品牌。而且這些品牌均為消費能力遠大於賺錢能力的年輕族群能夠消費得起的平價品牌，即使是設計師品牌，也以價錢較為平實的日本新銳設計師為主。然而，丸井百貨不賣貴婦紳士服、孕婦裝、童裝、和服、禮服，也沒有文具、家具、家庭園藝用品等雜貨。丸井百貨每個分館的定位都相當明確，若是你還不到 20 歲，就可以直接到 YOUNG 館去採購；購置行頭的男士請到 MEN' S 館；想買餐具就到 IN THE ROOM 館。為了吸引年輕的中國消費客群，丸井百貨還邀請中國明星黃曉明擔任其品牌的中國代言人。2009 年 6 月 15 日，丸井百貨更與支付寶達成合作，丸井百

貨旗下運營的中文購物網站（http://maruione.jp/cn/）正式透過接入支付寶的境外收單服務進軍國內市場。個性化、特色化經營模式使丸井百貨得以基業常青。目前，丸井百貨在日本擁有22家連鎖百貨商店，年銷售額5000億日元。丸井百貨的大紅招牌對於日本的時尚青年有著神奇的召喚力。

We may wish to learn from the case of Japan Marui Department Store (hereinafter referred to as Marui). Founded in 1931, Marui is a time-honored department store with more than 80 years of history, but it is also a favorite department store for young people in Japan. There are a variety of goods and brands chased after by young people today, such as the latest items of Japanese-made handbags SAZABY and G-SHOCK, the black-framed and blue-lensed sunglasses worn by Takashi Sorimachi in the Japanese TV series "Beach Boy", and Sponge backpacks of Hamlet Lang. There are also a lot of designer brands that emphasize creativity and subversion. Moreover, these brands are fair-priced brands that can be consumed by young people who are more able to spend money than make. Even the designer brands are mainly new brands with relatively fair prices. However, Marui does not sell ladies and gentlemen's clothing, maternity clothes, children's wear, kimonos, dresses, or groceries such as stationery, furniture, and home gardening supplies. Each pavilion of Marui Department Store has clear positioning. For those less than 20 years old, they can go directly to the YOUNG Pavilion. Men go to the MEN' S Pavilion to by clothes. To buy tableware, people can go to the IN THE ROOM Pavilion. In order to attract young Chinese consumers, Marui also invited Chinese star Huang Xiaoming as the Chinese spokesperson for its brand. On June 15, 2009, Marui reached the cooperation with Alipay. Its Chinese shopping website (http://maruione.jp/cn/) officially entered the domestic market by accessing

Alipay's overseas merchant service. The operation model of personalization and specialization has enabled Marui to keep long-term health. At present, Marui has 22 chain department stores in Japan with annual sales of 500 billion yen. The big red signboard of Marui has a magical appeal to Japanese fashionable youth.

近年來，連鎖行業的龍頭——麥當勞也在不斷強化個性化、餐廳的多樣化風格，在廣州市場，以 LIM 風格、Form 風格與 Allegro 風格升級主題餐廳。其中 LIM（Less Is More）設計理念源自歐洲，即化繁為簡，自 2006 年起由法國知名設計師 Philippe Avanzi 為麥當勞量身打造。LIM 設計理念有 4 種不同的設計主題：鋒尚前沿（Edge）、悅享美食（Food）、至潮體驗（Extreme）和炫彩活力（Fresh）。這種全新的設計強調色彩豐富的牆面語言，將現代藝術、風格獨特的燈飾與優雅舒適的沙發座椅完美融合。既巧妙地秉承麥當勞的風格，同時也為人們帶來更多驚喜——使麥當勞呈現出融個性、流行、時尚、友好、快樂等元素為一體的「新面孔」。[151]

In recent years, McDonald's, the leader in the chain industry, has also continuously enhanced the personalitzation and diversified style of restaurant. In the Guangzhou market, the theme restaurants have been upgraded in LIM style, Form style and Allegro style. The design concept of LIM (Less Is More) style originated from Europe, which is simplifying. Since 2006, the famous French designer Philippe Avanzi has been tailoring the concept for McDonald's. LIM

---

151 麥當勞 10 家主題餐廳全新亮相 到 2014 升級餐廳佔 7 成 . 中國吃網 . 餐飲資訊 . 2012-03-11 Brand New Appearance McDonald's 10 Theme Restaurants. Upgraded Restaurants will account for 70% by 2014. 6eat.com. cyzx.cn. March 11, 2012

has four different themes: Edge, Food, Extreme, and Fresh. This new design emphasizes the rich wall language and combines modern art and unique lighting with elegant and comfortable sofa seating. It not only cleverly adheres to the style of McDonald's, but also brings more surprises to people-elements like personality, popularity, fashion, friendship and happiness.[151]

2014 年 4 月 18 日，麥當勞在廣州推出全球第一家 EATERY 旗艦店，它是由老店翻新而成。EATERY 意為小食堂，強調輕鬆休閒、把人聚集在一起的體驗。在設計上，引入蒸籠、青磚、算盤、大圓桌等中國元素；在佈局上，按用餐需求進行區隔——丁字形吊燈籠罩下的圓形空間，透過大紅色算盤圍攏，適用於麥咖啡的消費者；能容納 10 多人的大長桌，適合聚會與聊天；而由彩色圓形光點鋪就的桌面，對小朋友有莫大吸引力。[152]

On April 18, 2014, McDonald's launched the world's first EATERY flagship store in Guangzhou, which was refurbished from an old store. EATERY means a small canteen, emphasizing the experience of relaxing and gathering people together. In the design, it introduces Chinese elements such as steamers, black bricks, abacuses, big round tables. In the layout, it separates areas according to dining needs-the circular space under the T-shaped hanging lantern, surrounded by a large red abacus, is prepared for McCAFE consumers. Large tables that accommodate more than 10 people each, are used for parties and chatting. The table tops with colored round spots are very attractive to children.[152]

---

152 麥當勞的新食堂 . 第一財經週刊 .2014-04-28 McDonald's New Canteen. CBNWeekly. April 28, 2014

圖 Figure 6-1 麥當勞 EATERY 餐廳 McDonald's EEATRY

連鎖速食業正不斷追求個性化、特色化，有想法的正餐企業、正餐餐館更是在個性化、特色化之路上不斷追求極致。我們可以分享上海一家建在蘇州河畔一個由破舊的老工廠改造的創意園區裡的感官餐廳 Ultraviolet 的感官盛宴。[153]

The chain fast food industry is constantly pursuing personalization and specialization. The dining enterprises and dining restaurants with ideas are constantly pursuing the acme all the way long. Let's enjoy the sensory feast of the Ultraviolet, a sensory restaurant in Shanghai, built in a creative park transformed by a dilapidated old factory on the banks of the Suzhou River.[153]

主餐廳是一個類似「駭客帝國」電影裡的科幻實驗室，「純白色，未來感」是首先蹦入我腦海的兩個詞。位於這個高聳白色空間的正中央，是一種類似「會議室」的擺設——一個巨大的白色長桌和 10 張白色工作椅，由多部投影機和鎂光燈組成穹頂，空氣中散發著某些未知香料的味道⋯⋯這一切彷

---

[153] 僅十個座位全球第一個感官餐廳 http://www.yoka.com/life/travel/2012/0605667039.shtml Only Ten Seats, the World's First Sensory Restaurant http://www.yoka.com/life/travel/2012/0605667039.shtml

彿是一個拍攝未來風格電影的製片廠。

The main restaurant is just like the science fiction lab in Matrix.「Pure white」and「futuristic」are the first two words that enter my mind. Located in the center of this towering white space, a huge white long table and 10 white work chairs, composed of multiple projectors and magnifiers, makes the space a 「conference room」. You can smell some unknown spices in the air... All these seem to be a studio for filming future style movies.

在餐廳經理 Fabien Verdier 的指引下，食客走到有自己名字的座位前。待所有客人坐下後，Fabien 隻身離開，燈光漸暗，好戲即將上演。

Under the guidance of restaurant manager Fabien Verdier, the diners walk to the seat with their own name. After all of them have sat down, Fabien leaves alone and the lights dim-the play is about to take place.

讓所有人大開眼界的一刻出現了——四面牆上的大螢幕突然亮起，讓人感到整個房間在迅速下沉 (原因是巨大投影給人的錯覺)，一個男人在急促地朗讀些什麼，類似那些美國科幻電影的開頭，房間下沉得越來越快，四面牆上的影像迅速轉變，人們不知所措但被強烈地吸引著，「5、4、3、2、1......歡迎來到 Ultraviolet」一個男人的聲音傳來。

The eye-opening moment comes-the big screen on the four walls suddenly light up, making people feel that the whole room is sinking quickly (due to the illusion effect). A man is reading something hurriedly, like the beginning of American science fiction movies. The room is sinking faster and faster and the images on the four walls are changing rapidly. People are overwhelmed but strongly attracted.「5, 4, 3, 2, 1… Welcome come to Ultraviolet,」said a man.

第一個前菜的形狀酷似雪茄，放在一個金屬質感的容器裡，容器裡還有一些類似菸灰的粉。這支「雪茄」的外殼是糖漿，裡面包裹著鵝肝慕斯。品嚐的方式是沾一下菸灰缸裡的粉，再搭配口感濃郁的西班牙雪莉酒 (Sherry)，二者的口感微甜，搭配得相得益彰。在上這道菜的時候，背景音樂是義大利菸槍作曲家 Ennio Morricone 的曲目，四面牆上投射出美國西部荒野的影像。上 Fish & Chips 的時候，牆面的場景變成了濕淋淋的英國。

The first appetizer is shaped like a cigar and placed in a metal container with some ash-like powder. The shell of this 「cigar」 is syrup, in which is wrapped foie gras mousse. The way to taste the dish is to dip the 「cigar」 in the powder and have it with the rich-flavored Spain Sherry, both of which have a slightly

sweet taste and complement each other. When serving the dish, the background music is the repertoire of the Italian smoker composer Ennio Morricone and images of the Wild West of the United States are projected on all four walls. After that, when serving the second dish Fish & Chips, the scene on the wall turns into the wet Britain.

圖 Figrue 6-2　感官餐廳 Ultraviolet 類似雪茄的前菜 Appetizer in the sensory restaurant Ultraviolet-「Cigar」

第三道菜是松露炙烤湯汁麵包 (Truffle Burnt Soup Bread)。這道菜登場的時候被罩在一個圓形的透明玻璃蓋中，裡面煙霧繚繞，一聞原來是雪茄煙霧。透明罩下的麵包浸在奶油醬汁中，上面是泡沫，另一面則是烤過的松露，搭

配來自美國加州 Francis Coppola 酒莊的 Chardonnay 白葡萄酒。背景牆上是森林的動畫，空氣裡散發著淡淡的松木香。同時，用餐者面前的一堵牆緩緩打開，出現在眼前的是一棵 300 年老樟樹的根部！活生生的、巨大的樟樹根基一清二楚展現在人們眼前，感覺相當的「震懾」──這簡直是感官料理和當代裝置藝術的結合。

The third dish is Truffle Burnt Soup Bread. The dish is covered in a round transparent glass cover when being served, filled with smoke, which is proved to be cigar smoke. The bread under the transparent cover is dipped in cream sauce with foam on top and roasted truffles on the other side, comes with Chardonnay white wine from Francis Coppola, California. The background wall is the animation of forest, and the air exudes a smell of pine wood. At the same time, a wall in front of the diners is slowly opening and the roots of a 300-year-old camphor tree appear in front of diners. The vivid and huge roots clearly come into view, giving a shock to diners-this is nothing but a combination of sensory cuisine and modern installation art.

上「海水浸龍蝦」這道菜的時候，用餐者被鏡頭帶到了海邊，被海浪環繞著，伴隨著海鷗的叫聲，晚宴的主人 Fabien Verdier 捧著蒸氣中帶有海水的鹹味兒的炊具在房間踱步。

When serving 「Seawater Dipped Lobster」, the diners are taken to the beach by the images on the screen and surrounded by the waves. With the sound of the seagulls, the host of the dinner party Fabien Verdier is stepping in the room holding utensils with seawater-smelled.

圖 Figure 6-3 感官餐廳 Ultraviolet 與「海水浸龍蝦」菜品相得益彰的海浪場景 The wave scene in the sensory restaurant Ultraviolet and the「Seawater Dipped Lobster」

Ultraviolet 餐廳還有自己的「遊戲規則」：1 天只接待 10 個客人，菜單定死，給啥吃啥，還要提前 3 個月預定才能排上號；而且只能透過餐廳官網預定，沒有其他管道；餐費 3000 元 / 人，且預定當天需支付訂金 1000 元 / 人，之後是漫長的 3 個月等待；餐廳沒有具體位址，客人不能私自前往，需統一集合後發車把客人送到用餐地點。也正是這家把個性化、特色化做到超乎想像的餐廳，2013 年、2014 年連續兩年獲得亞洲最佳 TOP50 餐廳稱號，排名國內第一。

Ultraviolet also has its own 「game rules」: it only accepts 10 guests a day with fixed menu. People can only make reservation three months in advance through the restaurant's official website. The meal fee is 3,000 yuan a person, and the down payment should be made at 1000 yuan a person on the day of the reservation. The restaurant does not have a specific address so that people can not go without reservation, they need to be gathered up and driven to the dining place. The restaurant, which has been personalized and specialized by beyond imagination, won the title of Best TOP50 Restaurant in Asia for two consecutive years of 2013 and 2014, ranking first in the country.

二

從經營商品轉變為經營生活方式

2. Shifting from commodity operation to lifestyle operation

　　傳統商業零售業經營的核心是商品，企業圍繞商品組織貨源、開展銷售，消費者為獲得某種商品或服務進入商店，選購心儀的商品。

The core of the traditional retail business is commodities, or goods. The enterprise organizes the supply of commodities and conducts sales centered on the commodities. Consumers enter the store to obtain certain goods or services.

　　傳統商業企業均圍繞商品組織經營，百貨店、超市、便利店、生鮮菜店等業態有明確區隔、實行專業經營、品類管理，商品少到幾十種，多至數萬種。人們去商店的主要目的是購物，採購生活中需要的商品，消費目的明確單一。

Traditional enterprises conduct sales centered on the commodities. Department stores, supermarkets, convenience stores, and fresh food stores have clear divisions, professional operation and category management. There are dozens or even tens of thousands of products. The main purpose of people going to the store is to shop and purchase the goods that are needed in daily life, the purpose of which is clear and single.

在「雲消費」時代，雖然購買標準化生活日用品的管道很多，方式更為便捷，但人們依然需要走出家門，逛街交友、親朋聚會，需要有空間，商業零售業承載的功能更為複合，不僅要滿足消費者的購物需求，更重要的是要適應現代生活方式，滿足消費者娛樂、休閒、社交等需求。

In the era of cloud consumption, although there are many convenient channels for purchasing standardized daily necessities, people still need to go out of their homes to go shopping and have parties, all of which requires space. Therefore, the functions of commercial retailing are more complex. Such functions not only need to meet the consumer's shopping needs, more importantly, to adapt to the modern lifestyle to meet consumers' needs for entertainment, leisure, social interaction and others.

因此，在「雲消費」時代，消費者走進商店，是為了體驗其所認同或嚮往的一種生活方式。為此，商業經營不應當以業態、專業、品類劃分，而要圍繞一類消費人群的共同的生活方式整合商品和服務。

In the era of cloud consumption, consumers enter the store to experience a lifestyle they agree with or yearn for. To this end, business operations should not be divided into formats, professions, and categories, but integrate goods and services centered on the common lifestyle of a group of consumers.

歐美快速發展的生活方式中心（Lifestyle Center）體現了這種經營思想的變革。生活方式中心是體驗式消費中心，它更關注的是消費者在消費過程中的身心感受，通常具有露天開放及良好環境的特徵，強調以舒適友好的環境設計，營造多元化的體驗式消費空間。

The rapidly growing Lifestyle Center in Europe and America reflects this change in business thinking. The Lifestyle Center is an experience consumption center, which pays attention to the consumer's physical and mental feelings during the consumption process. It usually has the characteristics of open air and good environment, emphasizing a comfortable and friendly environment design to create a diversified experience consumption space.

1990 年，在美國邁阿密開業的 Cocowalk，總營業面積 2 萬平方米，娛樂與餐飲所佔的比例大於購物，其中擁有 16 塊放映螢幕的 AMC 影劇院佔據購物中心總租賃面積的 25%，成為其主力店。Cocowalk 採取的開放式建築結構和偏重餐飲、娛樂的組合成為生活方式中心的重要標準。1996 年後，生活方式中心快速發展，現在已經成為美國發展最快的商業模式，每年設施規模增長率達 20%，已經超過了購物中心。

In 1990, Cocowalk opened in Miami, USA, with a total business area of 20,000 square meters. The proportion of entertainment and catering is more than shopping in the mall. The AMC theatre with 16 screens occupies 25% of the total rental area of the shopping mall, which plays a major role. The open architecture of Cocowalk and the combination of catering and entertainment have become important standards for the Lifestyle Center. Since 1996, the Lifestyle Center has developed rapidly and now become the fastest growing business model in the United States. The annual growth rate of facilities is 20%, which has already exceeded that of the shopping mall.

20 世紀 90 年代在美國田納西州孟菲斯 German Town 區誕生的 Saddle Creek 生活方式中心將 Mall 的專賣店和開放式的底層商鋪結合在一起，歐洲

式的典雅與傳統連同中心裡的瀑布、長椅和磚鋪人行道共同營造了一個優雅與舒適的購物環境。入駐的商戶均為著名的品牌店和特色餐館，其建築設計、環境氛圍和商品構成均迎合高收入群體的口味和追求。

Saddle Creek, a Lifestyle Center born in the German town of Memphis, Tennessee in the 1990s combines Mall's specialty store with the open floor store. Together with European elegance and tradition along with waterfalls, benches and brick-paved pedestrians, all create a graceful and comfortable shopping environment. The settled merchants are all famous brand stores or specialty restaurants, the architectural design, environment and commodity composition of which cater to the tastes and pursuits of high-income groups.

圖 Figure 6-4　Saddle Creek 生活方式中心 Saddle Creek, a Lifestyle Center

2002 年在美國洛杉磯開業的 The Grove 生活方式中心，租賃面積 5.5 萬平方米，擁有 1 萬平方米停車場。這裡仿照了一個美國風情小鎮，精美的雕塑，連通的街道，綠色的公園、池塘，精巧的噴泉與娛樂、餐飲、購物服務設施和諧共融，成為加州居民休閒享受的又一重要選擇。

The Grove, another Lifestyle Center, which opened in Los Angeles in 2002, has a rental area of 55,000 square meters and a parking lot of 10,000 square meters. It is modeled after an American-style town, with delicate sculptures, interconnected streets, green parks, ponds, and exquisite fountains that are harmoniously integrated with entertainment, dining, and shopping services, making it an important choice for California residents to enjoy.

圖 Figrue 6-5　洛杉磯 The Grove 生活方式中心 The Grove, a Lifestyle Center in Los Angeles

近年來，在圍繞消費者的生活方式、重新安排商業佈局、整合商品和服務等方面，中國的一些商家也進行了可喜的探索。

In recent years, some Chinese merchants have also made gratifying explorations on lifestyle of consumers, re-arranging business layout and integrating goods and services.

2013 年開業的上海 K11 購物藝術中心（以下簡稱 K11）就是一家為消費者帶來全新生活體驗的生活方式中心。這個新商業設施推崇「藝術 · 人文 · 自然」三大核心元素的融合，以藝術互動、舞臺感的購物體驗、最潮

的多元文化社區商業為定位，圍繞核心元素打造整體體驗。首先，連接地鐵通道的佈置就很有藝術氛圍，給人帶來良好的消費期待；中心的各個角落都有藝術品，讓消費者隨時都能感到驚喜。K11 相信消費者的藝術眼光，並致力於培養大眾的藝術鑑賞能力，讓更多的消費者去接近藝術、理解藝術和欣賞藝術。為此，K11 與很多當代年輕藝術家合作，也為年輕藝術家打造一個展示發佈平臺。其次，K11 彙聚了一批國際品牌，但無論是 BURBERRY、Chloé、DOLCE & GABBANA、MaxMara、Vivienne Westwood 等眾多國際一線品牌，還是相對小眾的 Axes Femme、Collect Point 等品牌，均結合 K11「藝術 • 人文 • 自然」的核心理念構建獨一無二的概念店，讓熟悉品牌的消費者在此充分感到品牌文化與 K11 文化融合的令人耳目一新的文化魅力。再次，K11 在業態的佈局上也全面體現出藝術性，每位消費者來到商場之後，如果對商場動線、佈局、藝術品不是很瞭解，客服會在最短時間內為有需要的消費者安排 30 分鐘的遊覽路線，並進行導遊，使其可以欣賞公共空間的每一個藝術作品，瞭解它們的故事及內涵。K11 為每一位 VIP 顧客提供「定做訂製服務」，根據其愛好、消費習慣、購物習慣為其訂製特別的服務模式。所有服務資訊都在客戶服務終端的電腦上，只要某貴賓進入 K11 的服務領域，其所有資料會第一時間在客服臺體現出來，而一對一的客服專員會在第一時間為其提供最需要的服務。這種服務效應透過朋友圈層的口碑傳播而後擴大。另外，K11 還引進「都市農莊」、ABC COOKING STUDIO 等類似業態，增加消費者的參與度。K11 的「都市農莊」位於其 3 樓美食區，近 300 平方米的室內生態互動體驗種植區被分割成一塊塊田地，種植著辣椒、甜菜、番茄等新鮮的蔬菜，不僅可以觀賞，而且也可以供給一些餐廳。每逢週末，「都市農莊」

還會舉辦各種互動種植活動，讓大人、孩子體驗從一顆種子開始到蔬菜滿園的樂趣。除了農莊之外，這裡還有一個小小的農場，小香豬、小乳牛等各類可愛的動物輪流來做客，讓周邊不少的白領和小朋友流連忘返……[154]

The Shanghai K11 Art Mall (hereinafter referred to as K11), opened in 2013, is a Lifestyle Center that brings new life experiences to consumers. The new commercial facility promotes the integration of the three core elements of 「art, humanity and nature」. Based on the art interaction, stage-sensed shopping and the most diverse multicultural community business, the mall creates an overall experience centered on the core elements. Firstly, the connection of the subway space is very artistic, giving people a good expectation of consumption. There are works of art in every corner so that consumers can be pleasantly surprised at any time. K11 believes in the artistic vision of consumers and is committed to cultivating the artistic appreciation of the public, allowing more consumers to approach, understand and appreciate art. To this end, K11 works with many contemporary young artists and creates a showcase platform for young artists. Secondly, K11 has gathered a number of international brands. Either the international first-line brands such as BURBERRY, Chloé, DOLCE & GABBANA, MaxMara, Vivienne Westwood, or relatively small brands such as Axes Femme and Collect Point, all have unique concept stores of their own, so that consumers who are familiar with the brand can fully feel the refreshing cultural charm of the

---

154 周思立‧李欣欣. 都市農莊入駐上海 K11 購物中心主打體驗元素 [N]. 新聞晨報‧2013-07-10.

combination of the brand and K11. Thirdly, K11 also fully reflects the artistry in the layout. For consumers who are not familiar with the pedestrian flow, layout, and artwork, the customer service staff will arrange a 30-minute guided tour for them in the shortest time, so that they can enjoy every piece of art in the public space and learn about their stories and connotations. K11 provides customized services for each VIP customer which means to customize special service models according to their hobbies, consumption habits and shopping habits. All service information is kept on the customer service terminal. As long as a VIP enters the service area, all the information will be showed in the customer service station in the first time, and the one-to-one customer service specialists will provide the best service. This kind of service effect is expanded by the word-of-mouth spread in the friends circle. In addition, K11 also introduced 「urban farms」, 「ABC COOKING STUDIO」 and other similar formats to increase consumer participation. The 「Urban Farm」 is located in the food section on the 3rd floor. The nearly 300 square meters of indoor ecological interactive planting area is divided into blocks, planted with fresh vegetables such as pepper, beet and tomato, which can not only be used as a landscape, but also be supplied to restaurants. On weekends, a variety of interactive planting activities will also be held on the 「Urban Farm」 to let adults and children experience the fun of planting a seed into a garden. In addition to the vegetable farm, there is also a small ranch where small fragrant pigs, small cows and other cute animals take turns to make guests, making many white-collar workers and children lingering...

圖 Figure 6-6　上海 K11 購物藝術中心──藝術與體驗的中心 Shanghai K11 Art Mall-the mall of art and experience

圖 Figure 6-7　上海 K11 購物藝術中心──消費者與小香豬親密互動 Shanghai K11 Art Mall-consumers and small fragrant pigs interact closely

北京甘家口百貨商場（以下簡稱甘家口百貨）圍繞所在社區消費者的生活方式，進行了改革創新。甘家口百貨是位於北京甘家口商圈的社區主力百貨店。隨著居民消費能力的普遍提升，人們的生活方式已出現質的轉變，已經從對商品價值的追求升級為對生活方式的追求，文化品位、群體認同、品質生活、快樂體驗將逐漸居於主導地位。做為一家社區型百貨店，甘家口百貨應該快速適應現代生活方式對零距離、「一站式」、休閒化、體驗化的社

區商業消費需求，由基本需求保障模式向集聚商業中心、休閒中心、生活服務中心、人際交往中心等多種功能於一體的社區商業發展模式轉型。為此，甘家口百貨提出了做「社區生活中心新模式的引領者」的戰略目標，致力於打造新一代社區休閒品質生活中心。

Beijing Ganjiakou Department Store has carried out reform and innovation based on the lifestyle of consumers in the community. Ganjiakou Department Store is the main department store in the business district of Ganjiakou. With the general improvement of residents' consumption power, people's lifestyle has undergone a qualitative change, from the pursuit of commodity value to the pursuit of lifestyle, enabling cultural taste, group identity, quality life, and happy experience gradually play a dominant part. As a community-based department store, Ganjiakou Department Store should quickly adapt to the modern lifestyle needs for zero-distance, one-stop style, leisure, and experience community consumption, and transform from the model of guaranteeing basic demand to the model that integrates multiple functions such as commercial centers, leisure centers, life service centers and interpersonal communication centers. To this end, Ganjiakou Department Store has proposed the strategic goal of 「leading the new model of community life center」 and is committed to creating a new generation of high-quality community leisure center.

甘家口百貨著手一系列改革調整，包括從以商業功能為主轉變為以休閒品質生活功能為主，從以購物功能為主轉變為以休閒、餐飲、生活服務等多元功能為主，從以大眾品牌為主轉變為以時尚休閒品牌為主，營造主題化、情境式的賣場環境，打造社區專屬時尚精品超市等。如在賣場環境營造方面，

甘家口百貨按不同人群特點打造不同的主題情境，在每個區域打造出不同的主題氛圍，在不同主題區域營造出不同的主題生活情境。據悉，甘家口百貨正以其新的戰略目標為核心，全力打造社區休閒品質生活中心，他們的發展值得期待。

Ganjiakou Department Store has embarked on a series of reforms, including shifting from commercial functions to leisure functions, from shopping functions to leisure, catering, and life services, from mass-market brand orientation to a fashion and leisure brand orientation, thus creating a themed and contextual store environment, and a community-specific fashion supermarket. For example, in the construction of the store environment, Ganjiakou Department Store creates different thematic situations according to the characteristics of different groups of people, different theme atmospheres in each area, and different theme life situations in different theme areas. It is reported that Ganjiakou Department Store is focusing on its new strategic goal and is committed to building a community leisure center. Their development is worth looking forward to.

三

從向消費者銷售轉變為消費者俱樂部的成員

3. Shifting from a seller to a member of the consumer club

　　在傳統零售業中，商家和消費者是對立的賣家和買家的關係，賣家向買家銷售商品，買家支付費用，透過賣家的銷售獲得商品，銷售是商品買賣的橋樑。

　　In the traditional retail industry, merchants and consumers are in a relation of opposition. Sellers sell goods to buyers and buyers pay for the goods. Sales is the bridge between the two parties.

　　在「雲消費」時代，以消費者為中心、以滿足消費者的個性化消費為價值核心，所以商家必須以消費者為中心開發產品，圍繞消費者期待整合資源，與消費者一起創造和分享價值，營造全面體驗。因此，商家一定要樹立我就是消費者的意識，把自身定位從商品的銷售者轉變為消費者俱樂部的成員。

　　In the era of cloud consumption, consumer orientation is the core value. Therefore, merchants must develop products and integrate resources centered on consumers, as well as creating and sharing value with consumers in order to create a comprehensive experience. To this end, merchants must realize that they are also consumers and transform their position from the sellers to the members of the consumer club.

　　這裡不妨借鑑紅星美凱龍為消費者免費提供家居顧問全程導購服務的案

例。紅星美凱龍是中國家居行業的領軍品牌，該企業以做「家居生活專家」為己任，為消費者提供「家居顧問全程導購」服務，以自身的家裝、家居專業設計人員為家居顧問，代入消費者需求，為消費者提供設計、預算、選材、施工及維修方面全程服務，滿足消費者「一站式」服務需求。[155]

Take the case of Red Star Macalline who provides free shopping guide services for consumers as an example. Red Star Macalline is the leading brand in China's home furnishing industry. The company is committed to being an expert on home furnishing and provides consumers with free full-time shopping guidance, with their own professional designers as consultants, who are eager to meet consumer demand and provide consumers with full service in design, budget, material selection, construction and maintenance to meet the "one-stop" service needs.[155]

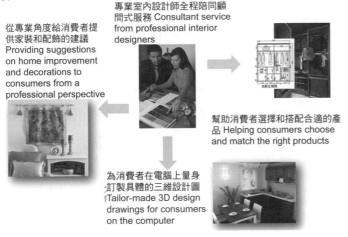

圖 Figure 6-8　紅星美凱龍家具顧問全程導購服務 Red Star Macalline's full-time shopping guidance

---

155 紅 星 美 凱 龍 從 規 模 體 量 到 文 化 內 核 的 全 面 超 越 . http://www.nbd.com.cn Comprehensive Transcend by Red Star Macalline from the Scale to the Cultural Core. http://www.nbd.com.cn

再看前文提到過的 NIKEID 線上平臺個人訂製時尚運動鞋。NIKEID 線上平臺的推出是建立在 NIKE 已經積累了足夠龐大的高黏性消費者群體和廣泛品牌認知以及無所不在的管道基礎上的。該平臺集合體驗行銷、個性化行銷、線上行銷等於一身，消費者可以任意選擇一款 NIKE 產品，透過 DIY 的形式改變每一個環節的顏色，設計出一款屬於自己的獨一無二的 NIKE 產品，還可以在這款產品上標明自己的個性化簽名。這種個性化的經營形式，契合了「雲消費」時代消費個人化、體驗化的特徵，商家從銷售方成為消費者俱樂部的成員，因此得到年輕族群的高度認可。

Let's return to NIKEID, the online platform for customized fashion sneakers mentioned above. The launch of the NIKEID is based on the fact that NIKE has accumulated a large and high-stickness consumer group and a wide range of brand awareness and ubiquitous channels. The platform integrates experience marketing, personalized marketing and online marketing. Consumers can choose a NIKE product, change the color of every detail through DIY at will to design a unique NIKE product. They can also mark their own signatures on the works. This kind of personalized business form is in line with the characteristics of consumer personalization and experience in the era of cloud consumption. The merchants become members of the consumer club instead of the sellers, so that they are highly recognized by the young people.

這裡轉引一位消費者在網路社區上的發言和曬的圖片，感受這種個人訂製化的經營方式給年輕人的生活帶來的快樂和分享 [156] ：

The following are a speech made by a consumer in the online community and a picture posted by him. Let's feel the joy this customized business brought to the young people[156]:

2010 年參觀世博會的時候，買了 1 雙耐克 AIR MAX 跑鞋。這時手裡現在已經有 4 雙耐克當季的旗艦跑鞋和 1 雙紅黑的 LEBRON 籃球鞋……但是，一直找不到我自己想要的那種鞋子！

When I visited the Expo in 2010, I bought a pair of Nike AIR MAX sneakers. At that moment, I already had four pairs of Nike's flagship sneakers and one pair of red and black LEBRON basketball shoes... But I couldn't find the kind of shoes I really wanted!

而現在，得益於 Nikeid 的線上訂製，終於可以製作出自己想要的那種感覺的鞋了！夢想終於實現了！

Now, thanks to NIKEID online customization, I can finally create the shoes I want! The dream has finally come true!

我一直想要一種鞋子：

What I really want is a pair of shoes that:

1. 可以在冬天穿（因為我那 4 雙跑鞋都是夏天才能穿的，不保暖）──所以需要皮製的。

a. can be worn in winter and made of leather(because another four pairs of sneakers are for summer whith no heat preservation).

2. 舒適、柔軟，長時間步行不累，甚至可以去旅遊穿。

---

156 http://www.chiphell.com/thread-1005588-1-1.html.【私人訂製之 NIKEid】第一雙訂製 Nike Max Air 跑鞋 [Customization-NIKEID] The First Pair of Customized Nike Max Air Sneakers

b. is comfortable, soft and energy-saving to wear even for a trip.

3. 萬能搭配！無論是穿休閒裝，還是正裝，都不用過多考慮，只需這 1 雙鞋子即可！

3. is all-match so that it matches whatever clothes.

正所謂，遠看像皮鞋，近看確實是皮鞋，但實質是跑鞋，呵呵！

That is what I mean by a pair of leather shoes which in fact is a pair of sneakers! Aha!

圖 Figure 6-9　網友在網路社區上曬的 Nikeid 訂製鞋 Customized NIKEID shoes posted by a netizen in the online community

從這個案例我們可以看到，透過 NIKEID 線上平臺個人訂製時尚運動鞋這種經營模式，消費者與商家的界限模糊了，消費者從中得到創造與分享的快樂，這何嘗不是商家的成功呢？

From this case, we can see that such business model narrows the boundaries between consumers and merchants, the former get the joy of creation and sharing which in the same, is the success of the latter.

四

從整合供應鏈轉變為整合需求鏈

4. Shifting from an integrated supply chain to an integrated demand chain

　　供應鏈的概念是從擴大的生產（Extended Production）概念發展來的，它將企業的生產活動進行了前伸和後延。哈里森 (Harrison) 將供應鏈定義為：「供應鏈是執行採購原材料，將它們轉換為中間產品和成品，並且將成品銷售到用戶的功能網鏈。」史蒂文斯（Stevens）認為：「透過增值過程和分銷管道控制，從供應商到顧客的流就是供應鏈，它開始於供應的源點，結束於消費的終點。」[157] 一般通常意義上，供應鏈是指商品到達消費者手中之前各相關者的連接或業務的銜接，是圍繞核心企業，透過對資訊流、物流、資金流的控制，從採購原材料開始，製成中間產品以及最終產品，最後由銷售網路把產品送到消費者手中的將供應商、製造商、分銷商、零售商，直到最終用戶連成一個整體的功能網鏈結構。

　　The concept of the supply chain is developed from the concept of Extended

---

157 唐納德·沃特斯（Donald Waters）. 供應鏈管理概論：物流視角 [M]. 高詠玲，譯. 北京：電子工業出版社，2011-1-1. Donald Waters. Introduction to Supply Chain Management: From The Perspective of Logistics [M]. Gao Yuling, Trans. Beijing: Publishing House of Electronics Industry, Jan 1, 2011.

Production, which extends the production activities of the company. Harrison defines the supply chain as 「the functional network chain that procures raw materials, converts them into intermediate products and finished products, and sells the finished products to users." Stevens regards the supply chain as 「the chain between suppliers and customers with the value-added process and distribution channel control, which starts at the source of supply and ends at the end of consumption.」157 In the general sense, supply chain refers to the interface or business relations among relevant poeple before the goods finally reach the consumers. Centered on the core enterprises, through the control of information flow, logistics and capital flow, it procures raw materials to produce the intermediate products and final products, and finally the products are sold to the consumers by the sales network including suppliers, manufacturers, distributors and retailers.

對企業而言，供應鏈就是透過計畫（Plan）、獲得（Obtain）、存儲（Store）、分銷（Distribute）、服務（Serve）等這樣一些活動在顧客和供應商之間形成的一種銜接（Interface），從而使企業能滿足內外部顧客的需求。優化供應鏈是傳統商業提升盈利能力的法寶，供應鏈每一個環節的優化，都能帶來明顯的直觀的效益。

For the enterprise, the supply chain is a kind of Interface between the customers and the suppliers through activities such as Plan, Obtain, Store, Distribute, and Serve, so that companies can meet the needs of internal and external customers. Optimizing the supply chain is a magic weapon for traditional business to improve their profitability. The optimization of each link in the supply chain can bring obvious and intuitive benefits.

需求鏈管理 (Demand Chain Management，DCM) 是按照從終端顧客到供應商的順序對整個供應鏈的管理和協調，由最終的顧客驅動整條供應鏈的運作，產品和服務由於需求的拉動 (不是推動)，實現從終端顧客到製造商、再到原材料供應商的一個過程。需求鏈管理擴展了傳統的供應鏈管理的概念，其核心是強調滿足顧客實際需求，創造卓越的顧客價值，其目標則在於獲取整個鏈條的協同效應，進而取得較之單個個體更好的效益。

Demand Chain Management (DCM) is the management model that manages and coordinates the entire supply chain in the order from the end customers to the suppliers. The operation of the entire supply chain is driven by the end customers. Products and services are driven (not promoted) by demand so that the whole process is from the end customers to the manufacturers and then to the raw material suppliers. Demand chain management expands the traditional concept of supply chain management of which the core is to meet the actual needs of customers and create superior customer value. The goal is to obtain the synergy effect of the entire chain, thus achieving better benefits than before.

需求鏈管理可以簡單地描述為以市場為導向的供應鏈管理。市場導向做為行銷觀念的體現，表現為顧客創造優良價值、為企業創造優異績效的組織文化。市場導向還被描述為行銷觀念的執行，即設計出比競爭者更好地滿足顧客的一套活動。不論市場導向的焦點在於組織文化還是行銷觀念的執行，學者們對於市場導向的要素存在著一致意見。這些要素包括：有關顧客和競爭者的持續和系統的資訊收集、資訊的跨職能分享和活動的協調，以及對競

爭者行動和變化的市場需要的快速回應。

DCM can be described simply as the market-oriented supply chain management. Market orientation is the embodiment of marketing concept, which manifests as the organizational culture in which customers create excellent value and performance for enterprises. Market orientation is also described as the implementation of marketing concept, that is, designing a set of activities that better satisfy customers than competitors. Whether the market focuses on organizational culture or the implementation of marketing concept, scholars have a consensus on the elements that determine the market orientation, including continuality of customers and competitors, systematic information collection, cross-functional sharing of information, activities coordination, and rapid response to competitors' actions and changing market needs.

與市場導向的 3 要素相對應,理想的需求鏈管理在資訊處理活動方面也存在 3 個維度:鏈中所有成員主動獲取消費者的需求資訊,分析顧客的結構與行為;相互傳遞、共用市場訊息和知識,使供應鏈成員在對市場的理解與認知上取得一致;基於市場訊息的獲取和傳遞,更好地協調供應鏈成員的行為,實現供應鏈管理的一體化。

Corresponding to the three elements of market orientation, the ideal DCM also has three dimensions in information processing. First, all members in the demand chain actively obtain consumer demand information and analyze customer structure and behavior. Second, all members communicate and share market information and knowledge with each other, so that supply chain members can achieve a consensus on understanding and of the market. Third, based on the acquisition and transmission of market information, all members better coordinate with others to realize the integration of supply chain management.

但在「雲消費」時代，商業盈利的關鍵是圍繞顧客需求提供商品和服務，所以研究顧客需求規律，發現需求、創造需求，並整合資源服務需求是企業經營的核心點。因此，在「雲消費」時代需求鏈意識應先於供應鏈意識。

However, in the era of cloud consumption, the key to profitability is to provide goods and services centering on customer needs. Therefore, researching customer demand patterns, discovering and creating demand, and integrating resource service requirements are the core points of business operations. That is why the demand chain awareness is prior to the supply chain awareness in the era of cloud consumption.

需求鏈尋求使顧客滿意並為顧客解決問題的共同目標，把管道成員聯合起來。其具體職能有：收集分析顧客未滿足的需求資訊；發現能夠執行需求鏈所需職能的夥伴；與鏈中其他成員分享有關顧客資訊及可利用的技術、物流機遇和挑戰等方面的資訊；開發解決顧客問題的產品和服務，以及開發並執行最優的物流、運輸和配送方法，以顧客期望的形式交付產品和服務。

The demand chain seeks common goals of satisfying customers and solving their problems. The specific functions include collecting and analyzing customer's unmet needs, finding partners who can perform the functions required in the demand chain, sharing information about customer information and available technologies, logistics opportunities and challenges with other members in the chain, and develop products and services that can solve customer problems, as well as develop and implement optimal logistics, transportation and distribution methods to deliver products and services in the form that customers expect.

下面我們共同分享日本零售企業 Family Mart 及美國亞馬遜需求鏈管理的案例：

Let's share the DCM case studies of Japanese retail company FamilyMart and Amazon from America.

（一）日本零售企業 Family Mart 的需求鏈管理實踐
DCM practice of Japanese retail company FamilyMart

隨著經濟的發展，日本消費者的消費方式開始轉變，不再單純追求便宜的商品，而是更加講究生活的品質和便利性。因此，Family Mart 於 2001 年設立 DCM(Demand Chain Management) 推進室，開始實施需求鏈管理計畫。這個計畫主要分為初期階段和完善階段。

With the development of the economy, the consumption patterns of Japanese consumers have begun to change. Instead of simply pursuing cheap goods, they are more concerned with the quality and convenience of life. Therefore, FamilyMart established the DCM Promotion Room in 2001 to implement the DCM plan, which is mainly divided into the initial stage and the polishing stage.

1· 初期階段 The initial stage

在初期階段，Family Mart 需求鏈管理的重心及 DCM 推進室的職責主要

集中在軟、硬體環境建設上，以達到能夠快速、準確、低成本地組織到與店鋪需求相一致的商品的目的。為此，Family Mart 主要從以下幾個方面進行了建設：

In the initial stage, FamilyMart's DCM and DCM Promotion Room's responsibilities are mainly focused on the construction of software and hardware environment, in order to quickly, accurately and cost-effectively organize goods that are consistent with the needs of the store. To this end, FamilyMart paid great attention on the following aspects.

■ 在資訊系統建設方面 Construction of information system

2001 年 6 月 Family Mart 在全部店鋪成功導入新店鋪系統，並且在店內開始應用移動式資料終端，實現了店鋪內商品資料獲取、處理的自動化。2001 年 8 月進一步導入新資訊分析系統，並在 2003 年 4 月在主要業務夥伴間實現了銷售資料共用。

In June 2001, FamilyMart successfully introduced a new store system in all its stores, and began to use mobile data terminals to automate the collection and processing of product data in the store. In August 2001, the new information analysis system was introduced. Moreover, sales data sharing was realized among major business partners in April 2003.

■ 在業務流程改造方面 Transformation of business process

2001 年 1 月，Family Mart 以實施新店鋪系統為契機，依託新店鋪系統的即時銷售資料處理機能，開始在全部店鋪引入單品管理機制，推行以訂貨分擔化為目標的 SST(Store Stuff Total) 體制，實現由店長、管理員訂貨向店

員參與訂貨的機制轉變。這樣不僅能及時捕捉到商品資訊動向，而且還進一步提高了訂貨的精確度。

In January 2001, FamilyMart took the opportunity of running the new store system to introduce a single item management mechanism in all stores, and implemented SST (Store Stuff Total) targeting order sharing based on the real-time sales data processing function of the new store system. In this way, staff can participate in restocking, which not only helps to capture the movement of product information in a timely manner, but also further improve the accuracy of the order.

■ 在物流規劃方面 Logistics planning

2001 年，Family Mart 與伊藤忠商社共同運營的物流中心進行了集約化改造，降低了物流成本。

In 2001, the logistics center operated by FamilyMart and ITOCHU Corp. was intensively transformed to reduce logistics costs.

2· 完善階段 The polishing stage

從 2006 年開始，Family Mart 的需求鏈管理進入了完善階段，主要從以下兩個方面採取了措施：

Since 2006, FamilyMart's DCM has entered the polishing stage where it takes measures mainly in the following two aspects.

■ 強化商品的需求管理意識 Strengthening the awareness of demand management of commodities

Family Mart 以與製造商、批發商資訊共用為核心，在實現商品流通合理

化、效率化的同時，透過開發、生產適銷商品追求顧客附加價值的最大化。2006 年 3 月，Family Mart 將原來的 DCM 推進室提升為 DCM 推進部，並將其置於商品本部之下。

FamilyMart focuses on the information sharing with manufacturers and wholesalers. While realizing the rationalization and efficiency of commodity circulation, it strives to maximize the added value of customers through the development and production of marketable products. In March 2006, FamilyMart upgraded the DCM Promotion Room to the DCM Promotion Department and placed it under the Commodity Headquarter.

■ 推行「結構改革」舉措 Promoting the structural reform

Family Mart 開展的「結構改革」舉措主要包括店鋪、商品改革，收益結構改革，成本結構改革和意識改革 4 項內容。[158]

The structural reform carried out by FamilyMart mainly include four aspects, they are store and commodity reform, income structure reform, cost structure reform and awareness reform.[158]

經過兩個階段的建設，到 2007 年，Family Mart 經過 6 年時間的探索已初步建立了自己的需求鏈管理系統（如圖 6-10 所示）。

After two phases of construction, by 2007, FamilyMart has initially established its own DCM system after six years of exploration (as shown in Figure 6-10).

---

158 劉振濱，田慧 . 日本零售企業需求鏈管理的經驗借鑑 [J]. 江蘇商論，2008（6）. Liu Zhenbin, Tian Hui. Experiences from the Demand Chain Management of Japanese Retail Enterprise [J]. Jiangsu Commercial Forum, 2008(6).

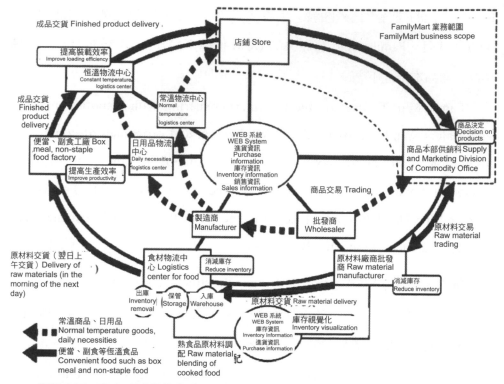

資料來源：社會·環境報告書（2007）.2007.06.P23 Source: Social and Environmental Report (2007).June, 2007.P23

（二）美國亞馬遜的需求鏈管理實踐
Amazon's practice on DCM

■ 以最低成本，提供最佳體驗 Providing the best experience at the lowest cost

亞馬遜創業之初是一家網路書店。亞馬遜以先進科技為依託，透過交易全流程掌控，建立精實、有效率的倉儲、配銷中心、存貨管理和訂單處理中心，精準掌握、預估甚至創造顧客的需求，提供便利的搜索功能和完整的產

品資訊，提供顧客從上網搜索、選購商品直到如期（甚至提早）收到貨品全過程的購物體驗。為此，亞馬遜主要開展了以下幾方面工作：

At the beginning, Amazon was an online bookstore. Relying on advanced technology, Amazon controls the entire process of transactions and establishes a lean and efficient center of warehousing, distribution, inventory management and order processing to accurately grasp, predict and even create customer needs, providing convenient search functions and complete product information as well as a shopping experience from the Internet search, purchase of goods until the receipt of the goods on time (or even in advance). To this end, Amazon has mainly carried out the following measures.

搜索：2003 年年底，亞馬遜推出「全文檢索」功能，隨後亞馬遜推出「A9 搜尋引擎」，可保留顧客的搜索及瀏覽紀錄。

Searching: At the end of 2003, Amazon launched the "full-text search" function, and then the "A9 Search Engine" to retain customers' search and browsing history.

選項：不但呈現書籍基本資訊，還提供專業編輯及讀者的評論和評等，做為顧客選擇時的參考。

Selection: Amazon provides not only the basic information of the books, but also the professional editors and readers' comments and ratings, as a reference for selection.

低價：亞馬遜堅持降價求售的模式，力求規模經濟，並透過有效率地管理訂單、倉儲與庫存來削減成本，以貼補由於提供優惠價格所招致的虧損。

Low price: Amazon insists on the model of price reduction and discounts to strive for economies of scale. It also cuts costs by efficiently managing orders, warehousing and inventory to offset the losses incurred by offering preferential prices.

免運費：在亞馬遜訂購款項超過一定額度，則運費全免。

Free shipping: If you order over a certain amount on Amazon, the shipping fee is free.

「一點靈」（1-Click）：亞馬遜開發出「一點靈」技術，顧客只要曾在網站進行交易，就可透過這項機制，輕點滑鼠鍵，交易即告完成，無須屢次重複填寫複雜的訂單。

1-Click: Amazon has developed the 1-Click technology. Customers who have made an order on the website before can use this mechanism by simply clicking the mouse button to complete the transaction without having to repeat the complicated filling-in.

■ 讓顧客找到產品，也讓產品找到顧客 Letting customers find products and vice versa

亞馬遜不斷致力於透過精密的技術，追蹤顧客的消費習慣，發掘出甚至連顧客自己都還不知道的需求。以著名的「推薦機制」為例，亞馬遜除了依據顧客實際購買的品類進行相關產品的推薦之外，還會透過「購買此書的顧客，也購買了……」的方式，營造出一種「電子同好」的氣氛，甚至顧客只是曾經在網站上搜索過某位作者的姓名，顧客很可能在下次進入亞馬遜網站時，就發現網頁上出現了該作者的著作。如果碰到健忘以致重複購買產品的顧客，亞馬遜也會根據以往的交易記錄，發出警訊。

Amazon is constantly striving to track customer consumption habits uncover needs that even customers themselves don't know with teh help of sophisticated technologies. Taking the famous recommendation mechanism as an example, in addition to recommending related products based in the same categories to the customers, Amazon will also create an atmosphere of 「virtual hobby friend」 through recommendation word like 「customers who bought this book also bought...」 What' s more, customers who just searched the name of an author on the website are likely to find the author's work on the webpage next time entering the Amazon website. For customers who incautiously repeats purchase, Amazon will also issue a warning based on transaction history.

■ 未下單，先發貨 Delivery in advance before placing the order

2013 年 12 月，亞馬遜獲得了一項名為「預測式發貨」的新專利，可以透過對使用者資料的分析，在他們還沒有下單購物前，提前發出包裹。雖然包裹會提前從亞馬遜發出，但在用戶正式下單前，這些包裹仍會暫存在快遞公司的轉運中心或卡車裡。[159]

In December 2013, Amazon received a new patent called 「Predictive Delivery」, which means packages are sent out in advance before customers place the order according to user data analysis. Although the packages will be sent out from Amazon in advance, they will remain in the courier's transshipment center or truck until the user officially places an order.[159]

---

159 亞馬遜開發新技術：未購買 先發貨 . 新浪科技 . 2014-01-18 Amazon Develops New Technology: Delivery in Advance Before Placing the Order. Sina Technology. Jan 18, 2014

■ 以客為尊，人人都是 VIP Customer-oriented, everyone is the VIP

在亞馬遜，無論是不是會員，永遠都享有折扣。而且，顧客每次造訪亞馬遜，網站都會自動辨識出訪客身分。亞馬遜還針對每位顧客打造了一間貴賓室，名為「XXX 的商店」，店裡所提供的推薦訊息，包括新產品、將要上市的產品、特價品等。在專屬商店裡，顧客不但可以針對亞馬遜所推薦的商品表達意見（評等、不感興趣、已擁有），還可以知道自己「為什麼會被推薦這項產品」。

In Amazon, whether you are a member or not, you will always get a discount. Moreover, every time a customer visits Amazon, the website automatically recognizes the identity of the visitor. Amazon has also created the VIP room for each customer, called "The Store of XXX". Recommendations provided by the store include new products, upcoming products, specials and so on. In the store, customers can not only make reviews (rated, not interested, owned) for the recommended products, but also the reason why they are recommended.

總之，亞馬遜透過掌握顧客的消費習慣與模式，預知甚至創造顧客內心深處的渴望，從而讓顧客的期待與產品能夠完美地結合，打造以顧客為中心的體驗。[160]

In short, Amazon grasps the customer's consumption habits and patterns, predicts or even creates the desire deep inside customers, so that customers' expectations can be perfectly combined with the products to create a customer-centric experience.

---

[160] 亞馬遜善用科技預知顧客需求. 中國證券報 Amazon Uses Technologies to Anticipate Customer Needs. China Securities Journal

五

善於「雲整合」資源 提升發展速度

5. The "cloud integration" of resources improves development speed

　　傳統零售業受到時間、空間的限制，商流、物流、資金流侷限於行業或企業內部封閉式運行，資源難以整合和共用，能夠收集的資訊有限。比如，顧客在一個門店何時來過一次，何時瀏覽過哪個產品，瀏覽時是什麼感覺，甚至包括表情、意願等。這些資料很難透過傳統的線下方式收集，也很難與每次光臨門店的資訊進行對比，很難快速在他下次光顧的時候，找到一個合適的產品推薦給他。同時單一企業的資料量有限，難以進行更深入的資料分析和比對。

　　The traditional retail industry is limited by time and space and the business flow, logistics flow and capital flow are limited to closed operations within the industry or enterprises so that resources are difficult to be integrated and shared, and the information that can be collected is limited. Data such as the time a customer visits a store and what products he has browsed as well as the feel or even facial expressions and willings, are difficult to collect through traditional offline methods, and it is difficult to compare with the information of each visit to the store. Besides, it is also difficult to quickly find a suitable product for the customer in his next visit. At the same time, the amount of data in a single enterprise is limited, making it difficult to conduct more in-depth data analysis and comparison.

在「雲消費」時代，商品流通的增值空間被壓縮，封閉化運行難以生存，而自由開放的市場和資訊資源平臺能夠為越來越多的交易提供更加低廉的產品或服務，因此企業必須轉變觀念，以「雲整合」意識為思維起點，在「雲」環境下，以「整合」為思想基礎，任我所需、任我所用，實現企業的跨越式發展。

In the era of cloud consumption, the value-added space of commodity circulation is compressed, and it is difficult for the closed operation to survive. The free and open market and information resource platform can provide more low-cost products or services for more and more transactions, so enterprises must change the business concept. They should take the cloud integration consciousness as the starting point of thinking to maximize the value of resources they need, thus achieving the leap-forward development of the enterprise.

我們認為，「雲消費」時代的「雲整合」，就是在「雲消費」環境下，轉變以我為中心、封閉運行的思維模式，利用專業技術手法，實現跨行業、跨時空、跨業態的整合資源、整合資訊、整合服務、整合交易、整合供應鏈……將一切企業和社會發展需要的內容整合在一起，發揮整合服務的優勢，創造新的價值。透過「雲整合」，在產業資源整合、供應鏈整合、網路商業構建等方面，發揮專業化分工的優勢，提升運行效率，發展速度將得以倍增。

We believe that the cloud integration in the era of cloud consumption is to transform the thinking mode that is self-centered and separated from the outside world in the cloud consumption environment, and use professional and technical means to integrate resources, information, services, transactions and supply chains in a cross-industry, cross-space and cross-industry manner as well as integrate all needed contents of business and social development and take advantage of integrated services to create new value. Cloud integration also brings into play the advantages of specialized division of labor, improve the operational efficiency to double the development speed in the aspects of industrial resource integration, supply chain integration, network business construction, etc.

　　淘寶網就是「雲整合」的典範。淘寶網的發展早已打破了百貨、超市、家電、建材家具、便利店等業態和業種的界限,甚至已打破線上線下的界限。淘寶網旗下主打團購概念的聚划算、主打拍賣概念的淘寶拍賣會、引導消費活動的一淘網等,不僅打破了零售與批發的界限,而且創造了更多的消費機會。隨著 2013 年 10 月 31 日淘寶網拿到證監會頒發的基金協力廠商電子商務平臺經營資質,11 月 1 日淘寶基金理財頻道上線,它甚至已經打破了資訊產業與金融產業的界限,使電子商務跨界、無邊界整合成為現實。

　　Taobao is a paragon of cloud integration. The development of Taobao has already broken the boundaries between department stores, supermarkets, home appliances, building materials and furniture, convenience stores and other industries, and even the boundaries between online and offline. Taobao's group purchasing platform Juhuasuan, the auction platform Paimai.taobao.com, and the Etao.com, which guides the consumption activities, not only break the boundaries between retail and wholesale, but also create more consumption opportunities. With the license of the third-party e-commerce platform issued by the China Securities

Regulatory Commission on October 31, 2013, Licai.taobao.com was launched on November 1st. It has broken the boundaries between the information industry and the financial industry, making the cross-border and borderless integration of e-commerce come true.

總部設在北京石景山區的易盟集團（以下簡稱易盟）對家政服務的整合，就是這樣一種「雲整合」。這家公司以給客戶帶來更加安全、便捷的家政服務做為重要職能。為此，他們對家政服務行業進行了規範整合：

The integration of domestic service by Yimeng Group (hereinafter referred to as Yimeng), headquartered in Shijingshan District, Beijing, is an example of cloud integration. The company focuses on bringing safer and more convenient domestic services to its customers and conducts the standardized integration of the domestic service industry.

一是整合人員。易盟整合了 10 餘萬人的家政服務大軍，相關服務人員的工作履歷均有詳細記錄，並為每一位客戶和家政服務人員上了保險。

First, integrating personnel. Yimeng has integrated the domestic service service troops of more than 100,000 in total. The work experience of relevant service personnel has all been recorded in detail, and insurance has been provided for every customer and domestic service personnel.

二是整合培訓資源。易盟對家政服務人員進行規範化專業培訓和指導。

Second, integrating training resources. Yimeng conducts standardized professional training and guidance for domestic service personnel.

三是整合評價系統。易盟讓客戶能在統一開放平臺上瀏覽所有家政服務

人員評價紀錄並進行評價。

Third, integrating the evaluation system. Yimeng allows customers to view and evaluate service evaluation of all personnel on a unified open platform.

四是整合支付體系。目前，易盟的盈利方式是與服務人員分成。可以是客戶直接付款然後易盟分成，也可以是客戶付款給家政人員所在公司，然後易盟與家政公司進行分成。

Fourth, integrating the payment system. At present, Yimeng's profit model is to divide commissions with service personnel. It can either be Yimeng to divide the direct payment by the customer, or Yimeng to divide the commissions with company where the service personnel are located.

五是統一平臺整合匹配服務資源。易盟 95081 家政寶在 PC 終端的實現上更類似於地圖搜索功能，極大提高了尋找匹配的家庭服務業員資訊的效率。在手機端上，95081 家政寶則是一款基於 LBS 地理位置服務的手機用戶端的應用軟體，為客戶提供小時工、鐘點工、保姆、月嫂、育兒嫂、保潔、管道疏通等「一站式」家政服務。客戶可以透過手機用戶端尋找離他最近的適合的服務人員，並與服務人員或服務人員所在企業取得聯繫。對於消費者，易盟利用十幾萬人的龐大的資料庫進行互聯網環境下的「碎片化」搜索，在資料庫內進行精準化匹配，這樣就能判斷這個活由哪個家政員去幹、由哪個公司去幹，在很大程度上解決了客戶尋找服務人員時面臨的資訊不對稱問題。對於服務人員，易盟個人終端可對服務人員進行精準定位和勞務資訊即時送達。用 O2O 把這些碎片化的勞動力資源同客戶需求整合起來，讓服務人員注明她在哪個時間段可以幹活，這樣就能對其進行合理安排，解決了資訊不對

稱的問題。

Fifth, unifying the platform, integrating the matching service resources. Yimeng 95081 Housekeeping is more similar to the map-search function on PC, which greatly improves the efficiency of finding the matching family service information. On the mobile phone, 95081 Housekeeping is a mobile application based on LBS geographic location service, providing customers with one-stop housekeeping services such as hourly employees, babysitters, maternity matrons, childcare, cleaning, and dredging. Customers can find the right service personnel closest to them through the mobile client and contact the service personnel or the company. For consumers, Yimeng uses a huge database of more than 100,000 people to conduct the fragmented searches in the Internet environment, and accurately do the matching in the database, so that it can be figured out which service personnel or which company is going to work. To a large extent, it solves the information asymmetry problem faced by customers when they look for service personnel. For the service personnel,the Yimeng terminal can accurately locate the service personnel and deliver the labor information immediately. Also, Yimeng uses O2O to integrate these fragmented labor resources with customer needs and lets the service staff indicate the time periods they can work, so that the resources can be reasonably arranged to solve the problem of information asymmetry.

家政服務「雲整合」的效果是非常明顯的，易盟集團 95081 的 APP 上線一年間，這家企業移動端的業務增量保持在平均每月 10% － 20%，每月在總業務量中的佔比保持在 15% 左右。北京地區日均接單量在 3000 單左右，其中移動端的接單量就有 300 － 400 單。易盟家政服務中心每天受理家庭服務

訂單 1.5 萬餘次，每天安排勞務人員就業 2 萬餘人。易盟集團已經成為「家政界的攜程」。

The effect of cloud integration of housekeeping service is very obvious. For the whole year since the 95081 APP has been online, business increment of the mobile terminal is maintained at an average of 10% － 20% per month, remaining at around 15% in the total business volume. The number of daily average orders in Beijing is about 3,000, of which 300 － 400 orders are available on the mobile terminal. The Yimeng Housekeeping Center accepts more than 15,000 home service orders and arranges more than 20,000 service personnel everyday. The Yimeng Group in the housekeeping service has the same status as Ctrip in travelling.

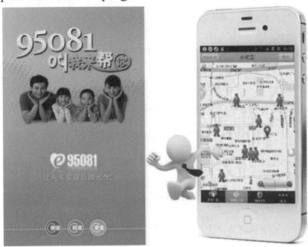

圖 Figure 6-11 易盟集團 95081 家政寶讓客戶輕鬆找到家政服務 Yimeng 95081 Housekeeping allows customers to easily find housekeeping services

六

從大企業化向輕量化轉型
6. Tansforming from a large enterprise to a light one

在「雲消費」時代，龐大的組織架構、繁瑣的運營流程都成為束縛企業發展的枷鎖，商業企業應該拋棄傳統大企業發展思維，充分借助社會分工，精簡人員，將力量集中於核心價值，輕量化小團隊作戰。

In the era of cloud consumption, the huge organizational structure and cumbersome operational processes have become the shackles that constrain the development of enterprises. Enterprises should abandon the development thinking of traditional large enterprises, fully utilize the social division of labor, streamline personnel, concentrate their efforts on core values and cut their teams into light weight.

我們知道，手機生產行業歷來是大企業馳騁的天地。隨著免費安卓系統的應用普及，以及更成熟開放的市場和廣泛的市場合作通路的打通，更多小企業甚至微企業可以生產自有品牌手機，並贏得大效益。如小米的成功就是如此。這家企業以小米品牌為核心，採取一系列外包方式，迅速佔領市場。如晶片主要用高通晶片，生產主要外包給聞泰等代工廠，物流倉儲利用凡客誠品，行銷推廣透過網路社區，同時利用運營商、天貓、微信等成熟的管道通路。2013 年小米手機的營業額已達到 316 億元，而小米員工不過 3000 人

左右，與當年諾基亞等企業動輒數萬人的企業規模不可同日而語。在小米內部，組織架構沒有層級，基本上是三級，7個核心創始人──部門 Leader──員工。而且它不會讓團隊太大，稍微大一點就拆分成小團隊。除了7個創始人有職位，其他人都沒有職位，都是工程師，晉升的唯一獎勵就是漲薪。小米沒有 KPI，沒有績效管理，沒有級別體系，但這並不妨礙公司形成獨到的文化，創造讓人驚嘆的業績。近年來，在國內風生水起的智慧手機生產行業中，小米還屬於大企業，更多數十人甚至一、二十人的小微企業已經成功地經營了自己的智慧手機品牌。

　　We know that the mobile phone production industry has always been a world where big companies strut their stuff. With the popularization of free Android applications, as well as the opening of more mature and open markets and extensive market cooperation channels, more small businesses and even micro-enterprises are able to produce their own branded mobile phones and win big profits. Xiaomi's success is the case. The company takes the Xiaomi brand as its core and adopts a series of outsourcing methods to quickly occupy the market. For example, Xiaomi phones mainly adopts Qualcomm chips, and the production is mainly outsourced to the foundries such as WingTech. In terms of the logistics and warehousing, Xiaomi cooperates with VANCL. Also, Xiaomi carries out the marketing and promotion through the online community, while using mature channels such as mobile operators, Tmall, and WeChat. In 2013, the turnover of Xiaomi's mobile phone has reached 31.6 billion yuan, while the employees of Xiaomi were about only 3,000, which could not be compared with Nokia's scale of tens of thousands of staff.Xiaomi does not have a hierarchical structure. Basically, it is divided into three levels- seven core founders, department leaders and employees. Xiaomi never lets the team become too big, a team is split into small ones even if it's a little bit big. Except for the seven founders who have positions, all the other employees are all engineers with no positions. The only reward for promotion is salary increase. There is no KPI, performance management or level system in Xiaomi, but this does not prevent the company from forming a unique culture and creating amazing performance. In recent years, Xiaomi is still a large enterprise in the smart phone production industry in China. Many small and micro enterprises with dozens or only a dozen of people have successfully operated their own smartphone brands.

在這裡，我們再看兩個讓人震驚的輕量化企業成功的案例：

Let's share two shocking examples of successful lightweight companies:

（一）WhatsAPP 的 55 人團隊
WhatsAPP's team of 55

WhatsAPP 創辦於 2009 年，專注於即時通訊，是一款跨平臺移動應用程式，用於智慧手機之間的通訊。在安卓應用下載排行榜當中，WhatsAPP 位居第五。目前用戶量已經超過 4.5 億人，其中日活躍用戶比例為 70%；每日新註冊用戶超過 100 萬人。而在背後支撐它的，僅僅是一個 55 人的工作團隊。其中，創始人 Jan Koum（CEO）和 Brain Acton 都曾為雅虎資深工程師，還有由 32 名最強悍的工程師組成的技術團隊。2014 年 2 月全球最大的社交網路 Facebook 斥資約 160 億美元（約 40 億美元現金外加約 120 億美元 Facebook 股票）收購 WhatsAPP，Facebook 將保留 WhatsAPP 品牌，

WhatsAPP 與 Facebook Messenger 兩大即時通訊應用將繼續獨立運行。Koum 和 Acton 分別握有公司 45% 和 20% 的股份，以 190 億美元收購價計算，前者的身價保守估計在 85.5 億美元左右，而後者的身價則在 38 億美元左右。Facebook 還將為 WhatsAPP 創始人提供約 30 億美元的限制股，將 4 年後授予。而 WhatsAPP 55 人的團隊，平均每人創造了約 3.46 億美元的價值。我們認為，WhatsAPP 的成功有兩條經驗值得借鑑：

Founded in 2009, WhatsAPP focuses on instant messaging and is a cross-platform mobile APP for communication among smartphones. In the Android app download rankings, WhatsAPP ranks fifth. At present, the number of users has exceeded 450 million, of which 70% are daily active users. There are more than one million new registered users per day. Behind the APP, there is a work team of only 55 people. Among them, Jan Koum (CEO) and Brain Acton, cofounders of the APP, have once been senior engineers in Yahoo, as well as a technical team consisting of 32 most resourceful engineers. In February 2014, Facebook, the world's largest social network, spent about $16 billion (about $4 billion in cash plus about $12 billion in Facebook shares) to acquire WhatsAPP. Facebook retained the Whats APP brand, and the two instant messaging apps WhatsAPP and Facebook Messenger continue to operate independently. Koum and Acton hold 45% and 20% of the company's shares respectively. Calculating by the purchase price of $19 billion, the value of the former is conservatively estimated at around $8.55 billion, while the latter about $3.8 billion. Facebook also provides restricted stock worth about $3 billion for the founders of WhatsAPP four years later. WhatsAPP's 55-person team created an average of about $346 million per person. We believe that we can learn from the success of WhatsAPP from two aspects.

一是 WhatsAPP 始終堅持「Do one thing and do it well」（只做一件事，但把它做好）的精神，將整個 APP 的重點放在「傳訊息」這個單一的功能上。由於時逢智慧型手機興起的階段，WhatsAPP 趕在即時通訊的熱潮前端，很快地在世界傳訊市場上佔有一席之地。

First, WhatsAPP always adheres to the spirit of 「Do one thing and do it well」, and focuses the entire APP on the single function of 「transmitting messages」. Due to the rise of smart phones, WhatsAPP has quickly gained a foothold in the world communications market.

二是「去管理層」的組織架構。據業內人士研究，在傳統商業模式下，由於足夠成熟，老闆 1 個月內真正的重要決策不超過 3 次；而在互聯網經濟下，由於機會稍縱即逝，需要臨機決斷的事宜太多，因此必須發揮每一位員工的核心職能，縮短決策半徑，實行扁平化管理。55 人就能創造龐大的業績。

Second, the organizational structure of de-management. According to industry insiders, in the traditional business model, the boss makes less than three important decisions within one month because the busniess is mature enough. However, in the Internet economy, because of the fleeting opportunities, there are too many things that need to be decided. Therefore, it is necessary to give play to the core function of each employee, shorten the decision radius, and implement flat management. Only 55 people can create huge achievements.

---

161 非典型性案例：一個只有 13 個人的上市公司. 品牌密碼 An Atypical Case Study: A Listed Company with Only 13 People. Brand Password

（二）13 人的上市公司
Listed company of 13

創建於 1970 年的日本微型製造商——A-one 精密（以下簡稱 A-one），包括老闆在內員工僅有 13 人，主要生產超硬彈簧夾頭，市場佔有率高達 60%，擁有 1.3 萬家國外用戶。並於 2003 年在大阪證券交易所上市。

Founded in 1970, Japanese micro-manufacturer A-one Seimitsu (hereinafter referred to as A-one) has only 13 employees including the boss. The company mainly produces super-hard spring collets, with a market share of 60% and 13 thousand foreign customers. It was listed on the Osaka Stock Exchange in 2003.

A-one 1 年開會的時間加在一起不超過 30 分鐘，最強的優勢是交貨的快速。A-one 從訂單到製造再到發貨這一系列流程都是自己在做，沒有中間環節，中間間隔不到 5 分鐘。這樣不但提高了發貨速度，而且不易於被銷售端控制價格。A-one 員工實行終身雇傭，不需要打卡，沒有組織和頭銜，可以分享公司盈利。[161]

The total time for A-one's meeting in the whole year is no more than 30 minutes. The strongest advantage of the campany is the fast delivery. A-one's series of processes from order to manufacturing to delivery are all done by themselves. There is no intermediate link, and the interval among every link is less than 5 minutes. This not only improves the delivery speed, but also is not easy to be controlled by the sales terminal. A-one employees are employed for life, thay do not need to clock in, have no organization and titles, and can share company profits.[161]

「雲消費」時代超市業的變革與探索

Chapter Seven
The Reform and Exploration of the Supermarket Industry in the Age of Cloud Consumption

第七章

「雲消費」時代超市業的
變革與探索

Chapter Seven The Reform and Exploration of the Supermarket Industry in the Age of Cloud Consumption

　　雲消費時代，傳統超市業對於消費者的價值——到店選擇購買的消費需求逐漸降低，生存根基發生動搖。消費者不用去實體店鋪，用手機端、電腦端、電話端，透過微信、淘寶、天貓、京東、當當、蘇寧、移動 APP 等智慧平臺，在任何時間任何地點都可以實現幾乎所有可以在超市得到的消費需求。從服裝、電器、日化、包裝食品到生鮮食品，從裝修裝飾到家政服務，涉及商品與服務的範圍越來越豐富。商品與服務提供者，從個體商戶，到廠商、經銷商、電商、店商等等，涉及的經營主體越來越多元。服務取得方式，從到店自提，自提點自提，到到家服務，服務提供方式據消費需求也呈現多元化趨勢。圍繞消費方式的變化，超市業對消費者的價值將重點體現在兩個方面，一是提供最後一公里的到家服務；二是提供與眾不同的消費體驗。

In the era of cloud consumption, the value of the traditional supermarket industry for consumers, that is, buying in the supermarket, is gradually reduced, and the survival base has been shaken. Consumers don't have to go to physical supermarkets. Through smart platforms such as WeChat, Taobao, Tmall, Jingdong, Dangdang, Suning, and Mobile APP on mobile phones and computers, almost all can be done in any place at any time. From clothing, electrical appliances, daily chemicals, packaged foods to fresh foods, from house decoration to housekeeping services, the range of goods and services is becoming more and more wide. Commodities and service providers from individual merchants to manufacturers, distributors, e-commerce, store operators are increasingly diversified. The ways to obtain services, from picking up in the supermarket or the self pick-up site to the home service, also show a diversified trend based on consumer demand. Focusing on the changes in consumption patterns, the value of the supermarket industry to consumers will be reflected in two aspects. One is to provide the last-mile home service. The other is to provide a different consumer experience.

超市業面臨系列變革
1. The supermarket industry faces a series of changes

（一）萬米大賣場將步入衰退期
Hypermarkets are headed for a recession

雲消費時代，消費者隨時隨地可以實現所需的消費，大賣場的消費者價值──「一站式購足」的優勢不再。大賣場的成本則越來越高昂，租金成本、運營成本、人工成本隨經營面積的增加而增大，隨著大賣場消費者價值的降低，經營業績會逐漸下滑，單位坪效逐漸減小，萬米大賣場將進入發展的衰退期。

In the era of cloud consumption, consumers can consume anytime and anywhere, and the consumer value, or the advantage of 「one-stop」 shopping of the hypermarket, disappears. The cost of hypermarkets is getting higher and higher. The rental cost, operating cost and labor cost increase with the increasing operating area. As the value of hypermarket consumers decreases, the operating performance and the unit efficiency will gradually decline. As a result, the hypermarkets are headed for a development decline.

近年各地大賣場業績普遍下滑，據聯商網統計，2014 年度，超市類門店共計關閉 178 家，其中沃爾瑪關閉 16 家門店，成為 2014 年關店最多的超市企業，2014 年沃爾瑪大規模裁員，裁員主要集中在中高層管理人員，涉及上百人。樂天瑪特關閉 6 家，人人樂關閉 5 家，華潤萬家、家樂福、樂購各關閉 4 家，世紀聯華、新華都各關閉 3 家，永輝超市關閉 2 家，麥德龍、北京華聯、中百倉儲、恆客隆、三江購物、胖東來、佳樂家各關閉 1 家。據家樂福相關資料顯示，家樂福 2014 年在中國市場的營業額下滑 6.4%，中國區平均單店業績下滑 15% 左右。截至 2015 年 6 月 30 日，2015 年主要超市企業在國內共計關閉 120 家，其中麥德龍、北京華聯、農工商、永旺、鄧尼斯關閉 1 家，樂購、新華都、世紀聯華分別關閉 2 家，家樂福關閉 3 家，華潤萬家關閉 6 家，永輝超市關閉 5 家。

In recent years, the performance of hypermarkets has generally declined. According to the statistics of Linkshop.com, in 2014, closed a total of 178 supermarkets were closed, of which Wal-Mart closed 16 stores, the most among all supermarket enterprises in 2014. Also in this year, Wal-Mart held a mass layoff, mainly focused on middle and senior administrative staff, involving hundreds of people. Lotte Mart closed 6. RenRenLe closed 5. China Resources Vanguard, Carrefour and Tesco closed 4 respectively. Century Lianhua and Xinhuadu closed 3 respectively. Yonghui Supermarket closed 2. Metro, Beijing Hualian, Zhongbai Warehousing, Hengkelong, Sanjiang Shopping, DL and Jialejia all closed 1. According to relevant data of Carrefour, its turnover in the Chinese market fell by 6.4% in 2014, and the average single-store performance in China fell by about 15%. As of June 30, 2015, the number of major supermarkets closed in China is

120, including Metro, Beijing Hualian, NGS, AEON and Dennis of 1, Tesco, New Hua Du and Century Lianhua of 2, Carrefour of 3, China Resources Vanguard of 6 and Yonghui Supermarket of 5 .

表 Table 7-1 2014 年度沃爾瑪關店一覽表 [162] List of Wal-Mart closed stores in 2014162

城市 City	門店 Store	關店時間 Closing date	開 業 時 間 Opening date	關 / 開店數 Number of stores closed/opened
江蘇江陰 Jiangyin in Jiangsu Province	新一城店 Xinyicheng Store	2014.4.8	2011.3.24	
江蘇鹽城 Yancheng in Jiangsu Province	迎賓南路 South Yingbin Road Store	2014.3.19	2009.9.24	
安徽蚌埠 Bengpu in Anhui Province	勝利路店 Shengli Road Store	2014.5.29	2010.10.14	
安徽馬鞍山 Ma'anshan in Anhui Province	雨山東路 East Yushan Road Store	2014.3.31	2010.11.19	
安徽合肥 Hefei in Anhui Province	曙光店 Shuguang Store	2014.12.9	2015.1	
安徽馬鞍山 Ma'anshan in Auhui Province	花雨路店 Yuhua Road Store	2014.3.31	2010.12.3	
湖南長沙 Changsha in Hunan Province	南門口店 Nanmenkou Store	2014.8.27	2003.7.17	
河南新鄉 Xinxiang in Henan Province	人民中路店 Middle People Road Store	2014.7.29	2010.11.26	
湖南常德 Changde in Hunan Province	水星樓店 Shuixinglou Store	2014.3.19	2009.1.17	16/25
湖南衡陽 Hengyang in Hunan Province	西渡店 Xidu Store	2014.7	2011.6.16	
湖南永州 Yongzhou in Hunan Province	道縣步行街店 Daoxian Walkway Store	2014.8.20	2012.8.23	
重慶 Chongqing	南濱店 Nanbing Store	2014.3.4	2009.8.25	
浙江杭州 Hangzhou in Zhejiang Province	朝暉店（好又多） Zhaohui Store (Trust-Mart)	2014.4.23	2003.11.8	

上海 Shanghai	殷高西路店 West Yingao Road Store	2014.3	2009.5.27	
廣東廣州 Guangzhou in Guangdong Province	東山口店 Dongshankou Store	2014.8.19	2000 年	
四川成都 Chengdu in Sichuan Province	府青店（好又多）Fuqing Store (Trust-Mart)	2014.4.23	2009.6.25	

表 Table7-2　2015 年上半年超市關店情況一覽表 [163] List of supermarkets closed in the first half of 2015

企業 Enterprise	城市 City	門店 Store	關店時間 Closing date	開業時間 Opening date
永旺 AEON	濰坊 Weifang	永旺濰坊店 Weifang Store	2015.1.11	2008.12.30
樂購 Tesco	台州 Taizhou	椒江店 Jiaojiang Store	2015.3.31	2009.2.28
	上海 Shanghai	盧灣店 Luwan Store	2015.3.31	2000.1.15
世紀華聯 Century	麗水 Lishui	聯華綜超麗水羅馬店 Lishui Roma Store	2015.3.3	2004 年
Hualian	柳州 Liuzhou	榮軍路店 Rongjun Road Store	2015.4	2007.4
家樂福 Carrefour	杭州 Hangzhou	余杭南苑店 Yuhangnanyuan Store	2015.3.23	2009.7.21
	珠海 Zhuhai	香洲店 Xiangzhou Store	2015.5.1	1998 年
	東莞 Dongguan	石龍店 Shilong Store	2015.5.30	2009.11.15

162 2014 年主要零售企業 (百貨、超市) 關店統計 . 聯商網 . http://www.linkshop.com.cn/web/archives/2015/315683.shtml
2014 Statistics of Closed Stores of major retailers (Department stores, Supermarkets). Linkshop.com. http://www.linkshop.com.cn/web/archives/2015/315683.shtml
163 2014 年主要零售企業 (百貨、超市) 關店統計 . 聯商網 . http://www.linkshop.com.cn/web/archives/2015/315683.shtml

華潤萬家 China Resources Vanguard	長春 Changchun	飛躍店 Feiyue Store	2015.4.1	2010.8
	宜賓 Yibing	財富廣場店 Fortune Plaza Store	2015.6	2013.9.29
	青島 Qingdao	九水路店 Jiushui Road Store	2015.6.1	2011.12.3
	九江 Jiujiang	九江店 Jiujiang Store	2015.6.5	2014.5
	北京 Beijing	北京西單店 Beijing Xidan Store	2015.6.24	2000 年
	惠州 Huizhou	河南岸店 Henan' an Store	2015.6.30	2012.9
永輝超市 Yonghui Supermarket	常州 Changzhou	茂業店 Maoye Store	2015.3	2011.12.31
	泉州 Quanzhou	東街店 Dongjie Store	2015.4.17	2009.7.18
	泉州 Quanzhou	鐘樓店 Zhonglou Store	2015.4.17	2009 年
	蚌埠 Bengpu	勝利路店 Shengli Road Store	2015.4.25	2011.12.1
	廈門 Xiamen	岳陽店 Yueyang Store	2015.4.17	2006.8.18
天天壹加壹超市 Everyday 1+1	銅州 Tongzhou	銅州約 30 家門店 About 30 stores in Tongzhou	2015.4.27	2000 年
麥德龍 Metro	佛山 Foshan	佛山南海店 Foshannanhai Store	2015.5.1	2011.12.15
雲南天順 Yunan Tianshun	雲南西雙版納等城市 Xishuangbanna and other cities in Yunnan Province	西雙版納等 40 家門店 40 stores including Xishuangbanna Store	2015.5.16	2001 年
農工商 NGS	寧波 Nongbo	寧波店 Ningbo Store	2015.5.6	2002 年
新華都 New Hua Du	寧波 Nongbo	四明店 Siming Store	2015.5.30	2011 年底
	寧波 Nongbo	興寧店 Xingning Store	2015.5.31	2011 年底
鄧尼斯 Dennis	新鄉 Xinxiang	新鄉平原店 Xinxiang Pingyuan Store	2015.6.1	2008.12.4
北京華聯 Beijing Hualian	惠州 Huizhou	BHG 華貿店 BHG Huamao Store	2015.6.15	2011 年

（二）千米中型超市以增加消費體驗為核心探索發展
Medium-sized supermarket explores development with the core of increasing consumption experience

　　傳統千米中型超市的消費者價值──「選擇性購買」，也在逐漸消失，但生鮮食品的現場加工，實地體驗式消費等需求需要到店實現。千米中型超市較大賣場規模小，運營成本也略低，鎖定目標客群，以生鮮食品為經營重點，以現場加工和體驗性功能為特色，強調與突出體驗性功能，給消費者與線上不同消費體驗，是綜合超市主要探索的發展方向。

　　The consumer value of the traditional medium-sized supermarkets, the selective purchases, is also gradually disappearing, but the on-site processing of fresh foods and on-site experience consumption and other needs can only be realized in the physical stores. The medium-sized supermarkets have small scales and low operating costs compared with hypermarkets. They focus on the target customers, put great emphasis on fresh food, and feature on-site processing and experience functions. By highlighting experience functions, they offer different consumer experiences from the online services, which is the main development direction of the comprehensive supermarkets.

　　安徽樂城超市為了給消費者更多的消費體驗做了很多探索，在千米綜合超市裡（樂城鄰里店）開設了免費的150多平方米的兒童樂園和幼童沙池，玩沙全部用中藥決明子；加重了生鮮熟食經營的比重，設置小型就餐區，滿足部分居民現場就餐的需求；闢出專門的園藝天地，提升效益的同時美化了

環境；在超市門口，擺放著一臺廢棄飲料容器回收機，顧客將一個礦泉水瓶放進回收機，回收機就會吐出一張樂城超市的一角錢的購物抵用券；賣場佈置體驗化，設計蛋殼形雞蛋銷售區、鮮肉櫃檯上擺著可愛的小豬等等，使顧客在購物中體驗到不一樣的樂趣。[164]

Anhui Lecheng Supermarket has made a lot of explorations to offer more consumer experiences. In the medium-sized supermarket (Lecheng Linli Store), a free children's playground and a children's sand pool of more than 150 square meters are opened, in which Chinese medicine cassia seeds are used as sand. The supermarket also increases the proportion of fresh cooked food operation and sets up a small dining area to meet some residents' needs of on-site dining. It also creates a special gardening world, which enhances the benefits while beautifying the environment. At the supermarket door, there is a beverage container recycling machine which spits out a dime-worth shopping voucher as a customer puts a mineral water bottle into the machine. Last but not least, the store layout is experience-oriented. The egg sales area is designed eggshell-shaped and there are cute pig dolls on the fresh meat counter, all making customers experience different kinds of fun while shopping.

美國全食超市（WHOLE FOODS）是美國著名的以有機食品為特色的連鎖超市，其以畫廊的燈光設計，渲染食品之美，超市內部進行主題化功能分區，針對分區設計主題化陳列，給消費者與眾不同的感觀體驗；以品類豐富

---

164 安徽樂城調研資料整理 Research Data of Anhui Lecheng

的各種速食食品，滿足消費者各種就餐需求；超市內設速食就餐區，有的門店內設酒吧臺，滿足消費者超市內即時就餐、休閒的需求，酒吧臺設置的嘗試增加了客流，WHOLE FOODS 很多門店都開始增設酒吧臺。

WHOOLE FOODS is a well-known supermarket chain featuring organic food in the United States. It uses the lighting design in the gallery to render the beauty of the food. The theme of the supermarket is functionally partitioned to offer customers different sensory experiences. Also, it satisfies consumers' various dining needs with a variety of fast foods. Every supermarket has a fast food dining area, and some even have a bar counter to meet the needs of instant dining and leisure in the supermarket. Such attempts have increased the passenger flow. Many WHOLE FOODS supermarkets have begun to add bar counters in them.

（三）中小型超市探索向食品超市轉型
Small and medium-sized supermarkets explore the transition into food supermarkets

在以即時購、隨時購為特徵的雲消費時代，緊張的城市生活節奏中，千米以下，貼近社區生活，以生鮮食品現場加工、綜合服務為特色，強調消費者生活圈服務功能的中小型食品超市，經營規模小、運營成本低、與快節奏的生活相契合，是超市企業探索發展的重點。

In the era of cloud consumption characterized by instant purchase and time-to-time purchase, as city life is much busy, small and medium-sized supermarkets with small scale of operation and low operating costs, fit the fast-paced life. They are close to the community life, feature on-site processing of fresh food and comprehensive services, and also emphasize on the service function of the consumer life circle. As a result, they are the focus of supermarket companies' exploration and development.

在這方面較早做出探索的是家樂福。2014 年 11 月家樂福在上海黃樺路開出家樂福中國的第一家小型食品超市──easy 家樂福，營業面積 300 平方米。easy 家樂福致力於打造便捷、休閒的生活方式，它比傳統便利店規模更大，經營品類更豐富，旨在透過豐富商品及優質服務滿足社區居民基本生活之需，把便捷的生活方式帶給周邊消費者。其裝修風格以橙色為主基調，店內主營食品，包括即食預製品如關東煮、現製飲品、盒飯；豬排飯、咖哩飯、意麵、飯糰等輕熟食便當，各種袋裝蔬菜沙拉、牛奶、軟飲、火腿、火鍋包裝食品、半加工食品；方便食品如薯條、餅乾、飲料；提供生鮮果蔬、酒類、基礎食品（米麵油）、嬰兒產品、日化用品及寵物食品等，還有一整個貨架的面積陳列家樂福法國之光的自有品牌系列食品，以及各國的進口食品，包括酒、調味醬、速食麵等。還用很大的面積陳列著包括洗衣粉、肥皂、清潔劑、沐浴露、洗髮露等日用品，以及鍋碗瓢盆、掃把拖把等各式日用廚房用品或生活雜貨用品。easy 家樂福注重環境品質的打造，營造咖啡店般的消費氛圍，店內設有簡易咖啡吧，並配備充電電源，供顧客小憩，並可享用飲料、簡餐等快速食品，店內覆蓋 WiFi，並配備便民生活服務終端，進門的牆上掛了一臺可以提供更多家樂福商品的觸控式螢幕，用手滑動螢幕就可以看到各類在售商品，每款商品都標明價格並附加了二維碼，用手機掃描可以完成預

定和到前臺結帳，門店提供預定和自取服務；門店收銀臺與支付寶錢包打通，在收銀台可以進行手機支付，店內還設有衛生間。2015 年 6 月 30 日上海第 2 家 easy 家樂福正式營業，營業面積 280 平方米，單品規模約 5000 種。2015 年 9 月 20 日上海第 3 家 easy 家樂福開業，營業面積 450 平方米，單品規模約 4500 種。

Carrefour made early explorations in this regard. In November 2014, Carrefour opened China's first small food supermarket, easy Carrefour, on Huanghua Road in Shanghai, with a business area of 300 square meters. Easy Carrefour is committed to creating a convenient and casual lifestyle. It is larger than traditional convenience stores and has a richer range of business categories. It aims to meet the basic needs of the community by enriching its products and services, and bringing a convenient lifestyle to the consumers nearby. The decoration style is based on orange. The main foods in the supermarket include ready-to-eat products such as Oden, fresh drinks and box lunches, light cooked foods such as pork rib rice, curry rice, pasta and rice balls, various bagged vegetable salads, milk, soft drinks, ham, hot pot packaged foods, and semi-processed foods, convenience foods such as French fries, biscuits and beverages. The supermarket also provides fresh fruits and vegetables, alcohol, basic foods (rice, noodles and oil), baby products, daily chemicals and pet food, etc. There is also an shelf displaying Light of France, Carrefour' s own brand of food, as well as imported food from various countries, including wine, sauces, instant noodles. It also uses a large area to display daily necessities such as washing powder, soap, detergent, shower gel, shampoo, as well as pots and pans, broom mops and other daily kitchen items or daily necessities.

Easy Carrefour pays attention to the creation of environmental quality to create a cafe-like consumption atmosphere. It has a simple coffee bar equipped with charging ports, where customers can enjoy fast foods like drinks, light meals. The supermarket is also wifi-covered and equipped with the life service terminal, a touch screen that provides more Carrefour products on the wall at the entrance, on which customers can swip their fingers to browse the items on sale. Every item is attached with a price and a QR code. Customers scan the QR code to complete the booking, and checkout at the front desk. The supermarket provides reservations and self-service services. The checkout counter is connected with Alipay wallet so that mobile payment can be made at the checkout counter. There is also a toilet in every supermarket. On June 30, 2015, the second easy Carrefour in Shanghai officially opened, with a business area of 280 square meters and a product amount of about 5,000. On September 20, 2015, the third easy Carrefour opened in Shanghai, with a business area of 450 square meters and a product amount of about 4,500.165

（四）便利超市成為今後發展重點
Convenience stores become the focus of future development

2015 年上半年超過 158 家大型超市、百貨店的「大店出現關閉潮」，而便利店 2014 年全國銷售額同比增長 25.12%，門店數同比增長 21.96%。

In the first half of 2015, more than 158 large supermarkets and department stores 「closed their big stores」, while the convenience store sales in 2014 increased by 25.12% year-on-year, and the number of stores increased by 21.96%.

雲消費時代，消費者需要的是「最後一公里」的生活和服務需求的滿足。深入社區、寫字樓的小型便利超市、便利店是實現「最後一公里」重要的終端節點。

In the era of cloud consumption, consumers need the satisfaction of the 「last-mile」 life and services. In the communities, small convenience supermarkets and convenience stores in office buildings are important terminal nodes for achieving the 「last-mile」 life.

業內各零售商都看到「最後一公里」的巨大商機。2014 年底，大潤發首家便利店在江蘇南通市開業，該店旨在打通實體店與網上商城，開拓新的便利店經營模式。2014 年德國最大的零售商麥德龍首家便利店「合麥家」在上海普陀開業，合麥家將以特許加盟的形式擴張，對於加盟商，麥德龍會為其留出 20%-25% 的利潤空間，並提供大部分商品。[165] 誕生於 2011 年的新銳便利店品牌「全時」，2017 年 9 月在北京的便利店已達 350 家，成為北京門店最多的便利店品牌，2017 年底，全時已全面佈局中國西南、華南、華中、華東、華北 5 大區域，實現在北京、天津、杭州、蘇州、武漢、成都、重慶等一、二線城市的快速擴張。全時便利店經營約 1500 種常規商品，其商品選擇主要基於兩點，首先是 25% 以上的高毛利商品；二是能夠與顧客進行感情溝通的

---

165 麥德龍正在進行入華 18 年來最大調整或開展特許加盟. 贏商網. http://sz.winshang.com/news-412785.html
Metro Is Undergoing the Largest Adjustment in China for 18 Years and about to Open the Franchise. Winshang.com. http://sz.winshang.com/news-412785.html

商品。其門店結合了中餐、飲品、便利、金融、服務等五項服務功能，很多門店將收銀臺改造成能夠銷售現磨咖啡、果汁、奶茶等十幾款飲品的咖啡吧。全時便利與中國鐵路總公司簽署了排他性協定，在部分門店中放置火車票互聯網取票／售票終端機，可以訂票取票。並投資千萬資金研發了一套獨特的管理系統「全時匯」，消費者可以在移動終端利用「全時匯」，進行點餐，然後到附近的門店去用餐，不用叫喊排隊，更不用催菜。[166]

Retailers in the industry have all seen huge business opportunities in the 「last- mile」 life. At the end of 2014, RT-Mart's first convenience store opened in Nantong, Jiangsu Province. The store aims to connect physical stores with online malls, and open up new business models of convenience store. In 2014, Germany's largest retailer Metro's first convenience store 「My Mart」 opened in Putuo, Shanghai. It will expand in the form of franchise. For franchisees, Metro will set aside 20%-25% of revenue as its profit, and offer most of its goods[165]. The new convenience store brand "Our Hours" was born in 2011. In September 2017, there were 350 its convenience stores in Beijing as it became convenient store brand with the most stores in Beijing. At the end of 2017, Our Hours has been fully deployed in the five major regions in China, Southwest China, South China and Central China, East China and North China, and achieved rapid expansion in first- and second-tier cities such as Beijing, Tianjin, Hangzhou, Suzhou, Wuhan, Chengdu

---

166 永輝等超 80 家零售商進軍電商 如何吸客是關鍵 . 中國經濟時報 . http://finance.sina.com.cn/roll/20140124/084518072448.shtml.　More than 80 Retailers Including Yonghui Entered the E-commerce, the Way to Attract Customers is the Key. China Economic Times. http://finance.sina.com.cn/roll/20140124/084518072448.shtml.

and Chongqing. Our Hours operates about 1,500 kinds of regular goods, and its product selection is mainly based on two points, goods with high-margin of 25% and goods that can be used to communicate with customers.The store combines five service functions of Chinese food, drinks, convenience, finance, and service. Many stores transform the checkout counter into a coffee bar that can sell more than a dozen kinds of drinks such as freshly ground coffee, juice, and milk tea. Our Hours signed an exclusive agreement with the China Railway Corporation. Some stores are placed with Internet ticketing terminal which can be used to book and pick up the train tickets. Our Hours also invested tens of millions of RMB to develop a unique management system called "D&N Hut" where consumers can order food on the mobile terminal before they go to the nearby stores to eat without being yelled at in the queue or asking about the order.[166]

（五）線上線下融合探索最後一公里服務方案解決
Online and offline integration to explore the solution of the last-mile service

線上線下融合探索最後一公里服務解決方案是業內一直關注和發展的重點。目前主要有幾種模式。

Online and offline integration to explore the solution of the last-mile service is the focus of the industry. Currently the main models are as follows.

1. 超市企業自建平臺，線上線下一體化，實現最後一公里服務
Supermarket enterprises build their own platforms, integrate online and offline, and realize the last-mile service

超市企業自建平臺，線上下單，線下體驗，並依託線下門店為消費者提供到家服務，線上線下一體化發展，是超市企業關注和探索發展重點。

Supermarket enterprises build their own platforms where orders are placed online, and experience is carried out offline. They rely on offline stores to provide home-to-home services. The integration of online and offline development is the focus of supermarket companies.

近幾年，超市企業紛紛進行線上業務探索。2011 年 12 月 16 日，屈臣氏中國正式進駐淘寶商城，開啟官方旗艦店，屈臣氏線上商店與線下商店不同，主要銷售其自有品牌和實體店的獨家品牌，促銷方面重點依據線民購物特點，制訂相應的促銷策略。麥德龍在試水天貓旗艦店後，2012 年 5 月上線了 B2B 商城，以自建物流和實體門店的方式進行全國配送。2013 年全國連鎖百強企業紛紛探索電商模式，全國大中型連鎖零售企業中涉足電商的已超過 80 家，其中 22 家為 2013 年首次進軍線上銷售，但不同企業探索的深度和成效不盡相同。「家樂福網上商城」於 2015 年 6 月在上海上線，充分融合線下門店資源，網上商城的貨品全都來自上海的 3 家門店，且價格和門店一致，消費者可憑訂單至上海任一家樂福大賣場辦理退換貨。永輝超市也率先在福州開設實現移動終端線上訂購、支付和超市門店線下提貨的「永輝微店」，試運行成熟後逐步向全國推廣[167]。

In recent years, supermarket companies have explored online business. On December 16, 2011, Watsons China officially entered Tmall and opened the official flagship store. Watson's online store is different from offline store, it mainly sells its own brands and exclusive brands sold in physical stores. For promotion, it focuses on the characteristics of online shopping of netizens and develops corresponding promotion strategies. After testing the Tmall flagship store, Metro launched the B2B Mall in May 2012, and delivered the goods nationwide through self-built logistics and physical stores. In 2013, the China top 100 chain enterprises explored the e-commerce model. Among the large and medium-sized chain retail enterprises in the country, there were more than 80 e-commerce companies, 22 of which entered the online sales for the first time in 2013, that set foot in e-commerce.[166] However, the depth and effectiveness of exploration by different enterprises are not the same. 「Carrefour Online Shopping Mall」was launched in Shanghai in June 2015, which fully integrated offline store resources. The goods in online store are all from three pgysical stores in Shanghai, of which the prices are the same as those in the physical store. Any physical store handles the return and exchange as long as customers bring along the order. Yonghui Supermarket also took the lead in setting up 「Yonghui Micro-Shop」in Fuzhou which realizes online ordering and payment and offline delivery by physical stores. The model is gradually promoted to the whole country as soon as it was mature.[167]

167 家 樂 福 官 網 資 料 整 理 . http://www.carrefour.com.cn/News/YearNews.aspx Information on Carrefour Official Website. http://www.carrefour.com.cn/News/YearNews.aspx

2. 電商企業的線上平臺提供最後一公里服務 Online platforms of the e-commerce enterprises provide the last-mile service

以 1 號店、中糧我買網為代表的電商企業一直是提供最後一公里服務的生力軍，近年得到快速發展。

The e-commerce enterprises represented by No. 1 Store and Zhongliang Womai have always been a new force to provide the last-mile service, and have also been rapidly developed in recent years.

1 號店，電子商務型網站，2008 年 7 月 11 日，「1 號店」正式上線，開創了中國電子商務行業「網上超市」的先河，成立以來，1 號店持續保持高速的增長勢頭，2013 年實現了 115.4 億元的銷售業績。1 號店已成為國內最大的 B2C 食品電商。至 2013 年年底，1 號店可銷售 SKU 已達 340 萬種，覆蓋了食品飲料、生鮮、進口食品、美容護理、服飾鞋靴、廚衛清潔用品、母嬰用品、數碼手機、家居用品、家電、保健器械、電腦辦公、箱包珠寶手錶、運動戶外、禮品等 14 個品類。2013 年年底，1 號店已擁有 5700 萬的註冊用戶，並擁有超過 1500 萬的移動端註冊用戶。1 號店在上海推出「準點達」，並推出「全國包郵」政策。[168]

No. 1 Store is an e-commerce website. On July 11, 2008, it was officially launched, which pioneered the online supermarket of China's e-commerce industry.

---

168 1 號店官網資料整理 . http://cms.yhd.com/cms/view.do?topicId=24183&ref=gl.1.1.1.[YHD_FOOTER_NAV_about].EhT`3y.EhT`ht　Data Collation from No. 1 Store Official Website. http://cms.yhd.com/cms/view.do?topicId=24183&ref=gl.1.1.1.[YHD_FOOTER_NAV_about].EhT`3y.EhT`ht

Since then, No. 1 Store has maintained a high-speed growth momentum. In 2013, it achieved sales of 11.54 billion yuan, which made it become the largest B2C food e-commerce company in China. By the end of 2013, No. 1 Store had reached 3.4 million SKUs, covering 14 categories of food and beverage, fresh food, imported food, beauty care, clothing and footwear, kitchen and toilet cleaning products, maternal and child supplies, digital products including mobile phones, household items, home appliances, health care equipments, computers and office supplies, bags, suitcases, jewelry and watches, sport outfits and gifts. At the end of 2013, the No. 1 Store had 57 million registered users among which more than 15 million were the mobile users. No. 1 Store launched 「On-time delivery」 service in Shanghai and the 「National free postage」 policy.[168]

中糧我買網是由世界 500 強企業中糧集團有限公司於 2009 年投資創辦的食品類 B2C 電子商務網站，致力於打造中國最大、最安全的食品購物網站。有 19 大類 6000 多種商品，每月新增 300 種，經營「中糧製造＋中糧優選」商品，不只賣中糧自己的產品，還設進口食品、地方特產、生鮮食品等專區。自有產品全產業鏈控制，保質期超過 1/3 不進庫，保質期超過 2/3 不出庫。市區 24 小時送達，郊區 48 小時送達，以京滬周邊區域為核心逐步發展。[169]

Zhongliang Womai is a food B2C e-commerce website invested by the world's top 500 enterprises COFCO Group Co., Ltd. in 2009, dedicated to building China's largest and safest food shopping website. There are 19 categories of more than 6,000

169 我買網官網資料整理. http://www.womai.com/Info/AboutUsDetail.do?id=60003
Information Compilation of the Official Website of Womai. http://www.womai.
com/Info/AboutUsDetail.do?id=60003

kinds of goods on the website, with 300 new arrivals every month. It operates both products of Zhongliang brand and other quality brand, not only selling self-owned products, but also setting up special sections of imported food, local specialties, fresh food and so on. The self-owned products are controlled by the whole industry chain. Products with the shelf life less than 2/3 are not permitted into the inventory, and products with the shelf life more than 2/3 are not permitted to be discharged. Orders in downtown are delivered in 24 hours, and orders in suburbs are delivered in 48 hours. The delivery service is developed taking the surrounding areas of Beijing and Shanghai as the core.[169]

3‧ 第三方資源整合平臺，透過整合多元商業資源，提供多元的最後一公里服務 A third-party resource integration platform that provides a diverse last-mile service by integrating diverse business resources

以愛鮮蜂、小 e 到家、生活圈 C、京東到家等為代表的綜合電商平臺是發展最後一公里服務的熱點。透過整合多元生活服務資源，提升多元的最後一公里服務，消費者的各種生活服務需求都可以找到相應的平臺資源隨時隨地得到滿足。

The comprehensive e-commerce platform represented by Beequick, xedaojia, C-Life, and Daojia.jd.com is a hot spot for the development of the last-mile service. By integrating multiple life service resources and enhancing the diversity of the last-mile service, consumers can find the corresponding platforms to meet the needs of any life service.

如愛鮮蜂定位為「掌上一小時速達便利店」，於 2014 年 5 月上線，其以眾包微物流配送為核心模式，基於移動終端定位的 O2O 社區電商平臺，利用社區便利店資源，透過店主來完成最後一公里的配送，主打一小時送達。愛鮮蜂的「鮮」強調食品的新鮮以及多樣性，「蜂」則表示配送人員的數量多、速度快，「愛」則是滿足用戶「即時消費需求」，讓用戶突然冒出來的消費意願在最快時間獲得滿足。公司成立後迅速在北京、上海、廣州、深圳、佛山、蘇州、杭州、南京等主要城市上線。愛鮮蜂平臺產品以生鮮為主，各種產地直採水果，各式海產鮮食，各類酒水飲料，各地特色滷味，以及麻辣小龍蝦、哈根達斯、星巴克等產品；還有生活必需品，如副食調料、電池、牙膏、蚊香等。[170]

For example, Beequick is positioned as the "one-hour arrival convenience store on the palm". Launched in May 2014, it takes the outsourcing micro-logistics as the core model. Its O2O community e-commerce platform based on mobile terminal makes the use of community convenience store resources to complete the last-mile delivery through the owner. Beequick emphasizes the freshness and diversity of the food. It has a large number of delivery personnel and the delivery speed is fast. It also tries hard to satisfy the 「instant consumption demand」 of the users, letting them finish the consumption intentions that suddenly pop up in their mind. After the establishment of the company, it quickly went online in major cities such as Beijing, Shanghai, Guangzhou, Shenzhen, Foshan, Suzhou, Hangzhou and

---

170 愛鮮蜂官網資料整理 . http://www.beequick.cn/show/info?tag=about Information on Beequick Official Website. http://www.beequick.cn/show/info?tag=about

Nanjing. Products on Beequick are mainly fresh foods including all kinds of direct-picked fruits, seafood, drinks, local flavors as well as spicy crayfish, Haagen-Dazs, Starbucks. Besides, there are other necessities such as non-staple food, seasonings, batteries, toothpaste, mosquito incense, etc.

結論：中小型超市和便利店是未來超市業發展的重點，也是最後一公里服務重要的節點，千米綜合超市以加強消費體驗為經營重點，爭得一席之地，線上線下結合，整合提供最後一公里服務是必然趨勢。

It can be concluded that small and medium-sized supermarkets and convenience stores are the focus of the future development of the supermarket industry, and also an important node for the last-mile service. They focus on strengthening the consumer experience, integrating online and offline, and providing the last-mile service, which is an inevitable trend of future development.

二

超市業發展的重點領域

2. Key developmental areas of supermarket industry development

（一）突出食品經營 Food business

　　城市生活節奏越來越快，針對不同的生活節奏和生活習慣，人們對於「吃飯」的消費需求越來越複雜，雲消費時代可以實現隨時隨地消費，但「吃飯」相關的消費，是一種有體驗傾向的消費，尤其是生鮮食品的挑選與現場加工，對速食食品等為解決吃飯問題到店消費有很大需求。突出並細分食品經營，將解決「吃飯」問題的餐桌食品做為經營重點將是超市業未來經營的重點。

　　The pace of urban life is getting faster and faster. With different life rhythms and living habits, people's consumption demand for 「dining」 is more and more complicated. In the era of cloud consumption, consumption can be done anytime and anywhere. However, the consumption related to dining is an experience consumption. Especially the selection and on-site processing of fresh foods pose a great demand for fast food to solve the problem of on-site dining. Highlighting and subdividing food business, and taking the fast food as the key points of business will be the focus of the future operation of the supermarket industry.

（二）突出服務功能 Service function

雲消費時代，商品消費不再是問題，但人們對生活服務的需求越來越多元，對高質便捷服務的需求越來越強烈，以居民生活為核心，圍繞居民生活方式，提供居民各種生活服務所需，使超市成為居民生活的一部分，突出服務功能，提供個人化專屬服務，因地制宜引進多元生活服務功能，尤其是提供最後一公里到家服務，也是未來超市業發展的重點。

In the era of cloud consumption, commodity consumption is no longer a problem. Nevertheless, people's demand for life services is becoming more and more diversified, and the demand for high quality and convenient services is becoming more and more intense. Supermarkets should provide all kinds of life services taking residents living as the core, and make themselves a part of the residents' life. Supermarkets should highlight the service function, provide personalized and exclusive services, and introduce multiple living services functions especially the last-mile service according to local conditions, which is the focus of the future development of the supermarket industry.

（三）突出消費體驗 Experience consumption

雲消費時代，體驗化是主流消費方式的四大特徵之一，消費者到店不再是為了消費，更多的是為了體驗，消費過程的體驗和樂趣有時重於商品和服務的內容。如果能帶給消費者特殊的、不同的消費體驗，超市就取得了制勝的先機。

In the era of cloud consumption, experiencing is one of the four characteristics of mainstream consumption. Consumers are no longer shopping for consumption, but more for experience. The experience and fun of the consumption are sometimes more important than the goods and services. If the supermarket can bring consumers a special and different consumer experience, then it has achieved the opportunity to win.

（四）發展商業智慧 Business intelligence

準確定位目標客群，針對需求隨時調整的商品品類與品牌，最適當的商品規格和數量，精準的針對消費者個人的行銷方案，提供消費者需要的服務功能等，是以大數據和商業智慧系統為基礎的，也是雲消費時代超市業生存與發展的必要條件。

Based on big data and business intelligence, necessary conditions for the survival and development of the supermarket industry in the era of cloud consumption include accurate location of the target customer group, adjustment of the product category and brand at any time, the most appropriate specifications and quantity of commodities, accurate marketing plan for the the individual, and providing the service functions that consumers need.

塔吉特是大數據應用的先驅，每位顧客初次到塔吉特刷卡消費時，都會獲得一組顧客識別編號，內含顧客姓名、信用卡卡號及電子郵件等個人資料。日後凡是顧客在塔吉特消費，電腦系統就會自動記錄消費內容、時間等資訊。再加上從其他管道取得的統計資料，塔吉特便能形成一個龐大數據庫，運用

於分析顧客喜好與需求。基於大數據採擷，塔吉特能提前預知消費需求，選取 25 種典型商品的銷售情況建立「懷孕預測指數」，針對一個女士幾個月的消費紀錄，比如懷孕頭三個月過後會購買大量無味的潤膚露，前 20 週大量的購買鈣、鎂、鋅，大量採購無味肥皂和特大包裝的棉球……塔吉特預測到該女士懷孕，並針對其孕期需求制訂專屬行銷方案，這是塔吉特比女士的父親更早知道女兒懷孕的經典案例。[171]

Target Corp. is a pioneer in big data applications. Each customer gets a set of customer identification numbers which contain customer names, credit card numbers and emails when he or she pay by card for the first time in Target. Later, whenever the customer consumes in Target, the computer system will automatically record information such as consumption content and time. Together with the statistics obtained from other channels, Target can form a huge database for analyzing customer preferences and needs. Based on big data mining, Target can predict consumer demand in advance, and select the sales of 25 typical products to establish a 「pregnancy prediction index」. Target studies women's consumption record in several months to tell whether she is pregnant or not. For example, a large amount of odorless body lotion will be purchased by a three-month pregnant woman. Also she will buy a large amount of calcium, magnesium and zinc tonic as well as a large amount of odorless soaps and extra large cotton balls in the first 20 weeks... Target tells the woman is pregnant and makes an exclusive marketing plan for her pregnancy needs. This is the classic case of Target knowing the woman's pregnancy earlier than her father.[171]

7-11 透過資訊系統，使用銷售資料和軟體改善企業的品質控制、產品定

價和產品開發等工作。7—11 可以一天三次收集所有商店的銷售資訊，並在 20 分鐘內分析完畢，可以更快地分辨出哪些商品或包裝吸引顧客。7—11 公司利用這些技能來增加有更高利潤的自有品牌產品的開發。7-11 系統每天收集氣象報告資料 5 次，供各地的門店參考，以免所訂的講求新鮮度的食物數量積壓或者不足。在商品自動訂貨的參數中加入天氣因素，該因素可能包括溫度、濕度、風力的高低以及緊急的天氣變化，例如暴雨、地震、颶風等等，這些影響都將對商品的訂貨和庫存計畫產生影響。礦泉水隨著溫度的上升，訂貨係數將會上升，具體上升的比例是根據以往銷售的最高峰和最低谷的變化程度來擬定天氣影響係數變化的範圍。第一天下雨和第二天還在下雨，對客戶的購物行為的影響是不同的，配比不同的訂貨數量。不同的天氣情況，不同的背景音樂以及問候語。突然下雨，門店的盒飯和一些速食的送餐訂單一定會大大增加，重新排班。**172**

7-Eleven uses sales data and software to improve quality control, product pricing, and product development through information systems. 7-Eleven can collect sales information of all stores three times a day, and analyze the information within 20 minutes, which helps to quickly identify the goods or packaging that attract customers. 7-Eleven uses these skills to increase the development self-owned

---

171 塔吉特讀心術——使用者資料分析的魔力.譯言網.http://select.yeeyan.org/view/147927/267161 Target's Mind Reading-the Magic of User Data Analysis. Yeeyan. http://select.yeeyan.org/view/147927/267161

172 向 7-11 便利店學「天氣管理」.外貿知識網.http://www.rfqy.net/infoFile/6/20071228203416.shtml Learn "Weather Management" from 7-Eleven Convenience Store. Foreign Trade Knowledge. http://www.rfqy.net/infoFile/6/20071228203416.shtml

products with a higher margin. 7-Eleven system collects meteorological report data five times a day as the reference for stores in various places, so as to avoid the backlog or shortage of foods that are ordered to be fresh. Add weather factors to the parameters of the automatic goods ordering, including temperature, humidity, wind level and weather emergencies, such as heavy rain, earthquakes, hurricanes, etc., which will affect the ordering and inventory planning of goods.. As the temperature of the mineral water rises, the order coefficient will increase, the ratio of which is based on the degree of change of the highest peak and the lowest valley of the previous sales to determine the range of the weather influence coefficient. Impact of raining for only one day and rainingg until the next day are different on customer's shopping behavior as well as the order quantitative proportion.There are different background music and greetings for different weather conditions. If it suddenly rains, the number of box meals in the stores and fast food delivery orders will definitely increase and should be rearranged.[172]

（五）構建高效配送體系
Building the efficient delivery system

　　做為雲消費時代超市業生存與發展的必要條件的針對需求隨時調整的商品品類與品牌，以及最適當的商品規格和數量，是以精準高效的物流為支撐的。

　　As a necessary condition for the survival and development of the supermarket industry in the era of cloud consumption, the categories and brands of goods should be adjusted at any time according to consumers' needs. The most appropriate product specifications and quantities are supported by precise and efficient logistics.

7-11 成功的關鍵要素之一，除了精準的大數據採擷與應用外，還在於其高效冷鏈物流體系的支撐。其冷鏈物流針對商品對溫度的要求，劃分四類，冷凍型（零下 20 攝氏度），如冰淇淋等；微冷型（5 攝氏度），如牛奶、生菜等；恆溫型，如罐頭、飲料等；暖溫型（20 攝氏度），如麵包、飯食等。對於有特殊要求的食品如冰淇淋，7 － 11 會繞過配送中心，由配送車早中晚三次直接從生產商門口拉到各個店鋪。對於一般的商品，7 － 11 實行的是一日三次的配送制度：早上 3 點到 7 點配送前一天晚上生產的一般食品；早上 8 點到 11 點配送前一天晚上生產的特殊食品如牛奶，新鮮蔬菜也屬於其中；下午 3 點到 6 點配送當天上午生產的食品。這樣一日三次的配送頻率在保持了商店不缺貨的同時，也保持了食品的新鮮度。碰到一些特殊情況造成缺貨，可以向配送中心打電話告急，配送中心則會用安全庫存對店鋪緊急配送。[173]

One of the key elements of 7-Eleven' s success is the support of its efficient cold chain logistics system in addition to accurate big data mining and application,. The cold chain logistics is divided into four categories according to the temperature requirements of commodities, they are the frozen type (minus 20°C) for ice-cream; the cool type (5°C) for milk, lettuce, etc.; the constant temperature type for canned food, beverage, etc.; the warm type (20°C) for bread, meals, etc. To deliver food

---

173 7 － 11 便利店的配送系統 . 物流天下 . http://www.56885.net/news/2008226/55702. html　　Distribution System of 7-Eleven Convenience Store.56885.net. http:// www.56885.net/news/2008226/55702.html

with special requirements such as ice-cream, the delivery vehicle will bypass the distribution center and deliver the food directly to each store three times in the morning, afternoon and evening. For normal food, 7-Eleven implements a three-times-a-day delivery system: normal food produced the night before the delivery day are delivered from 3 am to 7 am; special food such as milk and fresh vegetables produced the night before the delivery day are delivered from 8 am to 11 am; food produced in the morning of the delivery day is delivered from 3 to 6 pm. In this way, the delivery frequency of three times a day ensures that the store is not out of stock, and the freshness of the food is also guaranteed. If there are some special circumstances that cause the stockout, the storekeeper can call the distribution center to report the emergency, and the distribution center will deliver the emergency stock to the store.[173]

「雲消費」時代城市生活服務業的重構

Chapter Eight
Reconstruction of Urban Life Service Industry in the Era of Cloud Consumption

第八章

「雲消費」時代城市生活服務業的重構

Chapter Eight Reconstruction of Urban Life Service Industry in the Era of Cloud Consumption

　　生活性服務業是宜居城市的重要建設內容，是城市宜居水準的重要體現，它直接向居民提供生活所需的產品及服務，主要包括餐飲業、住宿業、家政服務業、洗染業、美髮美容業、沐浴業、人像攝影業、維修服務業等服務業態。本書重點研究「雲消費」時代與城市居民日常生活所需息息相關的生活服務業服務模式，其重點包括「菜籃子」、早餐、便利店、洗染、美容美髮、家政等門類。

　　The life service industry is an important construction content of a livable city and an important embodiment of the city's livability. It directly provides residents with the products and services needed for life, including catering, accommodation, housekeeping, dyeing, hairdressing and beauty, bathing, portrait photography and maintenance service. This book focuses on the service model of life service industry, which is closely related to the daily life needs of urban residents in the era of cloud consumption, including the 「food basket」, breakfast, convenience store, dyeing, beauty salon, housekeeping and so on.

「雲消費」時代對傳統城市生活服務業的挑戰

1. The challenges for the traditional urban life service industry in the era of cloud consumption

（一）「雲消費」時代對城市生活服務需求的變化

Changes in the demand for urban living services in the era of cloud consumption

　　如前所述，在「雲消費」時代，人們的消費需求逐漸向追求現代生活方式轉型，消費呈現體驗化、個人化、定位化、社群化四大特徵，隨之，對城市生活服務需求的多元化、便捷化、安全化也逐漸突出。

　　As mentioned above, in the era of cloud consumption, people's consumer demand is gradually transforming into the pursuit of modern lifestyle. Consumption presents four characteristics of experiencing, personalization, orientation, and socialization, followed by diversification, convenience and security needs for the urban life services, all of which have also become increasingly prominent.

1 · 多元化需求 Demand for diversification

生活服務需求的多元化主要體現在以下 3 個方面：①對社區生活服務品類的多元化需求。即希望所需所想的多品類的生活服務功能，包括就餐、洗衣、購物、洗染、修理等多種功能都能在社區內得以實現。②同類生活服務選擇的多樣化需求。即希望同類生活服務能有更多的選擇，比如早餐既可以選擇麥當勞、肯德基等西式速食，也可以選擇慶豐包子、老家肉餅等中式速食。③生活服務方式的多元化需求。即希望以多種形式提供社區生活服務，既可以選擇網路消費、上門消費（服務），也可以選擇現場消費，有多種消費方式可供選擇。

The diversification of life service demand is mainly reflected in the following three aspects. First, the diversified demand for community life service. That is to say, the multi-category life service functions including dining, laundry, shopping, dyeing, repair and other functions can be realized in the community. Second, diversified demand for the choice of life services. This means the hope for more choices of similar life services. For example, people can choose Western-style fast food such as McDonald's KFC, or Chinese fast food such as Qingfeng buns and meat pies for breakfast. Third, diversified demand for living services, which means people hope to use the community life services in various forms such as online consumption, door-to-door consumption (service) and on-site consumption. There are many ways for people to consume.

2 · 便捷化需求 Demand for Convenience

隨著城市經濟的發展、生活節奏的加快，人們對生活服務的便捷化需求

越來越強烈，即希望能在最短的時間內得到所需的優質的生活服務。尤其隨著移動終端的普及和大量服務應用的開發，越來越多的消費者已經適應了移動定位化消費，人們越來越希望可以隨時隨地滿足定位化的各種生活服務，包括購物、訂餐、訂票等。

With the development of the urban economy and the accelerated pace of life, people are increasingly demanding the convenience of living services, that is, they hope to get the quality life services they need in the shortest possible time. In particular, with the popularity of mobile terminals and the development of a large number of service applications, more and more consumers have adapted to mobile positioning consumption, and people are increasingly hoping to use various life services, including shopping, food ordering and tickets booking anytime, anywhere.

3・ 安全化需求 Demand for security

在「雲消費」時代，地溝油、轉基因等各種不安全資訊透過網路得以快速傳遞與發佈，人們越來越關注生活服務消費的安全性，希望得到安全和可靠的產品和服務。

In the era of cloud consumption, large amounts of insecurity news such as gutter oil and genetically modified food are quickly transmitted and released through the network. People are paying more and more attention to the security of life service consumption and hope to obtain safe and reliable products and services.

（二）「雲消費」時代對城市生活服務的發展要求 The development requirements of urban life service in the era of cloud consumption

為應對「雲消費」時代城市居民對消費的需求，城市生活服務業體系應按功能完善、服務便捷、流通安全的要求發展建設。

In order to meet the demand of urban residents for consumption in the era of cloud consumption, the urban life service industry system should be developed based on the requirements of perfect functions, convenient services and safe circulation.

1 · 功能完善 Perfect functions

要完善社區內生活服務體系功能，使便利購物、餐飲食品、便利服務等各項居民日常生活服務功能都在社區內得到高品質的滿足。

It is necessary to improve the functions of the life service system in the community, so that various daily life service functions such as shopping, catering, and convenience services can be met with high quality in the community.

2 · 服務便捷 Convenient services

社區生活服務業佈局合理、服務模式便捷高效，居民可以透過現代消費模式在最短時間內隨時隨地得到最完善的生活服務。

If the community life service industry has a reasonable layout and a convenient and efficient service model, residents can get the most complete life service at any time and any place in the shortest time through modern consumption model.

3 · 流通安全 Safe circulation

流通安全主要包括 3 方面內容：①流通的商品品質是安全可靠的；②流通商品在流通與加工過程中商品安全性有保障； ③社區生活服務的消費環境是現代安全的。

The safe circulation mainly includes three aspects. First, the quality of the goods in circulation is safe and reliable. Second, the safety of goods in circulation and processing is guaranteed. Third, the consumption environment of community life services is modern and safe.

（三）「雲消費」時代傳統城市生活服務業發展「瓶頸」
The bottleneck in the development of traditional urban life service industry in the era of cloud consumption

長期以來，政府各相關部門一直把民生問題做為關注的重點，把發展生活性服務業做為關注民生、保障民生的重要途徑，致力於提升生活服務業的品質和水準，並取得了明顯成效，居民滿意度不斷提升。但隨著「雲消費」時代的來臨，居民生活方式和消費習慣發生了重大變化，傳統城市生活服務

業的服務內容與服務方式難以滿足居民不斷發展變化的服務需求，還有待進一步創新與發展。

For a long time, all relevant government departments have always regarded the people's livelihood issue as the focus of attention, and regarded the development of life service industry as an important way to pay attention to and protect people's livelihood. They are committed to improving the quality and level of the life service industry and have achieved remarkable results as satisfaction continues to improve. However, with the advent of the era of cloud consumption, the lifestyle and consumption habits of the residents have undergone major changes. The service content and service methods of the traditional urban life service industry are difficult to meet the evolving and changing service needs of residents, and further innovation and development are needed.

1. 社區生活服務業規模化、規範化發展取得一定進展，但現代流通模式還沒有成為流通主管道 The scaling and standardization of community life service industry have made certain progress, but the modern circulation model has not become the main channel of circulation.

（1）「菜籃子」服務 The "Vegetable Basket" service

居民吃菜問題一直是政府關注的重點，自農業部 1988 年提出建設「菜籃子」工程至今，「菜籃子」工程建設從最初解決供需矛盾、以生產基地建設為主，發展為以生產基地與市場體系建設並舉。1997 年年底，全國農副產

品批發市場發展到約 4000 家。2000 年以來，「菜籃子」工程的發展建設日趨成熟，「菜籃子」的供求形勢從長期短缺轉向供求基本平衡；同時隨居民生活水準逐漸提升，「菜籃子」工程建設重點逐漸由規模化向品質化方向發展。2001 年，農業部組織實施「無公害食品行動計畫」，「菜籃子」工程關注重點逐漸由農產品大流轉、大流通轉移到蔬菜消費終端的發展建設。生鮮超市、連鎖超市等現代流通模式逐漸發展壯大，但菜市場等傳統流通模式仍是目前城市「菜籃子」的流通主管道，為此，「菜籃子」工程的連鎖化、規模化、規範化、現代化發展還有待提升。

The problem of residents buying agricultural products has always been the focus of the government. Since the Ministry of Agriculture proposed the construction of the 「Vegetable Basket」 project in 1988, the construction of the project has shifted from initially resolving the contradiction between supply and demand, and mainly constructing the production base to constructing both the production base and the market system. At the end of 1997, the national agricultural and sideline products wholesale market increased to about 4,000. Since 2000, the development and construction of the 「Vegetable Basket」 project has become increasingly mature. The supply and demand situation of the project has shifted from a long-term shortage to a basic balance between supply and demand. At the same time, with the gradual improvement of the living standards of the residents, the focus of the project has gradually shifted from development of scaling to quality. In 2001, the Ministry of Agriculture implemented the 「Non-pollution Food Action Plan」. From then the focus of the 「Vegetable Basket」 project was gradually shifted from the large circulation of agricultural products to the development of vegetable consumption terminals. Modern circulation models such as fresh supermarkets and chain supermarkets have gradually grown stronger whereas the traditional circulation model such as the vegetable market is still the main channel for the circulation of the project. For this reason, the chainization, scaling, standardization and modernization of the 「Vegetable Basket」 project need further improvement.

（2）早餐服務 The breakfast service

早餐關乎民生，與城市居民日常生活息息相關。所以政府相關部門一直把早餐服務做為關注的重點，以政策為引導的「早餐工程」取得了很大進展。以北京為例，1996 年，北京市開始恢復早點供應；2001 年，北京市政府決定採取以政策引導為主的「早餐工程」，正式進入組織實施階段；從 2002 年起，北京正式開始了「早餐工程」的探索。2009 年，商務部宣布以北京等城市為試點推動「早餐工程」。2009 年 8 月 19 日，市商務委啟動北京市早餐示範企業專案的招標工作，北京市「早餐工程」迅速展開。截至 2010 年，北京市 16 個區縣共有 2164 個早餐網點。同時餐飲企業也一直在進行早餐盈利模式的相關探索，早餐服務品質有了較大提升。但安全衛生的品牌早餐的服務人群規模還有待提升，正規網點覆蓋半徑內早餐服務解決度不夠、規範化早餐服務發展不均衡等問題在中國多數城市中普遍存在，同一社區內多樣化的品質早餐需求更是可望而不可即。

Breakfast is related to people's livelihood and daily life. Therefore, the relevant government departments have always regarded the breakfast service as the focus of attention, and the policy-oriented Breakfast Project has made great progress. Taking Beijing as an example, in 1996, Beijing began to resume the breakfast supply. In 2001, the Beijing Municipal Government decided to adopt a policy-oriented Breakfast Project and officially entered the implementation phase. Since 2002, Beijing officially began the exploration of the Breakfast Project. In 2009, the Ministry of Commerce announced the promotion of the Breakfast Project within cities such as Beijing as the pilot. On August 19, 2009, the Municipal Commercial Committee launched the bidding for the Beijing Breakfast Demonstration Enterprise Project, and the Breakfast Project in Beijing spread rapidly. As of 2010, there were 2,164 breakfast outlets in 16 districts and counties in Beijing. At the same time, catering companies have been exploring the profit model of breakfast, and the quality of breakfast service has been greatly improved. However, the size of the service group for safe and healthy brand breakfast needs to be improved. The problem of insufficient breakfast service settlement within the radius of regular outlets and the uneven development of standardized breakfast service are common in most cities in China, not to mention the variety of quality breakfast needs in the same community.

（3）再生資源回收服務 The renewable resources recycling service

再生資源回收服務做為人們生活中不可或缺的重要內容以及發展迴圈經濟和減輕環境污染的有效途徑，一直備受各領域的關注，同時也是政府著力解決的難點問題。2007 年，為促進再生資源回收規範化發展，商務部出臺《再生資源回收管理辦法》。《再生資源回收管理辦法》提出，「再生資源回收經營者可以透過電話、互聯網等形式與居民、企業建立資訊互動，實現便民、快捷的回收服務。」對再生資源服務形式提出明確方向。透過多年的整治，

中國再生資源回收服務的規範化程度有所提升。但目前多數城市非證照回收網站仍多於有證照回收網站，多數收購網點的組織化程度不高，各網站物資回收、流動、出售均各自為政，以個體收購運輸為主，沒有統一物流，在收購的過程中進行的一些簡易的分揀，存在各種污染隱患，影響城市環境。

As an indispensable part of people's life and an effective way to develop the recycling economy and reduce environmental pollution, renewable resources recycling service has always been concerned by various fields, and it is also a difficult problem that the government is trying to solve. In 2007, in order to promote the development of standardized recycling of renewable resources, the Ministry of Commerce issued the Regulations on the Management of Renewable Resources Recycling. It proposes that 「the renewable resource recycling operators can establish information interaction with residents and enterprises through telephone and the Internet to achieve convenient and fast recycling services,」 which is a clear direction for the form sof renewable resource services. Through years of rectification, the standardization of China's renewable resources recycling services has increased. However, at present, the number of non-licensed recycling sites in most cities is still more than that of the licensed recycling sites. The organizational degree of most purchased outlets is not high. Materials recovery, flow and sale of each site are completely separate, mainly the individual acquisition and transportation without the unified logistics. The individuals only carried out simple sortings, causing various pollution hazards and affectting the urban environment.

（4）洗染服務 The dyeing service

洗染業是對其周邊環境有一定污染的行業，所以衣服洗染所在地會嚴重影響周邊社區的環境品質。同時洗衣店運營需要洗衣設備的投入、人員的投入；而且規模越大、越先進、環境污染越低的設施，投入成本就越高。目前，大多城市的洗染企業以加盟連鎖運營模式為主，即將機器設備出售給加盟店，加盟店在社區內開店洗染一體化運營，有的加盟店在附近社區內多點收活，在社區大店內集中洗染。這種在社區內或附近直接洗染的模式一是直接產生污染，影響社區居民的生活環境品質；二是由各加盟店分別購入洗衣設施、投入人員，從社區洗染發展整體看，重複建設、成本高、效率低；三是加盟店一般只為周邊社區居民提供洗染服務，洗染服務半徑有限，一般不會引進污染小、效率高、投入高的先進設施。城市生態集約的現代洗衣模式仍需進行探索。

The dyeing industry is an industry that causes certain pollution to its surrounding environment, so the clothes dyeing factories will seriously affect the environmental quality of the surrounding communities. At the same time, operation of the business needs the input of the dyeing equipments and the personnel. The larger the scale of, the more advanced, the lower environmental pollution of the equipments, the higher the cost. At present, most of the city's dyeing enterprises mainly rely on the franchise chain, that is, selling the equipments to the franchise stores. Some franchise stores receive extra orders from the nearby communities and do the all the dyeing work in the main store. First, this mode of direct dyeing in or near the community directly produces pollution and affects the quality of the living environment of the community residents. Second, the franchise stores purchase dyeing facilities and input the personnel, from the perspective of overall development of the community dyeing business, is redundant and with high cost and low efficiency. Third, generally, franchise stores only provide dyeing services for residents in the surrounding communities. The radius of services is limited so the advanced facilities with low pollution, high efficiency and high cost will not be introduced. The modern laundry model of urban eco-intensive needs further exploration.

2 · 傳統城市生活服務業的發展「瓶頸」 The bottleneck in the development of traditional urban life service industry

(1) 生活服務業多是微利產業，單項服務功能進入社區難以盈利

The life service industries are mostly with low profit, and it is difficult for a single service to generate profits in the community.

生活服務企業以服務居民為宗旨，只有深入社區、深入居民才能得以生存和發展。盈利是企業生存發展的基礎，生活服務業普遍是微利產業，尤其單項社區生活服務功能利潤空間微薄，經營成本承受能力有限。目前，商業地產租金水準逐年上升，尤其是大城市中心城區，租金成本更為高昂，對於承租能力較低的生活服務業更難進入。同時單項生活服務功能各自獨立運營需要分別投入租金、設備設施、人工等各項運營成本，各自在社區內重複建設服務體系，服務企業投入成本較高，很難盈利。

Life service enterprises aim at serving residents, they can survive and develop only by involving in communities and residents. Profitability is the basis for the survival and development of enterprises. The life service industries are mostly with low profit. In particular, the profit margin of a single community life service is meager, and the operating cost tolerance is limited. At present, the rent level of commercial real estate is increasing year by year, especially in the central area of big cities, which is difficult for the life service industry with lower renting capacity to enter. Besides, each individual service needs various operating costs such as rent, equipments, facilities, labor, etc. Also, they have to repeatedly build a service system in the community, which requires a high input cost, making it difficult to make profits.

（2）很多城市可供發展的空間資源有限，難以支撐多樣化的生活服務需求 Many cities have limited space resources for development, and it is difficult to support diverse life service needs.

居民多元化的消費需求，需要多元化的生活服務組合來滿足。各類多元的生活服務功能進社區，需要可供開店的空間資源。單個生活服務業網點雖然需要的空間規模較小，但涉及的業種業態較多，所有網點需求的累加規模十分可觀。隨著城市的發展，優質的店鋪資源越來越稀缺，很多大中城市中心城區和老城區空間資源的使用相對飽和，可供開發利用的空間資源十分有限，社區很難拿出足夠的空間資源支持其發展。

Residents' diversified consumption needs require a diversified set of life services to meet. As all kinds of diverse life services enter the community, space resources for opening stores are needed. Although the space required for a single life service industry outlet is small, there are many industries involved, and the cumulative scale of demand for outlets is considerable. With the development of the city, store resources of high quality are becoming scarcer. The use of space resources in the central and urban areas of many large and medium-sized cities is relatively saturated, and the space resources available for development and utilization are very limited. It is difficult for the community to come up with sufficient space resources to support its development.

二

「零距離、雲服務」──「雲消費」時代城市生活性服務業的發展方向
2. Zero Distance and Cloud Service──The Development Direction of Urban Life Service Industry in the Age of Cloud Consumption

（一）發展目標 Development goals

根據「雲消費」時代居民生活服務需求的變化，以及城市生活性服務業發展現狀和發展「瓶頸」，「雲消費」時代城市生活性服務業發展應實現兩個目標：①要在有限的社區空間資源條件下，集約滿足居民多樣化的社區生活服務需求，尤其是滿足居民足不出戶，透過現代資訊手段即可以獲得所需的高品質的生活服務；②要保證為社區服務的優質生活服務企業有充足的利潤空間，實現企業健康可持續發展。

According to the changes in the needs of residents' living services in the era of cloud consumption, and the development status of urban life service industry and the bottleneck of development, the development of urban life service industry in the era should achieve two goals. First, under the condition of limited community space resources, the needs of residents' diverse community living services should be met, especially the needs of obtaining the high-quality living services by modern information approaches without leaving home. Second, ensure sufficient profit margins for the community service companies with high quality to achieve healthy and sustainable development.

（二）模式構建 Pattern construction

針對目前城市社區生活性服務業發展現狀、問題和發展瓶頸，需要探索既保障滿足居民高品質生活服務需求，又保證服務企業健康可持續發展的新的社區生活性服務業發展模式——社區生活「零距離、雲服務」體系。

In view of the current development status, problems and development bottlenecks of urban community life service industry, it is necessary to explore a new community life service industry development model that guarantees the high-quality living service needs of residents and ensures the healthy and sustainable development of service enterprises, and that is the "Zero Distance and Cloud Service" system.

1‧社區生活「零距離、雲服務」體系的內涵 The connotation of the "Zero Distance, Cloud Service" system of community life

「零距離」體現城市生活服務體系的便捷化特徵，指居民不出家門透過現代資訊手段即可以獲得絕大多數生活服務滿足。

「Zero Distance」 reflects the characteristics in convenience of the urban life service system, which means that residents can meet the majority of life service needs through modern information approaches.

「雲服務」體現城市生活服務體系強大的資源整合能力和現代高效的以資訊平臺為基礎的服務方式。一是結合區域內各社區不同的資源條件，跨行業、跨業態整合各類社區生活服務資源，使消費者在各種社區空間資源條件下，都能得到豐富多樣的社區生活服務需求滿足；二是運用現代高效的資訊平臺，使消費者透過簡單的平臺介面得到所需的各種生活服務。

「Cloud Service」 reflects the powerful resource integration capability of the urban life service system and the efficient modern information platform-based service. First, it integrates with resource conditions of various communities in the region in a cross-industry manner, so that consumers can receive rich and diverse community life services whatever community space resources they have. Second, it uses an efficient modern information platform that enables consumers to get the various life services they need through a simple platform interface.

在「雲消費」時代，商品交易更加自由開放，各類資訊資源平臺給消費者更多選擇空間，資源、資訊、服務等的「雲整合」是使「雲消費」時代商業體系運轉更有效率的重要手段。

In the era of cloud consumption, commodity transaction is more free and open, and various information resource platforms give consumers more choices. Cloud integration of resources, information and services is an important approach to make the commercial system run more efficiently in the era of cloud consumption.

2. 以整合資源為發展核心 Integrating resources as the core of development

在「雲消費」時代，城市生活性服務業的發展也應以資源整合為核心，結合城市各區域不同的資源條件，對生活服務資源進行跨行業、跨業態的整合，推進實現多種生活服務資源在城市所在區域內高效、集聚、整合發展。

In the era of cloud consumption, the development of urban life service industry should also focus on resource integration, combine different resource conditions in different regions of the city, integrate life service resources in a cross-industry manner, and promote the efficient, concentrated, integrated development of multiple life service resources in the region where the city is located.

以資源整合為核心，發展城市生活性服務業。首先，可以使消費者在不同區域、不同資源條件下，都可以實現滿足多樣的生活服務需求。消費者可以不出家門，透過電腦網路、歌華電視、電話等多種方式，透過一個平臺介面，實現零距離生活服務消費。消費者也可以到實體店親身體驗服務消費或手機二維碼消費，實現社區生活服務「一站式」全方位滿足消費需求。其次，透過多種生活服務企業資源的整合，很多生活服務項目不需要在社區內租用單獨的經營空間、不需要單獨雇傭員工，就可以為社區居民提供生活服務。在有效控制企業運營成本的同時，擴大企業的經營規模，提升企業的盈利能力，推進城市生活服務企業健康可持續發展。第三，可以使各類資源集約使用，節約成本，提升效率。最後，生活性服務業的整合發展可以提升區域生活服務品質。原來由於生活服務不方便，被迫選擇非正規網點的現象日趨減少，原來社區內髒亂差、不規範的服務將逐漸在市場競爭中被淘汰。安全、

便捷、舒適、高品質的生活服務體系將逐漸成為區域社區生活服務的主流，區域生活服務品質將得到很大提升。

With the integration of resources as the core, the development of urban life service industry should be further developed. First, consumers' various life service needs in different regions and different resource conditions should be met, so that they can stay at home and use a computer network, Gehua TV, telephone and other means to carry out the zero-distance life service consumption through a platform interface. Consumers can also experience service consumption or mobile phone QR code consumption in physical stores, meeting consumer demand for the 「one-stop」 community life services. Second, through the integration of various life service enterprise resources, many life service projects can be done without renting a separate business space and hiring employees in the community to provide life services for community residents. While effectively controlling the operating costs, enterprises expand the business scale and enhance the profitability to promote the healthy and sustainable development of urban life service. Third, the approach can make intensive use of all kinds of resources to save costs and improve efficiency. Finally, the integrated development of the life service industry can enhance the quality of regional life services. The phenomenon of being forced to choose informal outlets due to inconvenient living services in the past has been decreasing. Also, the messy and irregular services in the community will gradually be eliminated in the market competition. The safe, convenient, comfortable and high-quality life service system will gradually become the mainstream of regional community life service, and the quality of regional life service will be greatly improved.

（1）整合服務 Integration of services

　　每個城市都有很多各類優質的生活服務資源，但由於空間條件、租金水準、人員成本等多種條件的制約，眾多優質的服務資源如果孤軍作戰，難以充分發揮其能量，難以為居民提供更好的服務。在「雲消費」時代，可以將這些優質的生活服務資源整合起來，發揮各自優勢，集約地提供高品質的生活服務。

　　Each city has many kinds of high-quality living service resources. However, due to various conditions such as space, rental and personnel costs, energy of many high-quality service resources are difficult to be fully exerted if used alone. In the era of cloud consumption, these high-quality living service resources can be integrated to give full play to their respective advantages and provide high-quality life services intensively.

　　結合城市各區域空間資源條件，在限定的空間內對多種優質的生活服務資源進行整合，包括便利店、早餐店、果蔬店、洗衣洗染、家電維修、再生資源回收等，實現多樣的生活服務資源在有限的空間內整合提供。

　　Spatial resources of various regions of the city should be combined, and a variety of high-quality living service resources including convenience stores, breakfast shops, fruit and vegetable stores, dyeing, appliance repair, recycling of renewable resources in a limited space should be integrated to integrate and provide a variety of life service resources in a limited space.

（2）整合信息 Integration of information

　　「雲消費」時代是資訊爆炸的時代，企業發佈的資訊、政府發佈的資訊、

各類仲介組織發佈的資訊、消費者發佈的資訊等，同一條資訊會由不同的來源，從不同角度加以詮釋。身處眾多紛雜的資訊之中，消費者往往很難獲取適用的生活服務資訊。

The era of cloud consumption is the era of information explosion. Information released by enterprises, the government, various intermediary organizations and consumers is everywhere. The same piece of information from different sources is often interpreted from different angles. Surrounded by vast amount of information, it is often difficult for consumers to obtain useful information on life services.

將城市生活服務的各種服務資訊、企業資訊、店鋪資訊等資訊資源整合到統一的平臺上，可以提升資訊資源的使用效率，推進實現各種資訊資源的共用。

By integrating various life service information such as service information, enterprise information, and store information onto a unified platform, the use efficiency of information resources can be improved, and the sharing of various information resources can be promoted.

（3）整合平臺 Integration of platforms

在「雲消費」時代，網路成為資訊傳遞的主要手段，政府、企業、協力廠商機構都致力於平臺體系建設。政府主導的網路平臺（如 96156 線上服務平臺）、街道網路平臺（如北京廣內街道網路平臺）、門店網路平臺（如肯德基網路訂餐系統、唐久大賣場）與實體平臺（如 7-11 便利店）等各類平臺資源此起彼伏，消費者為獲取一項生活服務，可能會造訪多個平臺而不得其所。

In the era of cloud consumption, the Internet has become the main approach of information transmission. The government, enterprises, and third-party organizations are all committed to the construction of the platform systems. Government-led network platforms such as 96156 Online Service Platform, subdistrict network platforms such as Beijing Guangnei Subdistrict Network Platform, store network platforms such as KFC Food Ordering System, Tangjiu Hypermarket and physical platforms such as 7-Eleven Convenience Store and other types of platform resources come one after another. In order to obtain a life service, consumers may have too many platforms to visit among and feel at a loss.

進行平臺資源的整合，推進各類平臺資源間的有效對接，可以使居民透過簡單的方式從一個平臺介面得到所需的生活服務資訊與服務，提升居民獲取生活服務的便捷度。

The integration of platform resources and the effective docking of various platform resources enable residents to obtain the required life service information and services from a platform interface in a simple way, and improve the convenience of residents to obtain life services.

（4）整合物流 Integration of logistics

在「雲消費」時代，人們越來越習慣不出家門就獲得商品和服務的消費方式，由此催生了物流快遞業的快速發展。在城市中各類物流快遞走街串巷，人們一天中可能會接待幾個配送不同商品和服務的快遞人員。快遞服務在方便人們生活的同時，也增加了一定的安全隱患。同時，眾多配送人員在城市街巷中分揀、配送，也會對城市環境產生一定的不良影響。

In the era of cloud consumption, people are becoming more and more accustomed to the consumption of goods and services without leaving their homes, which has led to the rapid development of the logistics and express industry. In the city, all kinds of express couriers are coming and going street to street, people may receive several couriers who deliver different goods and services in one day. Express services make it more convenient for people's lives, but also increases certain security risks. At the same time, there are too many distribution personnel sorting and distributing packages in urban streets and lanes, which also causes certain adverse effects on the urban environment.

將各類生活服務物流功能，包括專業快遞公司（中通、圓通、申通、順豐等）、電子商務企業（京東商城、卓越、當當、1號店等）、餐飲公司的物流（肯德基、麥當勞、必勝客等）、家政公司等整合起來，使居民所需的生活服務可以透過一個配送人員一次性配送到戶，使各類生活服務物流能更安全、便捷、有序地在城市內運轉。

The logistics functions of various life services, including professional express delivery companies (Zhongtong, Yuantong, Shentong, SF, etc.), e-commerce enterprises (Jingdong Mall, Amazon, Dangdang, No. 1 Store, etc.), logistics of catering companies (KFC, McDonald's, Pizza Hut, etc.), and the housekeeping companies should all be integrated so that the living services required by the residents can be delivered to the household by distribution staff. In this way, all kinds of life service logistics can be operated safely, conveniently and orderly in the city.

（5）整合人員 Integration of personnel

日益增長的人工成本是制約生活性服務業發展的主要「瓶頸」之一。透過各類服務資源的整合，原來多個服務人員的工作，可以由一個服務人員實現，比如便利店的服務人員，因便利店服務功能的整合與擴展，同時承擔了餐飲服務、洗衣收活服務、快遞服務等多種服務職能，服務人員的使用更集約，用工成本降低，工作效率提升，進而生活服務的盈利空間也得到提升。

Increasing labor costs are one of the main bottlenecks that constrain the development of the life service industry. Through the integration of various service resources, work for multiple service personnel in the past can be done now by one person.For example, due to the integration and expansion of the convenience store service functions, the service staff of the convenience store can now undertake a variety of other service functions including the catering service and the laundry collection service and express service. In this way, the use of service personnel is more intensive, the cost of labor is reduced and efficiency is improved. As a result, the profitability of life services is also improved.

（6）整合網點 Integration of branches

在「雲消費」時代，網點資源的稀缺和租金成本的不斷增長，成為制約生活性服務業發展的主要「瓶頸」。整合使用城市內有限的網點資源，在有限的網點資源內將便利店、早餐店、洗衣店、蔬菜店、再生資源等各類生活服務功能整合集聚，使網點服務功能多元化，以此提升城市社區空間資源的使用效率。

In the era of cloud consumption, the scarcity of branch resources and the continuous increase of rental costs have become the main bottlenecks restricting the development of life service industry. The limited network resources in the city, and various convenience services such as convenience stores, breakfast, dyeing, vegetables, and renewable resources in limited branches should all be integrated to diversify the service functions of the branches. In this way, the efficiency of the use of urban community space resources will be improved.

（7）整合结算 Integration of payment

在「雲消費」時代，可以一切既便捷又安全的支付手段完成交易的「雲支付」，「雲支付」將逐漸成為一種消費習慣。城市生活服務的結算方式也應適用「雲支付」的需求，將各類支付結算方式進行整合，實現包括儲蓄卡、信用卡、智慧公交卡、手機儲值卡、支付寶網銀帳戶、消費儲值卡之間資金互通共用，各類結算方式在所在區域生活服務體系內無障礙使用、聯合支付，以此提升居民社區商業支付結算的便捷度。

In the era of cloud consumption, transaction can be completed by the cloud payment, a convenient and safe payment method, which will gradually become a consumption habit. The payment method of urban life service should also comply with the demand for cloud payment. Various payment methods should be integrated to realize Inter-funds sharing among deposit cards, credit cards, smart bus cards, mobile phone stored-value cards, Alipay and banking accounts and consumption stored-value cards. Also, various payment methods are used in the local life service system with accessibility and joint payment to improve the convenience of commercial payment in residential communities.

3· 探索發展多種整合模式 Exploring and developing multiple integration models

針對城市不同區域資源條件的不同，因地制宜地探索發展生活性服務業的多種整合模式。

In view of the different resource conditions in different regions of the city, it is necessary to explore and develop various integration models of the life service industry according to local conditions.

（1）以便利店為核心的整合模式

The integration model with convenience stores as the core

以具有整合能力的便利店為主體企業，整合其他服務資源，在便利店內實現多種生活服務功能的集聚，並鼓勵將各類商品與服務直接延伸至居民家中。

The model takes the convenience stores with integration capabilities as the main enterprises. It integrates other service resources to realize the agglomeration of various life services in the convenience stores, and encourages all kinds of goods and services to be directly extended to the residents' homes.

在空間資源有限的區域，可以以區域內現有便利店、菜店為核心，由便利店或菜店做為主體企業，將其他各種服務資源整合進來。

In areas where space resources are limited, existing convenience stores and restaurants in the region can be taken as the core, and convenience stores or restaurants as the main enterprises, to integrate other service resources.

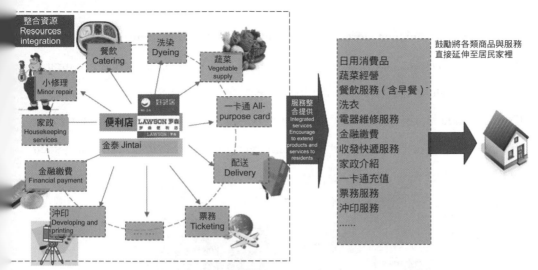

整合資源
Resources integration

餐飲 Catering
洗染 Dyeing
蔬菜 Vegetable supply
小修理 Minor repair
便利店 LAWSON 羅森
金泰 Jintai
一卡通 All-purpose card
家政 Housekeeping services
金融繳費 Financial payment
配送 Delivery
沖印 Developing and printing
票務 Ticketing

服務整合提供
Integrated services Encourage to extend products and services to residents

鼓勵將各類商品與服務直接延伸至居民家裡

日用消費品
蔬菜經營
餐飲服務（含早餐）
洗衣
電器維修服務
金融繳費
收發快遞服務
家政介紹
一卡通充值
票務服務
沖印服務
……

圖 Figrue 8-1　以便利店為核心的整合模式示意圖 Schematic diagram of the integration model with the convenience stores as the core

（2）社區服務中心整合模式

The integration model of community service center

社區服務中心一般有一定體量的商業空間，可以發展以便利店為核心店，以菜店、小型餐飲店、美容美髮店等為補充的整合模式。

The community service center generally has a certain amount of commercial space, where the convenience store can be developed as the core store, supplemented by a vegetable shop, a small restaurant, a beauty salon and the like.

擁有一定體量商業空間資源的區域，發展以社區便利店為核心店，針對區域的具體資源條件，由便利店整合部分服務功能，菜店、小型餐飲店以及美容美髮等其他服務功能可以針對可用資源規模因地設置，便利店和其他門店可透過招投標由不同的企業經營，也可以由一家主體企業整合實現所有功能，洗染、小修理、家政服務、金融繳費等服務功能並在便利店內搭載實現。

In the areas with a certain amount of commercial space resources, the community convenience stores can be developed as the core store, where some of the service functions can be integrated. Services like food stores, small restaurants and beauty salons can be set in the local area according to the scale of resources. Convenience stores and other stores can be operated by different enterprises through bidding or be integrated by a main enterprise to realize all functions, such as dyeing, minor repair, housekeeping services, financial payment, in the store.

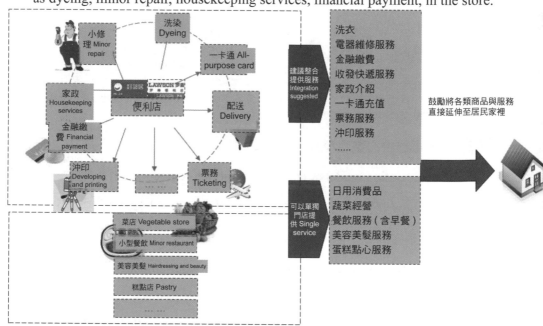

圖 Figure 8-2 社區服務中心整合模式示意圖 Schematic diagram of the integration model of community service center

（3）以網路服務商為核心的整合模式

The integratio model with network service providers as the core

在「雲消費」時代，隨著消費需求的變化，網路服務商隨之興起與發展。以網路服務商為核心，整合各類生活服務資源，實現消費者可以透過網路、電話等現代資訊手段，預約服務，由服務商以送貨上門或上門服務的方式為

消費者提供生活服務的「一站式」解決方案。

In the era of cloud consumption, wnetwork service providers have followed up the changes in consumer demand and developed. With network service providers as the core, various life service resources are integrated so that consumers can enjoy the one-stop solution, that is, making appointments through modern information approaches such as the Internet and telephone and then the service providers will provide service for consumers by means of home delivery or on-site service.

在沒有發展空間的區域，可以發展以網路服務商為核心的整合模式。

In areas where there is no room for development, an integration model centered on network service providers can be developed.

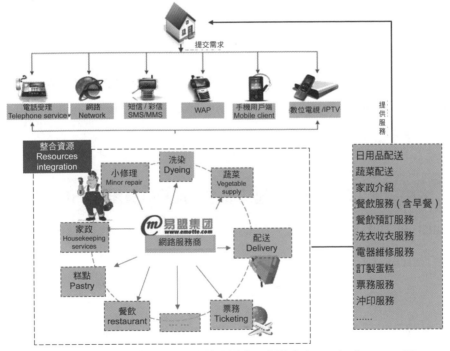

圖 Figure 8-3　以網路服務商為核心的整合模式示意圖 Schematic diagram of the integration model with the network service providers as the core

（三）模式功能 Model function

1‧ 一站式滿足居民多樣化的社區生活服務需求 Meeting the diverse needs of residents for community life services in a one-stop manner

透過各種生活服務資源的整合，居民既可以不出家門透過電腦網路、歌華電視、電話等多種方式，透過一個平臺介面實現零距離生活服務消費；也可以到實體店鋪親身體驗服務消費，或手機二維碼消費，實現社區生活服務一站式全方位消費需求滿足。

Through the integration of various life service resources, residents can not only use the Internet, Gehua TV, telephone and other means to achieve zero-distance life service consumption through a platform interface, but also experience service consumption in physical stores and QR code consumption on mobile phone, to meet consumer demand for community life services in a one-stop all-round manner.

2‧ 大幅降低生活服務企業運營成本，提升盈利能力 Significantly reducing the operating costs spent by life service companies and improving profitability

透過多種生活服務企業資源的整合，很多生活服務項目不需要在社區內租用單獨的經營空間、不需要單獨雇傭員工，就可以為社區居民提供生活服務，在有效控制企業運營成本的同時，擴大企業的經營規模，提升企業的盈

利能力。

Through the integration of various life service enterprise resources, many life service projects can provide life services for community residents without a separate business space in the community or special-hired employees. The model effectively controls the operating costs of enterprises and helps to expand the business scale to enhance the profitability of enterprises.

同時因各類生活服務資源的集聚，入戶服務可以將物流層級功能進行重新分割，最後一公里配送在社區內重新組織，為居民提供一站式生活服務配送，推進零距離服務效率提升，單位服務成本降低，服務範疇擴大，進而帶動企業盈利能力的提升。

At the same time, due to the accumulation of various life service resources, the household service can re-segment the logistics level functions, and the last-mile distribution is reorganized in the community to provide one-stop life service delivery for residents and promote the efficiency improvement of zero-distance service. The cost of services is reduced, and the scope of services is expanded, which in turn drives the profitability of enterprises.

3· 有效提升居民生活服務品質 Effectively improving the quality of life services for residents

「零距離、雲服務」體系透過生活服務資源的高效整合首先解決了區域社區空間有限、租金水準高，優質生活服務企業難以入駐的問題，尤其「零距離」的服務模式，是對原有社區生活服務的昇華，體現了安全、便捷、舒

適的服務標準，有效提升區域社區生活服務品質。

First, the 「Zero Distance, Cloud Service」 system solves the problem that the regional community has limited space, high rent level, and high-quality life service enterprises are difficult to settle in through the efficient integration of life service resources. In particular, the 「Zero Distance」 service model is the sublimation of original life services, which reflects the standards of safety, convenience and comfort, and effectively improves the quality of life services in the community.

其次，伴隨「零距離、雲服務」體系的構建與服務範疇的擴大，由於生活服務不方便，被迫選擇非正規網點的現象日趨減少，原來社區內髒亂差、不規範的服務將逐漸在市場競爭中被淘汰，安全、便捷、舒適、高品質的生活服務體系將逐漸成為區域社區生活服務的主流。

Second, with the construction of the 「Zero distance, Cloud service」 system and the expansion of the service categories, due to inconvenient life services, the phenomenon of forced selection of informal branches is decreasing day by day. The past messy and non-standard service in the community will gradually be eliminated in the market competition. Safe, convenient, comfortable, high-quality life service system will gradually become the mainstream of regional community life services.

三
城市生活性服務業創新與探索借鑑
3. Innovation and exploration of urban life service industry

事實上，基層政府和很多從事生活性服務業的企業已經進行的探索，代表了「零距離、雲服務」的發展。

In fact, the exploration carried out by the basic-level government and many companies engaged in the life service industry represents the development of「Zero Distance, Cloud Service」.

（一）「菜籃子」服務發展探索
Exploration of the "Vegetable Basket" service

如在「菜籃子」服務領域，上海市近年積極探索推進社區菜市場轉型升級的新模式，鼓勵發展四種模式：一是生鮮直投站模式，即生鮮產品 O2O 網訂店取模式。將傳統線下菜場零售與電子商務相融合，為市民提供 24 小時買菜服務，居民可以透過官方網站、自助終端、呼叫中心、手機應用、微信等五種途徑 24 小時不限時訂購生鮮產品，在社區生鮮直投站刷卡，自助領取生鮮包裹；二是恆溫自助售菜機模式。基地淨菜經分揀、包裝、全程冷鏈車輛

運輸到達社區內的恆溫自助售菜機，居民自助挑選生鮮產品後，透過投幣、刷會員卡付款。據瞭解，這一智慧終端機還將擴展至話費充值、信用卡還款、水電煤繳費、銀聯支付等功能；三是以線下實體交易連動線上服務。如康品匯社區生鮮直營加盟店推出的線上訂菜、線下一小時送菜上門服務；四是傳統菜市場的改造提升，如上蔬永輝結合傳統菜市場與現代超市的特點，從品質保障、安全可控、誠信服務、規模經營等方面提升社區菜市場，居民反響熱烈。[174]

For example, in recent years, Shanghai has actively explored a new model for promoting the transformation and upgrading of the vegetable market in the 「Vegetable Basket」 service, and encouraged the development of four models. First, the fresh products direct pick-up station model, that is, the O2O model of online ordering and pick-up in the store for fresh products. The model combines the traditional offline restaurant retail with e-commerce to provide citizens with the 24-hour grocery service, so that residents can order fresh products 24 hours a day in five ways as the official website, self-service terminal, call center, mobile application and WeChat. Customers swipe their card in the fresh products direct pick-up station to receive fresh food by themselves. Second, the constant temperature self-service vending machine model. After being cleaned in the base, vegetables are sorted, packaged, and transported all the way through the cold chain vehicles to reach the constant temperature self-service vending machine

---

174 上海推進菜市場轉型升級 [OL]. 國際商報 .2015-11-02 Shanghai Promotes the Transformation and Upgrading of the Vegetable Market [OL]. International Business Daily. Nov 2, 2015

in the community. After the residents have self-selected the fresh products, they pay by cash or swiping membership cards. It is understood that this intelligent terminal will also be further used in telephone recharging, credit card repayment, utilities payment and UnionPay payment. Third, physical stores interconnect with online services. For example, the Kangpinhui community fresh-selling franchise stores launched online ordering and offline one-hour delivery service. Fourth, the transformation of the traditional vegetable market. For example, Fmart combines traditional vegetable market and modern supermarket to improve the vegetable market in the aspects of quality assurance, safety and control, honest service, and scale management, to which residents responded fervently.[174]

（二）訂餐送餐服務發展探索
Exploration of the food ordering service development

「到家美食會」──整合餐飲資源，提供訂餐送餐服務「Daojia」– integrating catering resources and providing food ordering service

「到家美食會」成立於 2010 年，是中國國內較早出現的餐飲外賣平臺，專注於為城市中高收入家庭提供特色餐廳外賣服務。消費者足不出戶就可以享用到服務區域內多家餐飲店的美食服務。「到家美食會」運營重點有以下幾個方面：

Founded in 2010, 「Daojia」 is an early takeaway platform in China, focusing on providing specialty restaurant takeaway services for medium- and high-income families in the city. Consumers can enjoy the food service of many restaurants in the service area without leaving the house. The focuses of the Daojia are as follows.

（1）「到家美食會」將城市按一定規制劃分出若干服務區域，在各服務區域內簽約若干餐飲門店。在該服務區域內的消費者可以選擇這些門店的菜品進行訂餐。

a. It divides the city into several service areas according to certain regulations, and makes contract with several restaurants in each service area. Consumers in the service area can choose from the dishes in these stores and place an order.

（2）消費者可以透過網站、手機或「到家美食會」呼叫中心，從其周邊餐廳組合訂餐，由「到家美食會」的專業送餐員配送到戶。

b. Consumers can order food from the surrounding restaurants through the website, mobile phone or Daojia call center, and the food will be delivered to home by the professional delivery staff of Daojia.

（3）「到家美食會」由簽約餐飲企業訂餐返利獲取收益，專業送餐團隊依靠從消費者收取的送餐費支持。

c. Daojia makes profit from the rebate rewarded by the contracted catering enterprise, and the professional food delivery team earns the delivery fee charged from the consumers.

圖 Figrue 8-4「到家美食會」送餐網頁 Webpage of Daojia

2010 年前後，餓了麼、百度外賣、美團等一批企業相繼上線餐飲送餐服務並體現出良好的成長性。2008 年創立的本地生活平臺餓了麼經過不到 9 年的發展，截至 2017 年 6 月，其線上外賣平臺已覆蓋中國 2000 個城市，加盟餐廳130 萬家，用戶量達2.6億；2013 年11 月上線的美團外賣經過4年的發展，用戶數達 2.5 億，合作商戶數超過 200 萬家，覆蓋 1300 個城市，日完成訂單 1800 萬單；2014 年上線的百度外賣，到 2015 年 11 月已覆蓋 100 多個大中城市，平臺註冊用戶量達到 3000 多萬。據美團點評研究院發佈的《2017 年中國外賣發展研究報告》，2017 年中國線上餐飲外賣市場規模約為 2046 億元，較上一年增長 23%，線上訂餐使用者規模接近 3 億人。

Around 2010, a group of companies including Eleme, Baidu Takeaway, and Meituan have successively launched online catering service and demonstrated good growth. As of June 2017, after less than 9 years of development, Eleme, a local life platform started in 2008, has covered 2,000 cities in China, with 1.3 million restaurants and 260 million users. Meituan Takeaway, launched in November, 2013, has 250 million users and more than 2 million partners, covering 1,300 cities, with 18 million orders everyday after 4 years of development. Baidu Takeaway, started in 2014, has covered more than 100 large and medium-sized cities and more than 30 million registered users by November, 2015. According to the 2017 Research Report on China Takeaway Development released by the Meituan and Dianping Research Institute, the scale of China's online takeaway market in 2017 was about 204.6 billion yuan, an increase of 23% over the previous year, and the number of online ordering users was close to 300 million.

（三）便利店服務發展探索 Exploration of convenience store services development

1· 日本羅森便利店——搭載多項服務功能的運營模式 Lawson Convenience Store from Japan-the operational model with multiple service functions

在日本，羅森便利店的規模僅次於 7-11 便利店，分店遍及日本、中國、美國夏威夷、印尼和泰國。其中，羅森便利店在中國的分店分佈在上海、北京、重慶、杭州和大連 5 座城市。截至 2013 年 9 月，羅森便利店在日本的店數為11455家，在中國的店數為292家。羅森便利店出售的商品有雜誌、漫畫、飲品、藥物、零食及便當等，並以搭載提供多元服務功能而備受消費者推崇。

In Japan, Lawson is second only to 7-Eleven.It has branches in Japan, China,

Hawaii, Indonesia and Thailand. Among them, those in China are located in five cities, Shanghai, Beijing, Chongqing, Hangzhou and Dalian. As of September 2013, the number of Lawson Convenience Stores in Japan was 11,455, and the number of stores in China was 292. Lawson sells magazines, comics, drinks, medicines, snacks, box meals, etc. and is highly regarded by consumers for its multi-service capabilities.

（1）多樣化服務功能構建多元盈利模式

Diversified service functions to build a multi-profit model

羅森便利店搭載提供多元服務功能，為周邊消費者提供生活便利。其服務範圍包括郵政投遞、影印、資料處理設備、傳真、代收公用事業費、禮品包、快遞服務、電子錢包結算終端、接收網路購物商品並支付貨款、演出及娛樂、遊樂票務、預定住宿、支付機票、高速大巴車票、信用卡支付及信用卡還款、申請學校課程、支付考試費用、申請保險及支付保險費用、查看政府資訊、申請公共設施、訂購藝術產品、DVD 等。目前，羅森便利店的服務收入已經超出了其商品銷售的收入。

Lawson is equipped with a multi-service function to provide convenience for the nearby consumers. Its services include postal delivery, photocopying, data processing, fax, collection of utility fees, gift packages, courier services, e-wallet settlement terminals, receiving online shopping products and payment, performance and entertainment, amusement park ticketing, booking accommodation, payment for air tickets, high-speed bus tickets, credit card payments and repayments, application for school courses, payment for exams, application and payment for insurance, viewing government information, application for public facilities, ordering art products, DVDs, etc. At present, Lawson's service revenue has exceeded the its merchandise sales.

圖 Figure8-5 羅森便利店內景 Interior of Lawson

（2）完善的信息化體系支撐新型服務模式

Perfect information system supports new service models

羅森便利店有完善的資訊系統，適度改造、代碼對接即可使用。

Lawson has a complete information system, which can be used after moderate transformation or code docking.

Scan the code-cashier system displays all the information of commodity or service-Staff operates the system according to the instructions on the screen

圖 Figure 8-6 羅森便利店搭載服務模式操作示意圖 Operation diagram of service model in Lawson

2· 台灣 7-11 便利店的發展經驗

Development experience of 7-Eleven in Taiwan, China

從 1980 年在臺灣開設第一家店開始到現在，7-11 便利店已在臺灣開設 4861 家店。在這 33 年間，7-11 便利店逐漸滲入臺灣人的生活，而今已成為臺灣消費者生活方式的代表，在臺灣消費者的生活中不可或缺。

Since the opening of the first store in Taiwan in 1980, the 7-Eleven has already opened another 4,861 stores. During the 33 years, 7-Eleven gradually infiltrated into the lives of Taiwanese, and has now become the representative of Taiwanese consumers' lifestyles and are indispensable in the lives of Taiwanese consumers.

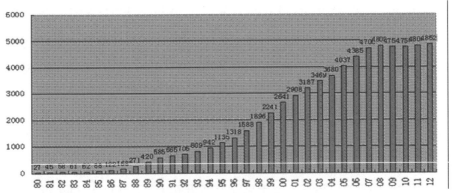

圖 Figure8-7 7-11 便利店在臺灣成長情況示意圖（單位：店）Diagram of the growth of 7-Eleven in Taiwan (Unit: store)

（1）推動物流和信息技術的革新升级，形成高效物流體系和強大的消費數據分析能力 Promoting innovation and upgrading of logistics and information technology to form an efficient logistics system and powerful consumer data analysis capabilities

1990 年，7-11 便利店不斷研發升級 POS 系統，成立自己的物流公司。經過 20 多年的演變，7-11 便利店資訊物流系統從早期的「今天訂貨，兩三天後到貨」，到「早上訂貨，晚上到貨」，甚至「一天配兩次」的快速運作系統，包括 1 日 3 次配送鮮食產品。而且到店準確率提升到 99%，前後只能有 50 分鐘的誤差。高效的資訊系統簡化了作業流程，提升了工作效率。

　　In 1990, 7-Eleven continued to develop and upgrade the POS system and set up its own logistics company. After more than 20 years of evolution, the information logistics system of 7-Eleven shifts from the past 「ordering today, arrival in two or three days later」 to 「ordering in the morning, arrival in the evening」 or even the rapid operation system of 「two deliveries in one day」, including the delivery of fresh food products for three times a day. Moreover, the accuracy of delivery has increased to 99%, and there can only be a 50-minute error before and after. The efficient information system simplifies the workflow and increases productivity.

　　高效的資訊系統是消費者資訊整合的強大後臺支援。結帳時輸入消費者的年齡、性別、商品價格等基本資料，經過系統測算後，可以更精確地掌握不同年齡層消費者的偏好，為產品開發與促銷方案的制訂提供依據。

　　The efficient information system is a powerful back-end support for consumer information integration. At the time of checkout, the basic data such as the age, gender, and commodity price of the consumer are input. After system measurement, the preferences of consumers of different ages can be accurately grasped, which provides a basis for development and promotion plan of products.

（2）整合提供各項生活服務，滲入民眾全方位生活 Integrating and providing various life services and infiltrating people's life

7-11 便利店整合各項生活服務資源，提供各項生活服務。消費者可以在住家附近的 7-11 便利店，選購各地名產、繳交各項費用、使用 ATM 提款機，也可影印、傳真、彩色列印各類檔、沖洗照片、收發快遞、部分再生資源回收等。

7-Eleven integrates various living service resources and provides various life services. Consumers can purchase local famous products, pay various fees, use ATM cash machines, color-print and fax various documents, develop photos, send and receive deliery, recycle some of the renewable resources, etc. at the 7-Eleven nearby.

24 小時不打烊的代收服務，從水電等基本民生消費到交通罰款、學費、信用卡等，在臺灣平均每人每年都要在 7-11 便利店繳交 7 張代收單據。7-11 便利店推出各地名產的預定服務，讓消費者隨地享受各地應季的名產、美食。並依照季節，提供不同的預定服務，可以預定北海道帝王蟹、中秋月餅，甚至可以把五星級飯店的年夜飯送到消費者家中。

7-Eleven has the 24-hour non-closing collection services, from basic consumption of utilities fees to traffic fines, tuition, credit cards, etc. In Taiwan, each person hands in 7 collection receipts at 7-Eleven per year. 7-Eleven launches a reservation service for famous products from all over the world, allowing consumers to enjoy the famous products and foods from all over the world. According to the season, different reservation services are provided, where people can book Hokkaido king crabs, mooncakes, and even the reunion dinner cooked by the five-star restaurants.

7-11 便利店運用其物流系統提供快遞服務，包括網路訂購商品快遞收發（類似淘寶購物），以及一般快遞收發。

7-Eleven uses its logistics system to provide the delivery services, including express delivery for online goods, which is similar to Taobao, as well as general express delivery.

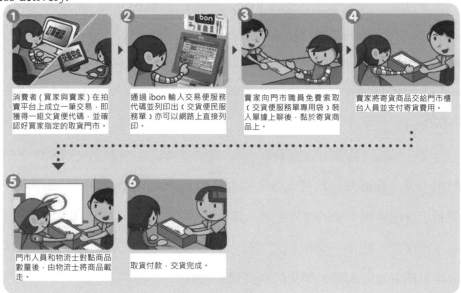

1. Consumers (including buyer and seller) make a deal on the auction platform and a set of code. Then the the buy decides the pick-up store.
2. Input the code on ibon or the website and print out the delivery order.
3. The seller asks for a delivery bag from staff and package in the certificate. Stick the bag on the parcel.
4. The seller sends the parcel to staff and pays for the fee.
5. The staff and courier make final check and the courier takes away the parcel.
6. The buyer receive the parcel and finish payment.

圖 Figure 8-8 臺灣 7-11 便利店網路訂購商品快遞收發流程示意圖 Schematic diagram of Taiwan 7-Eleven online goods express delivery process

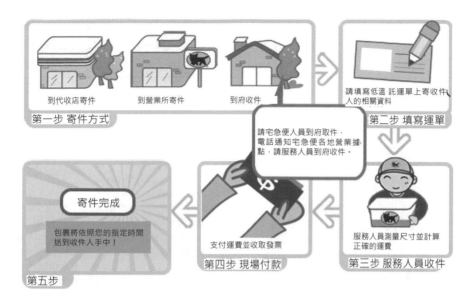

Step 1 Mail method: Send in the collecting store/Send in the business office/Receive at home---Step 2 Fill in the note Fill in the low temperature booking note---Step 3 Service staff picks up the parcel Service staff measures the parcel and gives the correct delivery charge---Step 4 On-site payment Pay and receive the invoice---Step 5 Finish The parcel will be delivered at the disignated time

圖 Figure 8-9 臺灣 7-11 便利店低溫快件收發流程示意圖 Schematic diagram of Taiwan 7-Eleven low-temperature express delivery process

　　7-11 便利店提供部分再生資源回收業務，包括廢筆記型電腦、廢手機、廢乾電池、光碟片、廢手機座充 / 旅充，回收再生資源換門店的購物抵用券。

　　7-Eleven provides the recycling business of some renewable resources, including waste laptops, mobile phones, batteries, CDs and mobile phone/travel chargers. The recycling can be exchanged for shopping vouchers.

表 Table 8-1　臺灣 7-11 便利店資源回收項目說明及商品抵用金額一覽表
Description of Taiwan 7-Eleven resource recycling project and voucher amount list

	回收種類 Tyep	回收方式 Method	抵用金說明 Amount
廢乾電池回收 Recycling of waste batteries（Q&A）	廢乾電池回收：1 號、2 號、3 號、4 號乾電池、水銀電池、鋰電池 (指手機或數位相機的鋰電池) 均可回收，但不包含工業用電池、汽 / 機車蓄電池、筆記型電腦電池 A、B、C and D dry batteries, mercury batteries, lithium batteries (referring to lithium batteries for mobile phones or digital cameras) can be recycled, but these do not include industrial batteries, automobile/ locomotive batteries and laptops batteries	將廢乾電池交由門市人員秤重回收，以每滿 0.5 公斤為單位，提供購物優惠折抵 8 元 (不包含代收、代售、香菸等商品)，限當次消費抵用 The waste dry batteries shall be handed over to the store staff for recycling, and the shopping discount is 8 yuan per 0.5 kg (excluding collection and commission sales, cigarettes, etc.), for the current consumption only.	例如，拿 0.5 公斤到門市回收，當次購物可折抵金額為 8 元，1 公斤可折抵 16 元，1.5 公斤可折抵 24 元，以此類推 For example, 0.5 kg equals a discount of 8 yuan. 1 kg for 16 yuan, 1.5 kg for 24 yuan, and the like.
廢光碟片回收 Recycling of CDs（Q&A）	一般的廢棄 CD、VCD、DVD 等光碟片，但不包含塑膠唱片 Normal discs like CDs, VCDs, DVDs, etc., excluding plastic records	將廢棄的光碟片交由門市人員秤重回收，以每滿 0.5 公斤為單位，提供購物優惠折抵 5 元 (不包含代收、代售、香菸等商品)，限當次消費抵用 The waste discs shall be handed to store staff for recycling, and the shopping discount is 5 yuan per 0.5 kg (excluding collection and commission sales, cigarettes, etc.), for the current consumption only.	例如，拿 0.5 公斤到門市回收，當次購物可折抵金額為 5 元，1 公斤可折抵 10 元，1.5 公斤可折抵 15 元，以此類推 For example, 0.5 kg equals a discount of 5 yuan. 1 kg for 10 yuan, 1.5 kg for 15 yuan, and the like.

廢筆記型電腦回收 Recycling of waste laptops （Q&A）	各廠牌的廢棄筆記型電腦（必須含有螢光屏、主機版、硬碟等零件），若電源線、電池遺失亦可回收 Waste laptops of all brands (must contain screens, mainframes, hard drives, etc.). If the power cord and battery are lost, they can still be recycled.	拿廢筆記型電腦至門市，在回收專用確認單上 1. 簽名及填寫回收日期。 2. 圈選回收物種類。 3. 門市人員蓋上店章。 之後將破壞袋封裝完成，即完成回收 Take the waste laptops to the store, and do the following on the recycling confirmation form. 1.Sign and fill in the date of recycling. 2.Circle the type of recycling. 3.Seal by store staff. After the damage bag is packaged, the recycling is completed.	每臺可折抵 120 元購物抵用金，2 臺可折抵 240 元，3 臺可折抵 360 元，以此類推 Each laptop can be discounted for 120 yuan for purchase, 2 for 240 yuan, 3 for 360 yuan, and the like.
廢手機回收 Recycling of waste mobile phones （Q&A）	各廠牌的廢棄手機（必須含有螢光屏、主機版等零件），若電池、電池蓋、充電器遺失亦可回收 Waste mobile phones of all brands (must contain parts such as fluorescent screen, host version, etc.), if the battery, battery cover, and charger are lost, they can still be recycled.	拿手機電至門市，在 回收專用確認單上 1. 簽名及填寫回收日期。 2. 圈選回收物種類。 3. 門市人員蓋上店章。 之後將破壞袋封裝完成，即完成回收 Take the waste mobile phones to the store, and do the following on the recycling confirmation form. 1.Sign and fill in the date of recycling. 2.Circle the type of recycling. 3.Seal by store staff. After the damage bag is packaged, the recycling is completed.	每支可折抵 12 元購物抵用金，2 支可折抵 24 元，3 支可折抵 36 元，以此類推 Each laptop can be discounted for 12 yuan for purchase, 2 for 24 yuan, 3 for 36 yuan, and the like.
手機座充 / 旅充回收 Recycling of waste mobile phone/ travel chargers （Q&A）	各廠牌的手機座充或旅充，必須含電源線完好才可回收 Power cord of Mobile phone chargers or travel chargers of all brands must be intact.	將廢棄的手機座充或旅充交由門市人員回收，每組提供購物優惠折抵 3 元（不包含代收、代售、香菸等商品），限當次消費抵用 The waste mobile phone chargers or travel chargers shall be handed in to the store staff, and each equals a discount of 3 yuan (excluding collection and commission sales, cigarettes, etc.), for the current consumption only.	每組座充或旅充可折抵 3 元購物抵用金，2 組可折抵 6 元，3 組可折抵 9 元，以此類推 Each charger can be discounted for 3 yuan for purchase, 2 for 6 yuan, 3 for 9 yuan, and the like.

（3）整合提供各色美食，運營滲入餐飲領域 Integrating and providing a variety of cuisines, infiltrating operations into the catering field

經營美食品類從御便當到小吃、從御飯糰到三明治，7-11 便利店將便利店的運營逐漸滲入餐飲領域。在影響餐飲模式創新的同時，促進便利店經營模式的升級與轉型。

By operating food from the box meal to snacks, from the rice balls to sandwiches, 7-Eleven gradually infiltrates the operation of the convenience store into the catering field. While influencing the innovation of the catering model, this also promotes the upgrading and transformation of the convenience store business model.

（4）部分門店經營生鮮蔬菜、水果，滿足民眾全方位生活消費需求 Some stores operate fresh vegetables and fruits to meet the needs of the people for all-round living.

3·唐久大賣場——便利店（唐久）與電商（京東）的線上線下合作探索 Tangjiu Hypermarket-exploration of online and offline cooperation between convenience store (Tang Jiu) and e-commerce (Jingdong)

2013 年 12 月 5 日，京東唐久網上大賣場正式啟動。京東—唐久的合作是便利店與電商線上線下合作的實踐探索，網上大賣場在試營業期間已經取得了日均近 1000 單、平均客單價 100 元左右、流量轉換率 7% 左右佳績。

On December 5, 2013, Jingdong Tangjiu Hypermarket was officially launched. Cooperation between Jingdong and Tangjiu is a practical exploration of online and offline cooperation between convenience stores and e-commerce. During the trial operation period, the online hypermarkets have achieved an average of nearly 1,000 orders everyday, an average customer consumption of about 100 yuan, and a traffic conversion rate of around 7%.

唐久與京東透過雙方系統與資源的深度對接與整合，實現唐久電商和門店的真正融合。

Tang Jiu and Jingdong realized the true integration of e-commerce and physical stores through the deep docking and integration of the two systems and resources.

（1）門店商品與網上商品實現對接

Docking of store goods and online products

唐久 800 多家便利店全部被整合到網上大賣場中。門店庫存可以在網上銷售，網上購買的商品可以到門店自提。

More than 800 convenience stores of Tangjiu have been integrated into online hypermarkets. Store inventory can be sold online, and online purchases can be picked up at the store.

（2）倉储物流體系的全面對接

Comprehensive docking of warehousing and logistics system

如果消費者購買的商品在門店有庫存，系統會根據 LBS 定位將商品交給距離消費者最近的門店，實現 1 小時達（甚至可以實現付費後 15 分鐘送達）；

如果門店無庫存，則由總倉發貨，次日送達。

If the goods purchased by the consumer are in stock at the store, the system will deliver the goods to the closest store to the consumer according to the LBS positioning, achieving 1-hour, or even 15-minute delivery. If the store is out of stock, then the goods purchased will be sent out from warehouse and arrive the next day.

（3）支付體系的全面對接

Comprehensive docking of the payment system

唐久與京東支付系統實現全面對接。京東擁有自建物流，在貨到付款方面一直是在電商領域最強的。網上賣場支持貨到付款、銀行轉帳、郵局匯款，並有京東信貸——京東白條支持。

Tangjiu and the Jingdong payment system achieve comprehensive docking. Jingdong has self-built logistics and has always been the strongest in the e-commerce field in terms of cash on delivery. The online store supports cash on delivery, bank transfer, postal remittance. There is also the support of Jingdong White Bar, or the Jingdong Credit.

（四）再生資源服務發展探索 Exploration of renewable resources services development

有代表性的是北京天天潔再生資源服務探索。天天潔公司是北京市第一批再生資源回收體系產業化試點企業，企業透過建立城市資源分類回收體系，

生產並推廣再生產品，宣導低碳綠色生活方式，推進有限資源的迴圈利用。

Beijing Tiantianjie' s exploration of renewable resources service is a typical example. Tiantianjie is one of the first pilot enterprises of industrialized renewable resource recycling system in Beijing. Through the establishment of urban resource classification and recycling system, the company produces and promotes recycled products, advocates a low-carbon green lifestyle, and carries forward the recycling of limited resources.

經營特色：

Its operation characteristics are as follows.

1· 規範收購 Normative acquisition

專業人員持證上崗。天天潔公司的工作人員必須經過嚴格的培訓上崗，並嚴格按照公司的規範提供服務。在社區內設置綠貓資源回收屋，派駐收購人員，用標準電子秤秤量，以統一的合理的價格實行定點回收。

Professionals are employed with certificates. The staff of Tiantianjie must go through strict training and provide services in strict accordance with the company's norms. The company sets up the Green Cat(Lumao.com.cn) resource recycling house in the community and sends the acquisition staff to it. The staff weigh the recyclable materials on standard electronic weigher to carry out fixed-point recycling at a uniform and reasonable price.

設立全新改版的綠貓網——北京再生資源回收利用服務平臺，居民透過網路或電話預約，天天潔公司可以提供上門服務。

The campany sets up a new version of the Green Cat, Beijing renewable resources recycling service platform, through which residents make network or telephone appointments and then Tiantianjie provides on-site service.

圖 Figure 8-10 天天潔公司再生資源回收網站 Tiantianjie website of renewable resources recycling

2· 統一物流，專業分揀 Unified logistics and professional sorting

天天潔公司擁有一個年處理能力 5 萬噸的分揀中心、45 輛專業物流車。回收網回收的物資，透過統一物流，直接配送至分揀中心，進行專業分揀加工。

Tiantianjie has a sorting center with an annual processing capacity of 50,000 tons and 45 professional logistics vehicles. The materials recycled by the recycling website are directly distributed to the sorting center through unified logistics for professional sorting and processing.

圖 Figure 8-11 天天潔公司專業分揀加工設備 Professional sorting and processing equipments of Tiantianjie

3． **廠商直掛** Direct channel to manufacturers

天天潔公司擁有再生產品研發設計中心，設計開發系列再生產品。經分揀加工的物資直接送到工廠回收利用，加工再生產品，形成完整的再生資源回收與迴圈利用的產業鏈條。

Tiantianjie has a R&D design center, where a series of recycled products are designed and developed. The sorted and processed materials are sent directly to the factory for recycling and processing into recycled products, forming a complete industrial chain for renewable resources recycling.

（五）洗染服務發展探索
Exploring the development of dyeing services

榮昌 e 袋洗—洗染服務 OTO 發展探索值得介紹。北京榮昌科技服務有

限責任公司（以下簡稱榮昌）成立於 1990 年，主要經營服裝、皮貨與毛貨清洗、洗衣設備代理等，旗下擁有榮昌 e 袋洗、榮昌洗染、伊爾薩洗衣、珞迪奢侈品養護等品牌。在全國擁有近千家店面，覆蓋 100 多個大中城市。歷年累計清洗衣物已逾 15 億件。從 2000 年開始，榮昌就開始探索與新浪共推網上洗衣；從 2004 年開始，榮昌開始探索 OTO 模式轉型；在 2007 年，建立了榮昌洗衣園；2010 年，實現商業模式轉型；2013 年，開始移動互聯網轉型。榮昌榮昌建立了業內首家居家服務網站和 400 呼叫中心，旗下 3 個品牌都已開通官方微信，消費者可以透過網站和微信實現洗衣預約、預定、購買、查詢、積分、投訴、支付等功能。

The OTO development exploration of Rongchang Edaixi, the washing and dyeing service, is worth introducing. Beijing Rongchang Technology Service Co., Ltd. (hereinafter referred to as Rongchang) was established in 1990. It mainly operates clothing, leather goods and wool cleaning, laundry equipment agents, etc.The campany owns Rongchang Edaixi, Rongchang Dyeing, Ilsa Laundry, and Lodevol Luxuries care and other brands. Also, it has nearly 1,000 stores across the country and covers more than 100 large and medium-sized cities. Over 1.5 million pieces of clothing have been cleaned over the years. Since 2000, Rongchang has begun to explore online laundry with Sina. Since 2004, Rongchang has begun to explore the transformation of OTO model. In 2007, Rongchang Laundry Park was established. In 2010, business model transformation was realized. In 2013, mobile Internet transformation began. Rongchang has established the first home service website and 400 call center in the industry. All the three brands have opened official WeChat account, where consumers can use the functions of laundry appointment, reservation, purchase, inquiry, points, complaints, payment.

2013 年 11 月 28 日感恩節，基於微信，榮昌推出榮昌 e 袋洗互聯網洗衣服務，將洗衣服務標準化。消費者將待洗衣物裝進指定洗衣袋裡，透過網路終端或移動終端預約取送時間和地點，然後由榮昌派專人上門取件，並按 99 元 / 袋收取清洗費用。榮昌 e 袋洗成功解決了消費者到乾洗店洗衣停車難、送洗衣物交接手續繁瑣、店面營業時間不能滿足消費者取送時間等一系列問題。取件時，取件人員不做衣物檢查，當面直接用 e 袋將衣物裝好鉛封。e 袋送到清洗中心後，在全程視頻監控下解除鉛封，對衣物進行洗前檢查和分類。

On November 28, 2013, the Thanksgiving Day, based on WeChat, Rongchang launched the Rongchang Edaixi Internet laundry service to standardize the laundry services. The consumer puts the clothes into the designated laundry bag, and makes an appointment for the time and place of the delivery through the online terminal or the mobile terminal, and then the staff from Rongchang drop in and pick up the clothes and charges the cleaning fee by 99 yuan per bag. Rongchang Edaixi successfully solves a series of problems such as the difficulty for the consumer to park car at dry cleaning shop, cumbersome delivery of the laundry and the store opening time can not meet the customers' needs. When staff picks up the bag, they do not check inside and directly seal the clothes with special bags. After the bags are sent to the cleaning center, the lead seal is released under full monitoring, and the clothes are inspected and classified before washing.

榮昌 e 袋洗的洗衣工序分為進廠掃描→衣物複檢→洗前分類→預去漬→

清洗→乾燥→整燙→質檢→包裝→送回等。衣物清洗完畢後，採用掛送的方式將衣物送回。使用的袋子也一同送回，供消費者下次使用。

Laundry process of Rongchang Edaixi is divided into: factory scanning → clothing re-examination → pre-classification → pre-stain reduction → cleaning → drying → ironing → quality inspection → packaging → return. After the laundry is done, the clothes are returned by hanging, together with the used bags for the consumer to use next time.

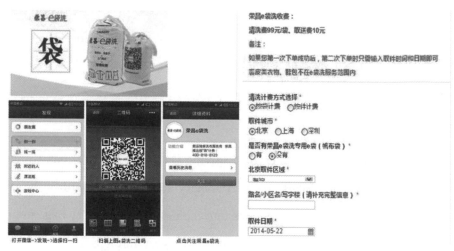

圖 Figure 8-12　榮昌 e 袋洗網頁 Rongchang Edaixi Website

（六）家政服務發展探索
Exploration of the domestic service development

易盟天地專業現代家政服務運營商（以下簡稱易盟天地公司）為專業現代家政服務業務的運營商，主要透過多網融合的手段為消費者提供全程閉環式服務。服務內容涉及家政、保潔、維修等，以滿足消費者對精細化服務的需求。易盟天地公司有全國統一的服務熱線：95081 以及相應的業務網址。

The Emotte Professional Modern Housekeeping Service Operator (hereinafter referred to as Emotte) is an operator of professional modern housekeeping service, which provides consumers with closed-loop service through multi-network integration. The service content involves housekeeping, cleaning, maintenance, etc. to meet consumer demand for refined services. Emotte has a unified national service hotline: 95081, and the corresponding business website.

95081 家庭服務中心成立了社區居家養老服務站，由民政局面向社會招聘社區居家養老工作者。透過 95081 家庭服務中心居家養老服務網路資訊化系統，即時對老人的健康等狀況進行監控，以社區居家養老的形式幫助老人度過健康安樂的晚年，解決了目前社會養老機構不足、服務人員緊缺的困難。

The 95081 Family Service Center has established a Community Homecare Service Station, and the Civil Affairs Bureau recruits community homecare workers to work there. Through the 95081 Family Service Center Homecare Service Network Information System, real-time monitoring of the health status of the elderly can be realized, helping the elderly to live a healthy and happy life in the form of community home care, and solving the current shortage of nursing institutions for the aged and service personnel

圖 Figure 8-13 易盟天地公司家政服務體系示意圖 Schematic diagram of Emotte housekeeping system

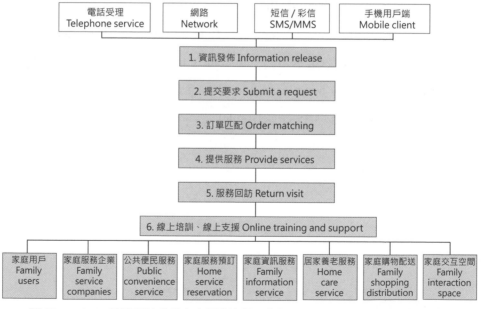

圖 Figure 8-14 易盟天地公司家政服務流程示意圖 Schematic diagram of Emotte housekeeping process

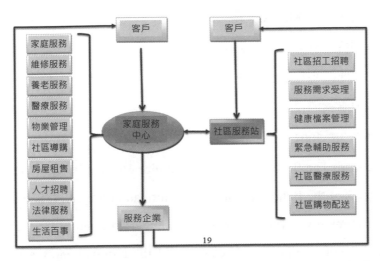

圖 Figure 8-15 易盟天地公司家政服務管理體系示意圖 Schematic diagram of Emotte housekeeping management system

（七）政府主導社區服務平臺建設發展探索
Exploration on the Construction and Development of Government-led Community Service Platform

1· 北京的 96156 在線服務平臺——政府主導的社區服務平臺建設探索 Beijing 96156 Online Service Platform—— Exploration on the Construction and Development of Government-led Community Service Platform

2002 年 12 月 26 日，北京市社區公共服務熱線 96156 正式開通。96156 社區服務熱線是北京社區資訊化工程的重點建設項目之一，由北京市社區服

務中心運營，實行 24 小時熱線服務，日接聽電話近萬次。全市範圍內的居民撥打服務熱線即可享受各種便民服務。北京市各區縣分別與 96156 北京市社區公共服務熱線對接，建設區縣社區服務資訊網為居民提供更便捷優質的服務。目前，96156 社區服務熱線能夠提供家政服務、為老服務、修理、租賃等 8 大類 200 多個服務項目。

On December 26, 2002, Beijing Community Public Service Hotline 96156 was officially opened. The 96156 Community Service Hotline is one of the key construction projects of the Beijing Community Information Project. It is operated by the Beijing Community Service Center, providing a 24-hour hotline service. It receives nearly 10,000 calls a day. Residents of the city can enjoy a variety of convenient services by calling the service hotline. Each district and county in Beijing is connected with the Hotline, and the district and county community service information network provides residents with more convenient and high-quality services. At present, the hotline can provide more than 200 service items in 8 categories including housekeeping services, senior services, repairs and leasing.

2014 年，96156 北京市社區服務熱線獲得了「4PS 聯絡中心國際標準四星級認證證書」。

In 2014, the 96156 Beijing Community Service Hotline won the 「4PS Contact Center Four-Star Certification of International Standard」.

圖 Figure 8-16 北京市社區服務資訊網網頁 Webpage of Beijing Community Service
Information Network

2・ 北京市西城區槐柏商圈社區便民服務網──街道主導的社區服務平臺建設探索 Business Circle Community Convenience Service Network in Huaibai of Xicheng District, Beijing－a street-based community service platform

2011 年 9 月，槐柏商圈社區便民服務網（www.dhbsq.com）開通。這是北京市首家社區服務資訊化電子商務平臺，可以為當地社區居民提供家政服務、購物、餐飲、家電維修等 11 大類服務資源資訊及網上預約服務，並在網站上為 150 家服務商開設電子門店，使社區居民不出家門，只要透過網路或電話就能得到購物、就餐、洗衣、廢品回收等便民服務，為轄區商戶和居民搭建了多方位雙向服務的「一站式」社區便民電子商務平臺。網上「跳蚤市

場」同時提供了以物易物的空間，宣導建設節約型社會。

In September 2011, Business Circle Community Convenience Service Network in Huaibai of Xicheng District, Beijing (www.dhbsq.com) was launched. This is the first community service information e-commerce platform in Beijing. It can provide local community residents with 11 categories of service resources such as housekeeping services, shopping, catering, appliance repair and online booking services. Also, it starts e-stores for 150 service providers on the website, so that community residents can get convenience services such as shopping, dining, laundry, waste recycling, etc. through the network or telephone without getting out of the house, as long as they can. In addition, it builds the one-stop community convenience e-commerce platform with multi-faceted and two-way service for merchants and residents in the area. The online 「flea market」 also provides space for bartering and advocates building a conservation-oriented society.

槐柏街道還與中國民生銀行合作推出「惠民興商卡」。透過該卡居民不出社區就可以在服務站繳納電話、水電、燃氣等公共費用，並在刷卡消費時享受轄區百家特約服務商的消費折扣，實現了惠民、便民、興商三項便利。

Huaibai Subdistrict also cooperated with China Minsheng Bank to launch the 「Huimin(discounts) Xingshang(prosperity) Card」. Through the card, residents can pay utilities fees such as telephone, water, electricity, gas, etc. at the service station, and enjoy the discount of 100 special service providers in the community, which realizes the convenience of discounts, convenience and prosperity.

四

實施建議 4. Implementation Suggestions

（一）實施原則 Principles of implementation

1 · 政策引導、市場運作 Policy guidance, market operation

發揮市場配置資源的核心作用，以政府的引導支持為助力，不斷推進安全、便捷、舒適、可持續發展的社區生活服務體系的形成。

Give play to the core role of market allocation resources, and use the guidance and support of the government as a boost to continuously promote the formation of a safe, convenient, comfortable and sustainable community life service system.

2 · 資源整合、獎優劣汰 Resource integration, award the superior and eliminate the inferior

鼓勵品牌企業跨行業、跨業態整合資源，鼓勵社區生活服務相關企業聯盟式發展，鼓勵品牌企業直營式連鎖發展，獎優劣汰，不斷提升品質、淨化市場。

Encourage brand-name enterprises to integrate resources across the industries, encourage alliance-based development of community life service-related enterprises, encourage brand-name enterprises to develop chain-oriented chains.

Award the superior and eliminate the inferior to continuously improve quality and purify the market.

3 · 嚴格標準、規範管理 Strict standards, standardized management

嚴格社區生活性服務業准入和退出標準，健全各項規章制度，提高社區生活服務水準和管理水準，推進社區生活性服務業整體健康發展。

Strictly meet the standards for access and exit of community life service industry, improve various rules and regulations, improve the level of community life service and management, and promote the overall healthy development of community life service industry.

4 · 以點帶面、試點先行 Point to area, pilot first

選擇試點企業、代表社區開展社區生活服務新模式試點，在取得成效基礎上，以點帶面逐步在全區、全市推廣。

Select pilot enterprises and carry out pilot projects for new models of community life services on behalf of the community. On the basis of the good results, gradually promoted the new model in the whole district and the whole city in a point-to-area manner.

（二）實施重點 Focus of implementation

1． 推進企業資源之間的有效對接
Promoting effective docking among enterprise resources

如前所述，居民生活服務需求趨於多樣化，需要在有限的空間資源內整合多品類多層次的生活服務資源，搭建「雲服務」平臺將各類企業的資源進行有效對接，但對接與整合需要將大量的企業資源統籌協調，一般企業沒有能力和實力進行資源之間的對接，需要政府發揮支援和引導作用，統籌、協調、推進企業資源之間的對接。

As mentioned above, residents' demand for life services tends to be diversified. It is necessary to integrate multi-category and multi-level life service resources in a limited space, and build a cloud service platform to effectively dock the resources of various enterprises. However, such docking and integration require a large amount of enterprise resources to be coordinated. Normal enterprises do not have the ability and power to carry out the docking among resources, which requires the government to play a supporting and guiding role to coordinate and promote the docking among enterprise resources.

重點組織協調「雲服務」平臺與「雲服務」體系上游企業——生活服務提供者與下游企業——零距離配送服務實施者之間的對接。在政府的組織引導下，使「雲服務」平臺與上下游企業之間形成思想認知的對接、盈利模式的對接、服務標準的對接、作業系統和資訊平臺的對接、設備設施的對接，以及合作方式的對接。

Docking between the cloud service platform and the cloud service system upstream enterprises; the life service provider and the downstream enterprises; the zero distance distribution service implementers should be coordinated as the key points . Under the guidance of the government's organization, the cloud service platform, and the upstream and downstream enterprises should form the docking of ideas and understanding, profit models, service standards, operating systems and information platforms, equipment and facilities, and the way of cooperation.

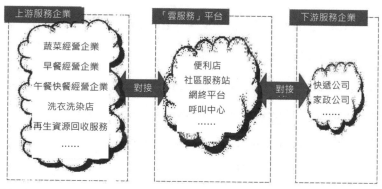

Upstream service companies(Vegetable companies / Breakfast companies / Lunch and fastfood companies / Laundry companies / Renewable resources recycling services)
Docking

「Cloud Service」 Platform(Convenience store / Community service station / Online platform / Call centre)
Docking

Downstream service companies(Express companies / Housekeeping companies)

圖 Figure 8-17　企業資源對接示意圖 Schematic diagram of enterprise resources docking

2． 推進「雲服務」平臺體系建設 Promoting the construction of the cloud service platform system

「雲服務」平臺的搭建是構建「雲服務」體系關鍵，政府應重點支援「雲

服務」平臺建設基礎性的投入，推進「雲服務」平臺體系建設。

The establishment of the cloud service platform is the key to building a cloud service system. The government should focus on supporting the foundational investment of the cloud service platform and promote the construction of the cloud service platform system.

（1）構建「雲服務」平臺體系

Building a cloud service platform system

「雲服務」平臺體系由實體平臺和電子平臺構成。

——實體平臺體系的搭建可以區域現有的便利店、社區服務站、新建社區服務中心為基礎進行。

——電子平臺以區裡現有的網路平臺為基礎開展。

The cloud service platform system consists of a physical platform and an e-platform.

—— The construction of the physical platform system can be carried out based on the existing convenience stores, community service stations and new community service centers in the region.

—— The e-platform is based on the existing network platform in the region.

（2）搭建「雲服務」後臺支持系統

Building a cloud service back-end support system

搭建保障社區「雲服務」體系順利運行的後臺服務支援系統。包括：資源系統建設，跨行業、跨業態整合有效資源，明確不同資源企業的核心優勢和支援方向，實現資源分享、優勢互補；資訊服務系統的建設和聯通；結算

系統建設；物流配送體系建設；專項配套扶持資金的立項和使用等。

The model establishes a back-end service support system that guarantees the smooth operation of the community cloud service system including the resource system construction that integrates effective resources in a cross-industry manner and identifies the core advantages and support directions of enterprises with different resources to achieve resource sharing and complementing each other's advantages; information service system construction and connectivity; settlement system construction; logistics distribution system construction; establishment and use of special supporting funds.

（3）搭建社區「雲服務」體系組織管理保障系統 Building a cloud service organization management support system for the community

包括：組織機構的建設；城市生活服務認證系統建設，透過標準化認證掛牌、嚴格獎懲等政策措施。

Including: the construction of organizational institutions; the construction of urban living service certification system; standardized certification and listing; strict rewards and punishments and other policy measures.

後　記

Afterword

　　2009 年筆者首次提出「雲消費」的認識，並隨著研究的深化而逐步系統化形成書稿。2015 年 7 月、2016 年 8 月中國經濟出版社相繼出版了《「雲消費時代」Ⅰ》、《「雲消費時代」Ⅱ》，就「雲消費時代」的基本特徵、主流消費模式、交易模式，零售業的變革與發展方向等加以探討，並提出系列觀點。本書在《「雲消費時代」Ⅰ》、《「雲消費時代」Ⅱ》兩書基礎上做了修訂，以英文版形式問世，希望將筆者關於「雲消費」的概念和認知介紹給更多朋友，並就教於海內外方家。

　　I proposed the concept of cloud consumption for the first time in 2009. With the deepening of research, I have been working on a systematic manuscript bit by bit. In July 2015 and August 2016, China Economic Publishing House successively published The Era of Cloud Consumption I and The Era of Cloud Consumption II, both of which discussed the basic characteristics of the era of cloud consumption, mainstream consumption patterns, transaction models, revolution and development direction of the teh retail industry as well as put forward a series of views. This book has been revised on the basis of The Era of Cloud Consumption I and The Era of Cloud Consumption II and is published in the English version. I hope to introduce the concept and understanding of the cloud consumption to more people, in the same time, beg for corrections from masters at home and abroad.

　　本書研究得到本人所在單位──北京財貿職業學院的支持，也得到了眾多業內學者、企業家的支持和指導，尤其是學院校長王成榮教授不僅對本項

研究給予大力支持，且對本書的最終成稿進行了指導，其諸多見解使本人受益頗豐。我院前任校長王茹芹教授也熱忱支持本專題研究，並在研究思路上給予很多啟發。在此，向所有指導、幫助本研究的學者、企業家表示衷心的感謝！

The research carried out in this book is supported by my work unit, Beijing College of Finance and Commerce, and many scholars and entrepreneurs. In particular, Professor Wang Chengrong, the president of the college, not only strongly supported the research, but also gave the guidance to the final version of this book. Many insights of his benefited me quite a lot. Professor Wang Ruqin, the former president of the college, also enthusiastically supported the research and gave a lot of inspiration to the research. I would like to express my heartfelt thanks to all the scholars and entrepreneurs who have guided and helped this research!

本書以筆者近年「雲消費」研究積累形成的觀點為核心，深化和系統化形成專著。本研究團隊之前完成的一系列以「雲消費」為核心的專題研究報告是形成本書的基礎。韓凝春副研究員承擔了全書的重要執筆和統稿工作；王春娟助理研究員承擔了本書的大量基礎研究工作；康健副研究員、黃愛光副研究員、李馥佳助理研究員、胡昕副研究員、趙挺助理研究員等均參與了「雲消費」的相關研究，為本書順利成書付出了努力。在此對他們一併表示感謝！

This book takes the viewpoints of the my research on the cloud consumption in recent years as the core, and forms into a monograph through deepening and systematizing. The series of research reports centered on cloud consumption completed by the research team is the basis for the book. Associate researcher Han Ningchun undertook the important writing and drafting work of the book. Assistant researcher Wang Chunjuan undertook a lot of basic research work in the book. Researchers including associate researcher Kang Jian, associate researcher Huang Aiguang, assistant researcher Li Fujia, associate researcher Hu Xin, assistant researcher Zhao Ting participated in the correlational research of cloud consumption and made efforts for the successful completion of this book. I would like to thank them all!

「雲消費」時代的零售革命是一個全新的事物，本書的研究僅僅是起點，需要深化的研究還有很多。服務商貿流通的發展與創新是本人及本團隊的責任和使命，誠請讀到本書的學者、企業家回饋真知灼見，指導本人及團隊繼續進步和提高，今後更好地以科研服務行業服務社會。

The retail revolution in the era of cloud consumption is a brand new thing. The research in this book is only the starting point, and there are still many studies that need to be deepened. The development and innovation of service commerce circulation is the responsibility and mission of my team. I sincerely beg the scholars and entrepreneurs who read this book to give precious feedback to guide my team to further improve and better serve the industry and the society with research work in the future.

賴 陽

Lai Yang

2017 年 8 月

August, 2017

國家圖書館出版品預行編目（CIP）資料

雲消費時代 / 賴陽著.
-- 第一版. -- 臺北市：樂果文化出版：紅螞蟻圖書發行,
2020.03
　面；公分. -- (樂經營；16)中英對照
ISBN 978-957-9036-24-5(平裝)

1. 電子商務

490.29　　　　　　　　　　　　　108020535

樂經營 016

雲消費時代

作　　　　者／賴陽

總　編　　輯／何南輝

行 銷 企 畫／黃文秀

封 面 設 計／引子設計

內 頁 設 計／沙海潛行

出　　　　版／樂果文化事業有限公司

讀者服務專線／（02）2795-3656

劃 撥 帳 號／50118837　樂果文化事業有限公司

印　刷　　廠／卡樂彩色製版印刷有限公司

總　經　　銷／紅螞蟻圖書有限公司

地　　　　址／台北市內湖區舊宗路二段121巷19號（紅螞蟻資訊大樓）

電　　　　話／（02）2795-3656

傳　　　　真／（02）2795-4100

2020年3月第一版　定價／450元　ISBN 978-957-9036-24-5